数据科学与工程技术丛书

WEB AND NETWORK DATA SCIENCE

MODELING TECHNIQUES IN PREDICTIVE ANALYTICS

Web与网络数据科学

建模技术在预测分析中的应用

[美] 托马斯 W. 米勒（Thomas W. Miller）著
何泾沙 等译

图书在版编目（CIP）数据

Web 与网络数据科学：建模技术在预测分析中的应用 /（美）托马斯 W. 米勒（Thomas W. Miller）著；何泾沙等译．—北京：机械工业出版社，2017.1
（数据科学与工程技术丛书）
书名原文：Web and Network Data Science: Modeling Techniques in Predictive Analytics

ISBN 978-7-111-55844-6

I. W… II. ① 托… ② 何… III. 网络数据库 IV. TP311.132

中国版本图书馆 CIP 数据核字（2017）第 002923 号

本书版权登记号：图字：01-2016-2918

本书以作者在美国西北大学开设的“Web 网站分析学”课程为基础，介绍了可用性测试、网站性能、使用分析、社交媒介平台、搜索引擎优化（SEO）等方面的知识。同时，书中在涵盖实际应用与介绍社交网络分析和网络科学领域中现有的最新知识之间取得了一个良好的平衡，清楚地展示出如何将所涉及的理论知识应用于解决实际的商业问题。

本书可供计算机及相关专业学生阅读，也可供数据研究人员和分析师学习参考。

出版发行：机械工业出版社（北京市西城区百万庄大街 22 号　邮政编码：100037）
责任编辑：关　敏　　责任校对：董纪丽
印　　刷：北京瑞德印刷有限公司　　版　　次：2017 年 2 月第 1 版第 1 次印刷
开　　本：185mm × 260mm　1/16　　印　　张：16.75
书　　号：ISBN 978-7-111-55844-6　　定　　价：79.00 元

凡购本书，如有缺页、倒页、脱页，由本社发行部调换
客服热线：（010）88378991　88361066　　投稿热线：（010）88379604
购书热线：（010）68326294　88379649　68995259　　读者信箱：hzjsj@hzbook.com

译者序

当今社会是一个快速发展的社会，科技发达、信息流通，人们之间的交流越来越密切，居民的生活越来越方便，大数据就是这个高科技时代的最新产物，近年来迅速成为全球IT行业中的热门词汇。大数据中所隐含的理念以及潜在的发展前景与价值已经得到越来越广泛的认可，影响着政治和经济社会中的各个方面，被认为是各类组织和机构乃至国家层面的重要战略资源，成为提高核心竞争力的有力武器，也理应得到我们每一个用户、每一个消费者的高度重视。大数据具有数据量大、种类繁杂、实时性强、蕴藏的潜在价值大等特征，公开与分享已经成为大势所趋。然而，如何鉴别数据的真伪？如何从价值密度稀疏的大数据中获取隐藏在其中的真正价值？这些疑问给人们提出了技术上的巨大挑战。Web是大数据的一个重要来源，我们每一个人通过个人电脑、手机或各类移动终端敲击的每一个字、选择的每一条词语、录制的每一段语音留言、浏览的每一个网页，都成为大数据的组成部分，进入巨大的数据海洋中，成为被提取、分析、使用的基本元素，成为形成各种商业决策的依据以及通过分析对未来可能发生的事件进行预测的基础支撑。社会信用体系中政务诚信、商务诚信、社会诚信的构建也将建立在大数据的基础之上。近年来日渐流行的社交网站在给广大用户提供即时沟通交流工具以及形成在线社区平台的同时，更是成为大数据的一个重要来源。在当今社会的线上线下，数据无处不在、持续产生，然而，众多的数据纷繁杂乱，数据之间存在的关系复杂而不明朗，我们需要去搜索、处理、分析和归纳，以挖掘出数据的深层次规律以及数据之间存在的相互关系。

大数据的价值不仅仅在于数量巨大，通过建立新的模型、提出新的方法、构建新的系统、开发新的工具对大量、动态、持续产生的结构化、半结构化和非结构化数据进行分类、融合、分析与挖掘，以获得具有实际应用价值以及能够预测未来事件与行为的结果，这才是大数据的真正价值所在。虽然数据的迅速膨胀将决定企业、机构的未来发展，然而很多企业和机构并没有意识到数据爆炸性增长所带来的机遇以及潜在的隐患。但是随着时间的推移，随着大数据分析技术的进一步成熟与完善，大数据将得到越来越多的应用，实现越来越大的价值，人们也将越来越多地意识到大数据在企业和机构运作中所起的重要作用。在未来的商业、经济及其他领域中，关键的决策将越来越多地基于大数据与数据分析而做出，而越来越少地基于经验和直觉。

大数据将在观念上给我们带来一些颠覆性的转变。首先，我们面对的将是全部数据，而不再是随机抽取的样本；其次，大数据提供给我们的是混杂性，而不再是精确性；再次，大数据之间存在的是相关关系，而不再只是因果关系。因此，对大数据的分析需要我们提出和构建新的模型、方法和工具。本书就是为了满足这些新的需求和新的要求而撰写的，是将数据科学与网络科学相结合形成的“Web与网络数据科学”。本书不但包含了大数据分析与应用所需的理论知识与建模技术，还提供了大量的应用实例，并通过提供建模技术方面的资料和参考指南对研究人员及分析师的工作提供进一步的帮助，同时面向实际应用向编程人员展示如何使用目前在数据分析与应用领域中得到广泛应用的Python和R语言编写能够正确运行并解决实际商业问题的计算机软件，还提供了大量的Python和R语言代码实例。全书涵盖了Web与网络数据科学领域中的若干主要问题，如网站设计与用户行为、网络路径与通信、社区与影响、个体与群体行为、信息与网络等，具体分12章，在对开展Web与网络数据科学方面的研究所需的相关技术进行概述后，分别对Web在线消息传递技术、Web爬行与抓取技术、Web链接测试以及体验与外观改进技术、在线竞争性情报搜集与分析技术、网络可视化技术、社区发现与分析技术、情感度量技术、基于文本的共同主题发现技术、推荐技术、网络博弈行为的建模技术进行了深入浅出的介绍，最后对未来Web的发展进行了展望。此外，本书使用较大篇幅，以附录的方式对目前数据建模与分析中常用的技术进行了简要介绍，包括数据库与数据准备、数据统计学、回归与分类、机器学习、数据可视化以及文本分析学，对开展在线研究的流程与方法进行了系统的归纳，最后通过提供若干实用案例为本书中介绍的理论知识和应用技术画上了一个完美的句号。本书还向读者提供了在应用预测分析学的建模技术中常用的代码与共享程序、常用术语以及丰富的参考文献，为读者进一步学习提供专业帮助与技术指南。因此，本书对于从事基于Web的数据搜集、分析和应用的技术人员以及在相关领域中从事科学探索和技术研发的科研人员具有较重要的参考价值。

本书由三峡大学“楚天学者计划”主讲教授、北京市特聘教授何泾沙博士负责翻译，三峡大学贺鹏教授、中国科学院软件研究所朱娜斐博士协助了全程的翻译工作，中国航天科技集团公司第九研究院第十三研究所张玉强博士、清华大学徐晶博士、中国科学院信息工程研究所徐菲博士、北京工业大学博士生赵斌、研究生朱星烨、方静、刘畅、黄辉祥参与了部分章节的翻译工作。何泾沙博士对全书进行了最终统稿及全文校验。由于译者的水平有限，再加上时间方面的限制，译文中难免存在不够准确之处，敬请广大读者批评指正，译者在此深表谢意。

何泾沙

2016年12月于北京

前　言

“斯考特，把我弹射出去。”

Captain Kirk（William Shatner 饰）
电影《星际旅行 4：抢救未来》（1986 年）

Web 是一个由众多网页相连接而形成的网络，是一个通信媒介，是一个覆盖全球的信息来源。人们花费大量的时间在 Web 上进行搜索，获取有用的数据与信息，并对它们进行分析。有效使用 Web 给人们的生活带来了很多的便利。本书将告诉你以上这一切是如何实现的。

本书是根据我在西北大学（Northwestern University）讲授的一门课程的内容撰写而成的。此课程从介绍 Web 网站分析学入手，主要关注在 Web 搜索中使用数据的统计与性能。之后，我又在此课程中增加了来自网络科学和社交媒体的概念。在讲授此课程两年后，我认识到从 Web 上收集信息可以成为一个独立的话题，有太多关于 Web 与网络数据科学方面的知识可以学习。本书就像我讲授的课程那样，是关于以上这些知识的指南。

Web 与网络数据科学是数据科学和网络科学相结合而形成的，关注的是将 Web 看成一个提供信息的来源。因而，最好的学习方法就是通过实例进行讲解。因此，本书中包含大量的实例，通过提供建模技术方面的资料和参考指南给研究人员与分析师提供帮助。我们也会向编程人员展示如何基于基础代码编写能够正确运行并用于解决真实商业问题的软件。

我们想要做的事情都会通过所编写的代码体现出来。本书中包含的这些代码将作为参考资料提供给每一位读者，当然会有部分读者对这些代码进行进一步调试。为了鼓励学生学习，每一段程序代码都包含详细的注释以及如何进一步分析的建议。所有的数据集以及计算机程序代码都可以直接从本书的网站 http://www.ftpress.com/miller/ 下载。

Python 这个名字来源于 Monty Python。大家会看到有些软件包的名称比较奇特，如 Twisted 或 Scrapy。R 语言拥有自己的 lubridate 与 zoo 软件开发包。好的结果来源于辛勤工作并热爱工作的人们。那些追求快乐而不是名利的人们为开源软件做出了贡献，而我很高兴自己能够成为开源软件 Python 和 R 语言社区中的一员。那就让我们一起开始这段快乐的旅程吧！

对于 Web 和网络中存在的问题，使用 Python 可以有效便捷地解决某些问题，而使用

R 语言可以有效便捷地解决其他一些问题。常常还会出现两种语言都适用的情况，这时就需要进行权衡。总体来说，Python 和 R 语言能够用于对 Web 及网络数据进行有效的收集与分析。

在本书中，我们还会提到编程时会使用到的很多工具。对网站的正常运行负有责任的 Web 专业技术人员还会使用很多其他语言和技术，如 JavaScript、Apache、.Net Web 服务，以及数据库系统。本书的讨论将会涉及这些技术，但不会提供任何编程代码。

本书中大多数数据来源于公共域数据源。用于支持案例的数据来源于加利福尼亚大学尔湾分校的机器学习信息库（Machine Learning Repository）和斯坦福大学的大型网络数据集（Large Network Dataset Collection）。所获取的影视方面的数据得益于互联网影视数据库（Internet Movie Database）所给予的使用许可。IMDb 影视评价数据由斯坦福大学的 Andrew L. Mass 及同事整理完成。安然（Enron）案例数据由卡耐基 – 梅隆大学的 William W. Cohen 维护。Quake Talk（地震谈话）案例数据由 Maksim Tsvetovat 维护。我们对以上这些学者为我们的研究提供了丰富的数据表示深切的感谢。

很多人对我这些年来的知识积累都产生过重大的影响。他们中有出色的思考者，有善良的同仁，还有我会永远感激的老师以及导师。不幸的是，尤西纽斯学院（Ursinus College）哲学系的 Gerald Hahn Hinkle 和语言系的 Allan Lake Rice 以及明尼苏达大学（University of Minnesota）哲学系的 Herbert Feigl 已经永远离开了我们。在此，我还要感谢明尼苏达大学心理测验学系的 David J. Weiss 以及曾经在俄勒冈大学（University of Oregon）经济系任教的 Kelly Eakin。好的老师（没错，他们都是伟大的园丁）终身都将得到人们的尊重。

感谢 Stan Narusiewcz 给了我职业生涯中的第一份工作，那是一个网络工程师的岗位。感谢 Tom Obinger 指导我成为一个成功的计算机系统和网络销售人员。还有 Bill JoBush 和 Brian Hill，在我作为信息系统专业人员整个职业生涯的各个阶段，他们曾经是我的直接上司或同事。

感谢 Michael L. Rothschild、Neal M. Ford、Peter R. Dickson 和 Janet Christopher 在威斯康星大学麦迪逊分校（University of Wisconsin–Madison）伴我一起度过几年美好的时光并给予我无私的帮助。特别感谢 A. C. Nielsen Center for Marketing Research 的学生和顾问委员会的专家以及 Jeff Walkowski 和 Neli Esipova，后两位在我组织在线调查与专题讨论小组期间曾经同我一起工作，我们所使用的方法那时才开始在重要的研究中得到应用。

我很有幸参与了西北大学成人教育学院开展的研究生远程教育的课程教学活动。感谢 Glen Fogerty 给我提供了讲授课程的机会，并让我负责西北大学预测分析学项目。感谢所有参与这个很有特色的研究生项目的同事和管理人员。最后，感谢帮助过我的众多学生们和老师们，你们令我受益匪浅。

ToutBay 是数据科学领域中一个快速成长的公司。与公司的共同创始人 Greg Blence 一样，我对公司的未来发展抱有很大的信心。感谢 Greg 让我有这样一个参与创业以及面对商业活动中的现实而能够更加脚踏实地的机会。学术以及数据科学模型毕竟有其局限性，为了能够真正产生影响，我们必须实现我们的想法和模型，并且与他人进行共享。

我的家在加利福尼亚州，道奇体育馆（Dodger Stadium）以北四英里⊖，但是我在位于伊利诺伊州埃文斯顿市（Evanston, Illinois）的西北大学任教，同时在位于佛罗里达州坦帕市（Tampa, Florida）的一个名叫 ToutBay 的数据科学公司指导产品研发。这样的工作和生活方式充分体现出了互联网带给我们的巨大便利。

TEXnology 公司的 Amy Hendrickson 使本书的编排、文字、图表看上去都是那么出色和完美，这是开源软件的又一个成功实例。感谢 Donald Knuth 以及TEX/LATEX整个社区对这个出色的系统在编排和出版方面做出的贡献。

本书中包含的内容主要源于在西北大学讲授的 Web 与网络数据科学这门课程。参与课程学习的学生提出了很多想法和启示。Lorena Martin 对本书进行了评阅，提供了许多宝贵意见。Candice Bradley 不但评阅了本书，还是本书的文字编辑。我对他们给予的帮助和鼓励表示衷心感谢。最后还要感谢我的编辑 Jeanne Glasser Levine 以及本书的出版商 Pearson/FT Press，是他们使本书的成功出版成为可能。在此特别声明，我个人对所有写作方面的事宜、存在的错误与问题以及不足负全部责任。

我的好朋友 Brittney 和她的女儿 Janiya 总是抽空陪伴我。我的儿子 Daniel 总能与我同甘共苦，是我一辈子的朋友。我对于他们给予的信任致以崇高的敬意。

Thomas W. Miller

美国加利福尼亚州格伦代尔市

⊖ 约 6.4 公里。——编辑注

目　录

第 1 章

相关技术概述

你为什么不抽时间来看我？

Lady Lou（Mae West 饰）
电影《侬本多情》(1933 年)

我的职业生涯起始于明尼苏达州罗斯维尔市的一名网络工程师。从明尼苏达大学统计专业研究生毕业后，我具备了很好的数学和建模方面的基础，但是缺少商业方面的知识。我很快就意识到要在职业发展上取得成功就需要掌握更多管理方面的知识。

20 世纪 70 年代末是一个通过拨号或专线上网的时代，异步通信、半同步通信及同步通信是当时的主流。我们将网络协议通过轮询方式和信息比特位来表达，并标注出每一条通信线路能够承载的传输速率。排队理论（Queuing Theory）及离散事件模拟是开展以上分析的理论基础。

银行里的柜台职员会提交一个请求，然后按下终端上的回车键。终端连接到一个控制器，控制器再连接到远端的一个汇聚处理器。汇聚处理器再通过专线连接到一个前端处理器，从而建立起一个连接到大型计算机的通道。以上就是那个年代网络中的节点和连接。排队理论用于估算银行的柜台职员需要等待多久才能得到大型计算机的响应。

40 年飞快地过去了。我们已经远离了拨号和专线的时代。今天通信协议的基础是数据包的交换以及移动通信。使用网络的用户已经遍及各个角落，而不再只局限于银行、公司和研究机构。绝大多数大型计算机也被小型计算机群所取代。我们的口袋里装着最小的计算机。如果愿意，我们还可以将计算机穿戴在身上。然而，当我们向远端的系统提交请求时，我们仍然需要等待响应，不同的是我们现在可以在任何地方等待，同时可以做些其他的事情。

随着计算机硬件方面的差别逐渐消失以及软件走向开源化，现有的科技公司都在寻找商业智能与数据科学方面的商机。IBM 从一个以生产硬件为主的公司转型成为一个软件开发商，然后又转型成为一家咨询公司。HP（惠普公司）已经一分为二，一部分专营硬件，另一部分则以提供商业服务和工具为主要经营范围。同时，苹果（Apple）与亚马逊（Amazon）和谷歌（Google）在多媒体发布领域中展开激烈竞争，并对三星公司（Samsung）违反软件著作权提起法律诉讼。

今天主要的商业竞争都涉及信息以及信息的在线发布。知识产权、专业知识、竞争智能、专长及艺术都给在线市场增加了很多新的价值，不考虑以上因素的话，在线获取信息基本上都可以免费实现。

人们很难抵挡 Web 的诱惑，因为它为用户提供了拥有无穷无尽信息的空间，还可以提供能够到达所有这些信息的连接。Web 是一个巨大的数据仓库，是一条通向知识的道路，更是一个开发新知识的研究媒介。

Web 与网络数据科学由多项科学技术以及建模技术所组成，其中某些技术已经相当成熟，还有些新技术仍然处在发展与完善当中，这些技术能够帮助我们了解生活中的 Web 及网络。Web 也有多项技术得到应用，Alexa Internet 公司（2014）、W3Techs 公司（2014）等企业专门对各项技术的市场占有率进行跟踪。

为了更有效地开展 Web 与网络数据科学方面的研究，需要具备相应的技术背景，至少需要对 Python、R 语言和 JavaScript（Java 脚本语言）有一定的了解。Python 是一款进行数据预处理的必选工具（有时也称为数据改造）。R 语言是一款专门用于数据建模和可视化的工具。JavaScript 是 Web 客户端开发中使用的一种语言，主流的 Web 浏览器都可以支持。在解决 Web 和网络方面的问题中，以下方面的知识和技能也非常有用：HTML5、CSS3、XPath、各种文本和图像文件格式、Java、Linux、Apache、.Net Web 服务、数据库系统以及服务器端开发中使用的语言，如 Perl、PHP。虽然需要具备上面提到的相关专业知识，但是一本书能够涵盖的内容毕竟有限。因此，我们将会在本书的最后增加一个有关术语表的附录。

自 Brendan Eich 于 1995 年在网景公司（Netscape）工作期间开发出 JavaScript 以来，作为一种开发语言，JavaScript 迅速成为 Web 客户端开发所使用的语言，是一款基于浏览器对用户交互进行管理的引擎。JavaScript 在客户端占有垄断地位，88% 的网站使用该技术，另外 11.8% 的网站仍然使用静态的纯 HTML 技术，无法为客户端提供任何编程能力（W3Techs 2014）。

2008 年，Crockford 总结了 JavaScript 的优缺点，而更多的人则告诉我们如何在实际系统中去使用它（Stefanov 2010; Flanagan 2011; Resig and BearBibeault 2013）。近年来，随着 Node.js 的出现，JavaScript 在服务器端也开始得到应用（Hughes-Croucher and Wilson 2012; Wanderschneider 2013; Cantelon, Harter, Holowaychuk, and Rajlich 2014）。有些人大力推动端到端的 JavaScript 应用程序，即在客户端和服务器端都使用 JavaScript 程序以及文件数据库（Mikowski and Powell 2014）。JavaScript Object Notation（JSON）提供了一种数据交换格式，可读性比 XML 更好，并可以很容易地与任何 MongoDB 文件数据库进行集成（Chodorow 2013; Copeland 2013; Hoberman 2014）。如果具有作为一个建模与分析语言所需要的功能，JavaScript 肯定也能在 Web 领域发挥主导作用。不幸的是，JavaScript 并没有做到。

今天的数据科学领域吸引了能够熟练使用 R 语言的统计专家和能够熟练使用 Python 的信息技术专家。这两类研究人员还有很多需要相互学习之处。对于想要真正开展实际工作的数据科学家来说，熟悉以上两种语言会使他们具备更多的优势。

R 语言由 Ross Ihaka 和 Robert Gentleman 在 1993 年设计并推出，是一款可扩展、面向对象的开源脚本语言，用于通过编程对数据进行处理。此语言在数据统计界得到广泛应用，其语法、数据结构及编程方法与它的前身 S 语言和 S+ 语言相似。此语言的贡献者为广泛应用提供了超过 5 000 个软件开发包，主要提供传统的数据统计、机器学习及数据可视化方面的功能。R 语言目前是数据科学界使用最广泛的编程语言，但它并不是一个通用目的的编程语言。

3

Guido van Rossum（Monty Python 的一个追随者）在 1994 年设计并发布了 Python 1.0 版本。这个通用目的编程语言在随后的年代里慢慢流行起来。很多系统编程人员都从先前使用 Perl 改为使用 Python，Python 尤其获得从事数学和自然科学的研究人员的青睐。很多高等院校将 Python 作为介绍面向对象程序设计语言基本概念的一种手段。一个非常活跃的开源社区贡献了超过 15 000 个 Python 软件开发包。

Python 时常会被誉为一种"胶合语言"，它为科学编程与科研提供了一个非常丰富的开源环境。对于需要占用大量计算机资源的应用程序来说，Python 提供了调用通过正确编译的 C、C++ 和 Fortran 子程序的功能。我们也可以使用 Cython 将 Python 代码转换成优化后的 C 语言代码。对于没有在 Python 中实现的建模技术或图像来说，我们可以在 Python 程序中调用 R 语言程序。

有些问题可以使用 Python 很容易地解决，其他一些问题则可以使用 R 语言很容易地解决。Python 作为一个通用目的编程语言给我们提供了很多便利，并能通过使用 R 语言编程软件包得到传统的数据统计、时间序列分析、多元方法、统计图表制作以及丢失数据处理方面的功能。因此，本书将包含 Python 和 R 语言的代码实例，是一部在 Web 与网络数据科学领域同时使用这两种语言的指导书籍。

浏览器的使用在发展过程中经历了很大的变化，随着 Google（谷歌）Chrome 浏览器市场份额的增加，微软（Microsoft）Internet Explorer（IE）的使用呈现下降趋势。表 1-1 和图 1-1 展现的是 2008 年 10 月～2014 年 10 月全球浏览器的使用情况，图表中的数据来源于 StarCounter (2014)。熟练使用浏览器以及由浏览器提供的查阅文本元素和网页结构的工具非常有帮助。

表 1-1　全球 Web 浏览器占有率统计（2008～2014）

年	IE	Chrome	Firefox	Safari	其　他
2008	67.68	1.02	25.54	2.91	2.85
2009	57.96	4.17	31.82	3.47	2.58
2010	49.21	12.39	31.24	4.56	2.60
2011	40.18	25.00	26.39	5.93	2.50
2012	32.08	34.77	22.32	7.81	3.02
2013	28.96	40.44	18.11	8.54	3.95
2014	19.25	47.57	17.00	10.95	5.23

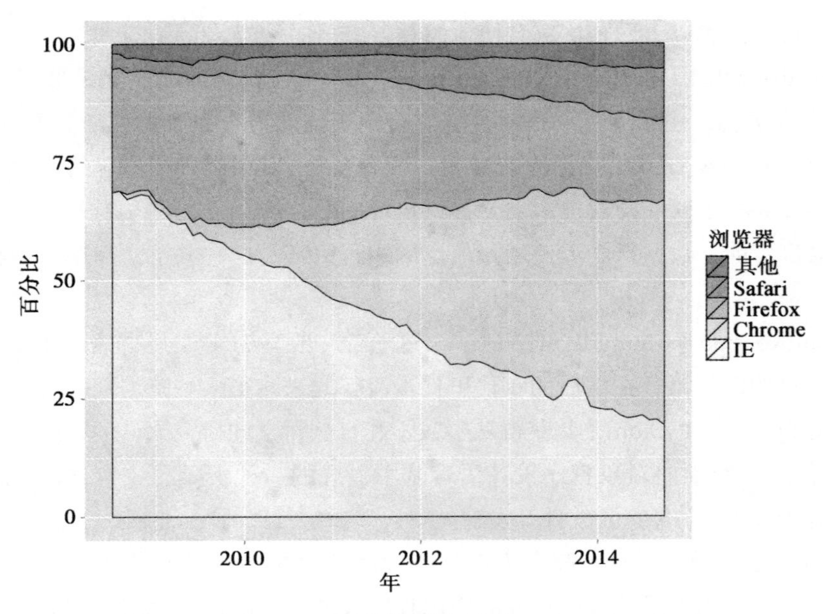

图 1-1 全球 Web 浏览器使用情况（2008.7～2014.10）

“大数据”带来的挑战不仅仅是数据的数量，而是我们对数据产生的来源缺乏足够的了解，特别是 Web 以及社交媒体。数据存在于 Web 的各个角落。我们需要能够找到有关数据并获取这些数据的高效方法。

应用程序编程接口（API）是一个从 Web 获取数据的方法。Russell（2014）对社交媒体 API 做了一个整体回顾。不幸的是，调用 API 对语法、参数及授权码都有要求，数据提供者可以随意进行修改。我们采取的方法有所不同，关注重点是从 Web 自动获取数据的通用技术。

4

图 1-2 对在线研究过程进行了总结。数据采样、收集及准备会耗费大量的时间，使二次研究超过初步研究而处于主导地位。在线二次研究以 Web 中已经存在的数据为基础。我们将在第 3 章介绍二次研究方法，这些方法将在随后的章节中应用。在线初步研究得到了 Web 的支持，我们将会在附录 B 中对相关的方法进行介绍。

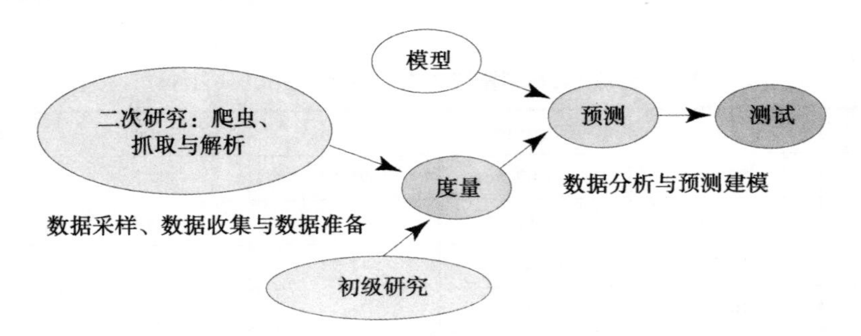

图 1-2 Web 与网络数据科学：在线研究过程

Web 与网络数据科学的范围很大，有众多问题亟待解决，主要问题如下。

- **网站设计与用户行为**。正如很多人理解的那样，Web分析学涉及对某个特定网站的用户数据进行收集、存储与分析。这就带来了很多需要解决的问题。如何设计和实现网站（达到易用、可视性好、市场宣传、高效搜索、将访问转化为销售等目的）？如何从Web中高效获得信息？如何将半结构化和非结构化的文本转换为可以用于分析和建模的数据？哪些方法适合对网站及社交媒体进行度量？谁在使用网站，他们如何使用网站？网站提供的服务是否满足用户的需要？如何将一个网站与其他网站进行比较？

5~6

- **网络路径与通信**。Web与网络数据科学远远超出网站分析学的范畴。我们对网站的审视或评价都是相对于Web中的其他网站。我们围绕着网络进行思考，即提供信息的节点之间相互连接、用户之间相互通信。那么，两个节点之间的最短路径、最快路径或成本最低路径在哪里？通过网络传递消息最快的方法是什么？在完成一个项目的关键路径上都有哪些活动？从服务器得到响应需要多长时间？
- **社区与影响**。电子社交网络从社交媒体可见一斑。在这个领域中，存在很多关于社交网络分析方面的问题。在某个社区中是否存在可识别的用户群体？哪些用户是群体中关键或最重要的参与者？拥有声望、影响力或权利的用户在哪里？哪些用户在成长为群体领导者的过程中处于最有利的位置？
- **个体与群体行为**。作为一个研究数据的科研人员，要求做到的不仅仅是描述现象而已，而是要对将来的行为或性能进行预测。因此，有更多的问题需要我们解决。已知一个用户与其他购买者或非购买者之间的关系，那么这个用户也会购买吗？已知一个用户与其他投票人的关系，这个用户也会投票给某位候选人吗？已知个人动机，一个群体接下来会做什么呢？已知网络过去的发展，我们预测网络将来会怎样发展呢？
- **信息与网络**。作为信息的来源，Web是无可比拟的。在线信息属性方面的问题由而产生。哪些网站是获取某方面信息的最好来源？谁是提供信息的最可靠来源？如何归纳出领域知识的特征？如何使用Web获得具有竞争力的智能？如何像使用数据库一样将基于Web的信息用来回答问题（专业问题及通用问题）？

本书的目的就是对Web与网络数据科学领域进行概括性介绍。我们将介绍为了回答问题而需要使用的度量与建模技术，并提供进一步学习的参考资料。介绍的技术从基础到超前，然而它们对于数据科学研究来说都非常重要。

7

某些人认为所谓的数据科学其实就是一些新的统计方法。在一个由数据主宰的世界里，数据科学同时似乎开始提供新的商机以及新的信息技术。在需要解决Web和网络问题时，后面这一点显得尤为重要。由Web产生和传输的无穷无尽的数据将会使这些研究持续相当长一段时间。

作为本书的首个软件程序，代码1-1是一段展示Web浏览器使用统计数据的Python程序。代码1-2是使用R语言编写的与之相对应的程序，其中用到了由Wickham和Chang（2014）编写的制图软件。

8

代码 1-1 浏览器使用情况分析（Python）

```
# Analysis of Browser Usage (Python)

# prepare for Python version 3x features and functions
from __future__ import division, print_function

# import packages for data analysis
import pandas as pd  # data structures for time series analysis
import datetime  # date manipulation
import matplotlib.pyplot as plt

# browser usage data from StatCounter Global Stats
# retrieved from the World Wide Web, October 21, 2014:
# \url{http://gs.statcounter.com/#browser-ww-monthly-200807-201410
# read in comma-delimited text file
browser_usage = pd.read_csv('browser_usage_2008_2014.csv')
# examine the data frame object
print(browser_usage.shape)
print(browser_usage.head())

# identify date fields as dates with apply and lambda function
browser_usage['Date'] = \
    browser_usage['Date']\
    .apply(lambda d: datetime.datetime.strptime(str(d), '%Y-%m'))
# define Other category
browser_usage['Other'] = 100 -\
    browser_usage['IE'] - browser_usage['Chrome'] -\
    browser_usage['Firefox'] - browser_usage['Safari']

# examine selected columns of the data frame object
selected_browser_usage = pd.DataFrame(browser_usage,\
    columns = ['Date', 'IE', 'Chrome', 'Firefox', 'Safari', 'Other'])
print(selected_browser_usage.shape)
print(selected_browser_usage.head())

# create multiple time series plot
selected_browser_usage.plot(subplots = True,  \
     sharex = True, sharey = True, style = 'k-')
plt.legend(loc = 'best')
plt.xlabel('')
plt.savefig('fig_browser_mts_Python.pdf',
    bbox_inches = 'tight', dpi=None, facecolor='w', edgecolor='b',
    orientation='portrait', papertype=None, format=None,
    transparent=True, pad_inches=0.25, frameon=None)

# Suggestions for the student:
# Explore alternative visualizations of these data.
# Try the Python package ggplot to reproduce R graphics.
# Explore time series for other software and systems.
```

9

代码 1-2 浏览器使用情况分析（R 语言）

```
# Analysis of Browser Usage (R)

# begin by installing necessary package ggplot2

# load package into the workspace for this program
library(ggplot2)  # grammar of graphics plotting

# browser usage data from StatCounter Global Stats
# retrieved from the World Wide Web, October 21, 2014:
# \url{http://gs.statcounter.com/#browser-ww-monthly-200807-201410
```

```
# read in comma-delimited text file
browser_usage <- read.csv("browser_usage_2008_2014.csv")
# examine the data frame object
print(str(browser_usage))
# define Other category
browser_usage$Other <- 100 -
    browser_usage$IE - browser_usage$Chrome -
    browser_usage$Firefox - browser_usage$Safari

# define time series data objects
IE_ts <- ts(browser_usage$IE, start = c(2008, 7), frequency = 12)
Chrome_ts <- ts(browser_usage$Chrome, start = c(2008, 7), frequency = 12)
Firefox_ts <- ts(browser_usage$Firefox, start = c(2008, 7), frequency = 12)
Safari_ts <- ts(browser_usage$Safari, start = c(2008, 7), frequency = 12)
Other_ts <- ts(browser_usage$Other, start = c(2008, 7), frequency = 12)

# create a multiple time series object
browser_mts <- cbind(IE_ts, Chrome_ts, Firefox_ts, Safari_ts, Other_ts)
dimnames(browser_mts)[[2]] <- c("IE", "Chrome", "Firefox", "Safari", "Other")
# plot multiple time series object using standard R graphics
pdf(file="fig_browser_mts_R.pdf",width = 11,height = 8.5)
ts.plot(browser_mts, ylab = "Percent Usage", main="",
    plot.type = "single", col = 1:5)
legend("topright", colnames(browser_mts), col = 1:5,
    lty = 1, cex = 1)
dev.off()

# define Year as numeric with fractional values for months
browser_usage$Year <- as.numeric(time(IE_ts))

# build data frame for plotting a stacked area graph
Browser <- rep("IE", length = nrow(browser_usage))
Percent <- browser_usage$IE
Year <- browser_usage$Year
plotting_data_frame <- data.frame(Browser, Percent, Year)

Browser <- rep("Chrome", length = nrow(browser_usage))
Percent <- browser_usage$Chrome
Year <- browser_usage$Year
plotting_data_frame <- rbind(plotting_data_frame,
    data.frame(Browser, Percent, Year))
Browser <- rep("Firefox", length = nrow(browser_usage))
Percent <- browser_usage$Firefox
Year <- browser_usage$Year
plotting_data_frame <- rbind(plotting_data_frame,
    data.frame(Browser, Percent, Year))

Browser <- rep("Safari", length = nrow(browser_usage))
Percent <- browser_usage$Safari
Year <- browser_usage$Year
plotting_data_frame <- rbind(plotting_data_frame,
    data.frame(Browser, Percent, Year))

Browser <- rep("Other", length = nrow(browser_usage))
Percent <- browser_usage$Other
Year <- browser_usage$Year
plotting_data_frame <- rbind(plotting_data_frame,
    data.frame(Browser, Percent, Year))

# create ggplot plotting object and plot to external file
pdf(file = "fig_browser_usage_stacked_area_R.pdf", width = 11, height = 8.5)
area_plot <- ggplot(data = plotting_data_frame,
    aes(x = Year, y = Percent, fill = Browser)) +
    geom_area(colour = "black", size = 1, alpha = 0.4) +
```

10

```
    scale_fill_brewer(palette = "Blues",
        breaks = rev(levels(plotting_data_frame$Browser))) +
    theme(legend.text = element_text(size = 15))  +
    theme(legend.title = element_text(size = 15)) +
    theme(axis.title = element_text(size = 15))
print(area_plot)
dev.off()
```

11

第2章
在线传递消息

“你的问候打动了我。”

Dorothy Boyd（Renée Zellweger 饰）
电影《甜心先生》(1996年)

我个人的管理方式很简单：雇佣最优秀的人，让他们独立做事。由于不需要花时间告诉别人该做些什么，我就有更多的时间做我自己喜欢的事情。要使这种粗犷的管理方式产生好的效果，就需要将公司的发展远景清楚地传递给公司的员工，并且需要员工对发展远景充满信心。

消息传递对于在线工作来说尤为重要。在缺少面对面会议的情况下，我们完全依赖电子通信方式。我们将信息置于 Web 之中，希望让更多的人收到并了解其内容。我们等待来自于在线社区的响应。

我们可以对服务器端或客户端消息传递的成功与否进行测量。服务器端保存着用户请求的记录，包含的信息可以完整地显示出请求的来源（IP 地址的位置及用户使用的浏览器）、请求访问的网页以及访问请求的状态。时间戳指明了请求的到达时间和服务提供时间。服务器端记录能够反映出的是以服务器为中心的网站性能评价。

如果想要从用户的角度去了解一个网站，我们只能求助于网络服务提供商。当前，Google（谷歌）是对用户端进行网络监控这个领域的市场领导者[⊖]。

对客户端进行性能监控要求网站提供者给予明确的配合以及网站访问者的隐式配合。选择使用基于 Google 客户端监控的机构还必须将 JavaScript 代码嵌入在需要进行监控的每一个网页中。

网站管理员可以选择是否使用 Google 网站监控。大多数网站管理员都选择使用免费提供的基本服务。目前全球约有 49.8% 的网站使用 Google 网站监控服务，涵盖了 81.5% 的网

⊖ 截至 2014 年 10 月，Google 的 Web 监控产品包括 Google Analytics 和 Google Universal Analytics。Google 免费提供这些产品的基本版本。Google 同时也提供收费的高级版本。我们对客户端 Web 监控的讨论将局限在 Google Analytics 基本版本范围内。

站流量（W3Techs 2014）。

当用户或客户端收到从网站返回的一个网页时，客户端的浏览器将执行嵌入在网页中的跟踪代码，其结果是在客户的计算机上放置一个 cookie，并将此 cookie 随同其他信息（包括用户的 IP 地址）一起发送给 Google。用户的计算机将保存这个 cookie 文本文件以备将来使用。因而，当用户再一次访问同一个网站时，Google 可以识别出该用户是一个重复访问者。cookie 将一直保存在客户的计算机上，直至用户将其删除。

要正确运行对客户端 Web 的监控，用户必须允许将多个 cookie 存储在客户端的设备上，这些设备可以是计算机或平板电脑，也可以是智能手机。大多数 Web 用户都同意使用 cookie。也许他们对 cookie 本身并不理解，也许他们不知道如何禁用 cookie，而更有可能的是他们喜欢 cookie 带来的个性化的和良好的 Web 体验。

对于参与的网站和用户来说，Google 对每一个网站的每一次访问都进行跟踪。使用 Google 客户端监控服务后，网站管理员能够获得大量的数据。

附录 C 中的“开启 ToutBay 之旅”案例包括 Google 提供的网站性能数据。这些数据总结了 ToutBay 网站在最初 23 周运行期间的网站流量情况（2014.4.12～2014.9.19）。通过对这些数据进行分析，我们可以知道网站访问者使用浏览器和操作系统的情况，分别显示在

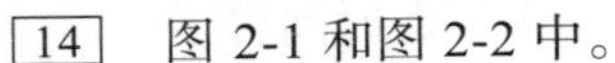

14 图 2-1 和图 2-2 中。

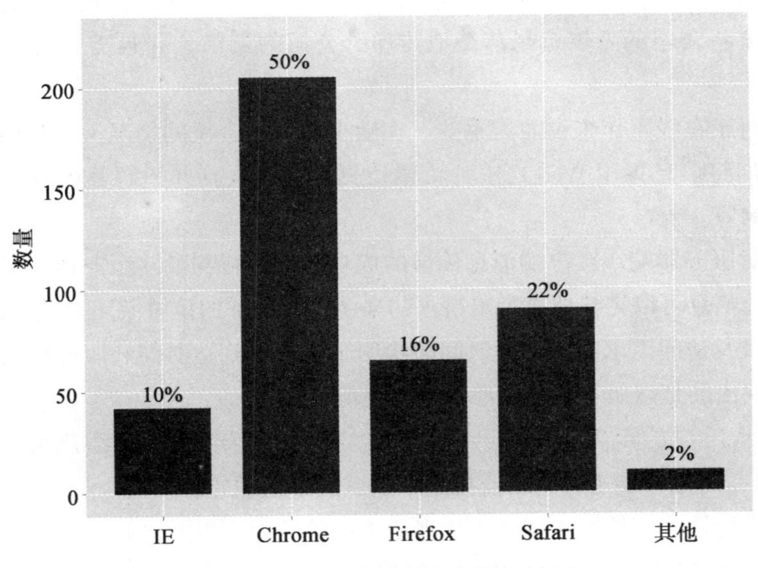

图 2-1 网站访问者使用浏览器情况

在对网站流量的监控中，我们可以特别标示出会话个数、网页浏览量以及会话平均持续时间（以秒计），如图 2-3 所示。这些数据显示出，在国际 R 语言用户大会（International R Users’ Conference（UseR!））期间，对 ToutBay 网站的访问量达到了一个峰值。ToutBay 是该次国际会议的一个赞助者，并在会场设置了展台。

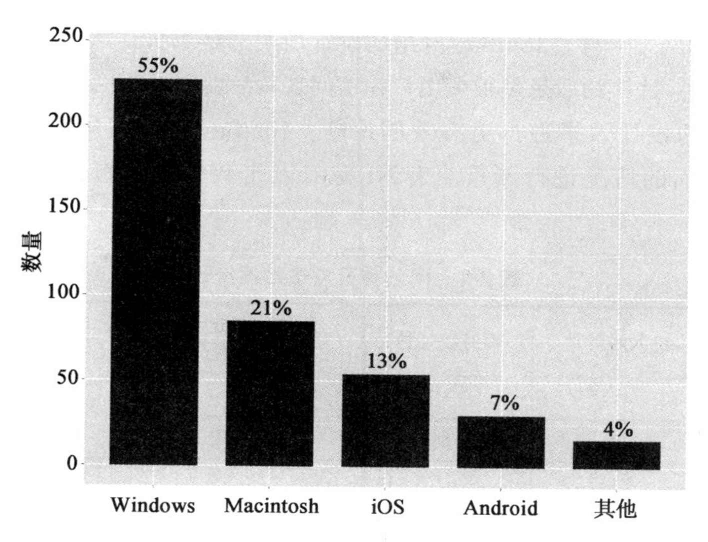

图 2-2 网站访问者使用操作系统情况

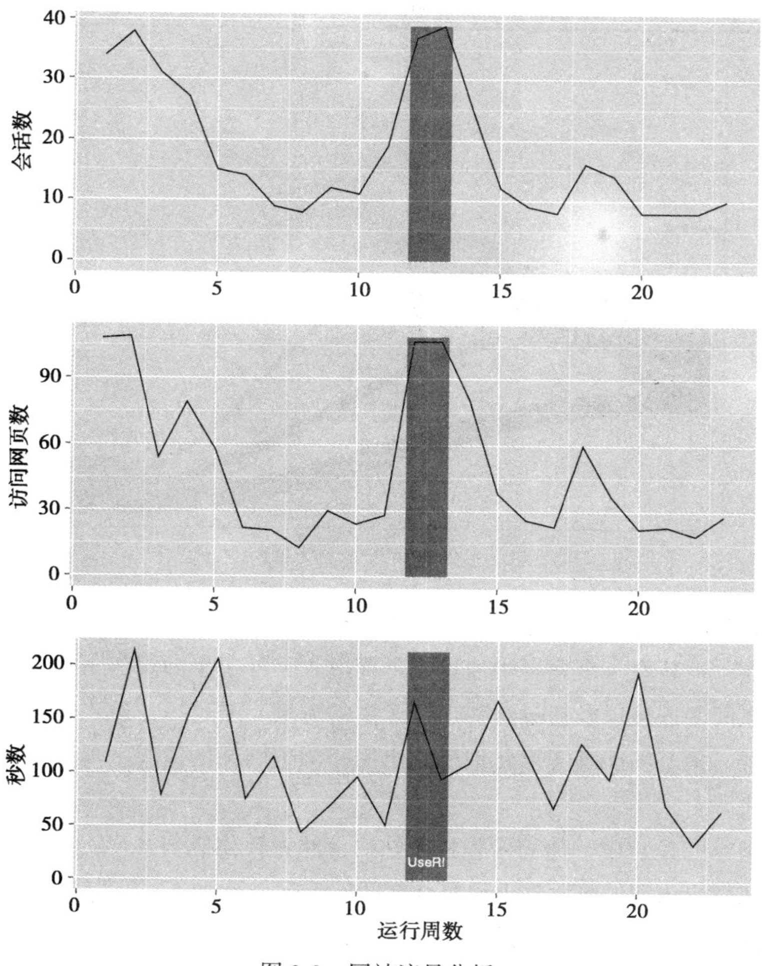

图 2-3 网站流量分析

与许多时尚的网站一样，ToutBay网站的设计使用的是单页式风格，有关公司的大量信息都提供在首页上。这样的后果是很多用户在访问首页后就离开了网站。

为了对网站的设计效果进行更有效的诊断，ToutBay编写代码对用户在首页上点击进入其他网站页面的情况进行测量。表2-1是对点击数据的分析，图2-4是相应的桑基（Sankey）图。

表2-1 网站首页点击数据统计

点击处	进入统计数	离开统计数	从前一点击点进入的访问者百分比数	访问者会话总数的百分比数
首页顶端	412			
ToutBay Video	296	116	72	72
What's ToutBay?	234	62	79	57
How does it work?	193	41	82	47
FAQ	174	19	90	42
News	125	49	72	30

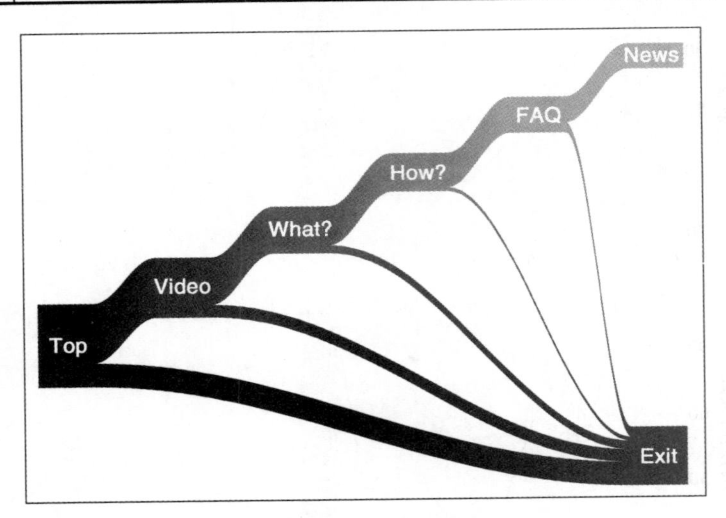

图2-4 首页点击的桑基图

为了绘制桑基图（通常也称为过程图或河流图），我们将网站上的网页以及网页内的网页作为网络的节点，把从在一个网页中通过点击而进入另外一个网页作为节点之间的连接，以此来识别出节点间的连接以及通过这些连接产生的流量。通过每一条连接的流量利用此连接的粗细度来体现。桑基图能够图形化地表达出用户浏览网站的情况，可以用于对网站的结构及性能进行评估。

在线电商追求的是订单，他们希望将网站的访问者转化成为客户。我们因此相信在线电商会非常关心从用户访问的页面到下单页面之间的连接。

在Web监控讨论中较少涉及的另外一个话题就是用户隐私。只要允许cookie在计算机、平板电脑或智能手机中存在，像Google这样的网站服务提供商就可以看到用户发往被监控网站的每一个请求。同时，不论是否使用 cookie，网站管理员也可以通过服务器上的记录看到用户发往此网站的每一个访问请求。用户的个人隐私取决于能够对 Web 活动进行控制、管理及监控的个人或机构。

某项研究表明，99% 的 Web 用户允许 JavaScript 代码运行，85% 的 Web 用户允许 JavaScript 代码以及直接交互方和其他第三方设置 cookie（Priebe 2009）。这些统计数据也许会随着用户对 cookie 及 IP 地址有了更多的了解而发生变化。

15 ~ 18

欧盟电子隐私指令（European Union's e-Privacy Directive）要求只有事先得到用户的同意才能在用户的计算机上设置 cookie，除非所提供的服务必须要使用 cookie（European Parliament 2002; Wikipedia 2014b）。IP 地址可以被认为是一个能够对个体进行识别的信息。在某些国家（包括德国），在没有得到用户的明确允许下收集 IP 地址信息属于违法行为。因此，在这些国家开展商务活动，必须要对 IP 地址进行匿名化处理（Clifton 2012）。

代码 2-1 是用于对 ToutBay 网站的客户端数据进行分析而编写的 R 语言程序。这段代码来源于由 Auguie (2014), Grolemund and Wickham (2014), Neuwirth (2014), Weiner (2014) 和 Wickham and Chang (2014) 提供的 R 语言软件代码包。

19

代码 2-1　网站流量分析（R 语言）

```
# Website Traffic Analysis (R)

# begin by installing necessary packages

# load packages into the workspace for this program
library(gridExtra)  # grid plotting utilities
library(ggplot2)  # grammar of graphics plotting
library(lubridate)  # date and time functions
library(riverplot)  # Sankey diagrams
library(RColorBrewer)  # colors for plots

# user-defined function to convert hh:mm:ss to seconds
make_seconds <- function(hhmmss) {
    hhmmss_list <- strsplit(hhmmss, split = ":")
    3600 * as.numeric(hhmmss_list[[1]][1]) +
        60 * as.numeric(hhmmss_list[[1]][2]) +
        as.numeric(hhmmss_list[[1]][3])
    }

# read in data from ToutBay Begins case
toutbay_begins <- read.csv("toutbay_begins.csv", stringsAsFactors = FALSE)

# examine the data frame object
print(str(toutbay_begins))

# set date as date object
toutbay_begins$date <- parse_date_time(toutbay_begins$date, "mdy")

# convert ave_session_duration to ave_session_seconds, total_session_seconds
toutbay_begins$ave_session_seconds <- numeric(nrow(toutbay_begins))
for (i in seq(along = toutbay_begins$ave_session_duration))
    toutbay_begins$ave_session_seconds[i] <-
        make_seconds(toutbay_begins$ave_session_duration[i])
# compute total seconds across all sessions in the day
toutbay_begins$total_session_seconds <-
    toutbay_begins$ave_session_seconds * toutbay_begins$sessions
```

```
# 161 days = 23 weeks so we can index by weeks
week <- NULL
for (i in 1:23) week <- c(week, rep(i, times = 7))
toutbay_begins$week <- week

# compute other_browser browser counts
toutbay_begins$other_browser <- toutbay_begins$sessions -
    toutbay_begins$chrome - toutbay_begins$safari -
    toutbay_begins$firefox - toutbay_begins$internet_explorer

# compute other_system operating system counts (Linux and others)
toutbay_begins$other_system <- toutbay_begins$sessions -
    toutbay_begins$windows - toutbay_begins$macintosh -
    toutbay_begins$ios - toutbay_begins$android
# extract daily counts and totals
toutbay_daily <- toutbay_begins[,
    c("date", "week", "sessions", "users", "pageviews", "scroll_videopromo",
      "scroll_whatstoutbay", "scroll_howitworks", "scroll_faq",
       "scroll_latestfeeds", "internet_explorer", "chrome", "firefox",
       "safari", "other_browser", "windows", "macintosh", "ios", "android",
       "other_system", "total_session_seconds")]
# examine the daily data frame
print(str(toutbay_daily))
print(head(toutbay_daily))

# aggregate by week using sum()
toutbay_weekly <-
    aggregate(toutbay_daily[, setdiff(names(toutbay_daily), c("date","week"))],
    by = list(toutbay_daily$week), FUN = sum)
names(toutbay_weekly)[1] <- "week"  # rename first column of data frame

# compute average session duration in seconds
toutbay_weekly$ave_session_seconds <-
    toutbay_weekly$total_session_seconds / toutbay_weekly$sessions

# examine the weekly data frame
print(str(toutbay_weekly))
print(head(toutbay_weekly))

# create browser data frame for plotting
Browser <- "IE"
Count <- sum(toutbay_weekly$internet_explorer)
browser_data_frame <- data.frame(Browser, Count)
Browser <- "Chrome"
Count <- sum(toutbay_weekly$chrome)
browser_data_frame <- rbind(browser_data_frame,
    data.frame(Browser, Count))
Browser <- "Firefox"
Count <- sum(toutbay_weekly$firefox)
browser_data_frame <- rbind(browser_data_frame,
    data.frame(Browser, Count))
Browser <- "Safari"
Count <- sum(toutbay_weekly$safari)
browser_data_frame <- rbind(browser_data_frame,
    data.frame(Browser, Count))
Browser <- "Other"
Count <- sum(toutbay_weekly$other_browser)
browser_data_frame <- rbind(browser_data_frame,
    data.frame(Browser, Count))

pdf(file = "fig_toutbay_begins_user_browsers_R.pdf", width = 11, height = 8.5)
browser_bar_plot <- ggplot(data = browser_data_frame,
    aes(x = Browser, y = Count)) +
    geom_bar(stat = "identity", width = 0.75,
        colour = "black", fill = "darkblue") +
```

20

```
    ylim(0, 225) +
    theme(axis.title.y = element_text(size = 15, colour = "black")) +
    theme(axis.title.x = element_blank()) +
    theme(axis.text.x = element_text(size = 15, colour = "black")) +
    annotate("text", x = 1, y = browser_data_frame$Count[1] + 5,
        label = paste(as.character(round(100 * browser_data_frame$Count[1]/
            sum(browser_data_frame$Count), digits = 0)), "%", sep = "")) +
    annotate("text", x = 2, y = browser_data_frame$Count[2] + 5,
        label = paste(as.character(round(100 * browser_data_frame$Count[2]/
            sum(browser_data_frame$Count), digits = 0)), "%", sep = "")) +
    annotate("text", x = 3, y = browser_data_frame$Count[3] + 5,
        label = paste(as.character(round(100 * browser_data_frame$Count[3]/
            sum(browser_data_frame$Count), digits = 0)), "%", sep = "")) +
    annotate("text", x = 4, y = browser_data_frame$Count[4] + 5,
        label = paste(as.character(round(100 * browser_data_frame$Count[4]/
            sum(browser_data_frame$Count), digits = 0)), "%", sep = "")) +
    annotate("text", x = 5, y = browser_data_frame$Count[5] + 5,
        label = paste(as.character(round(100 * browser_data_frame$Count[5]/
            sum(browser_data_frame$Count), digits = 0)), "%", sep = ""))
print(browser_bar_plot)
dev.off()

# create operating system data frame for plotting
System <- "Windows"
Count <- sum(toutbay_weekly$windows)
system_data_frame <- data.frame(System, Count)
System <- "Macintosh"
Count <- sum(toutbay_weekly$macintosh)
system_data_frame <- rbind(system_data_frame,
    data.frame(System, Count))
System <- "iOS"
Count <- sum(toutbay_weekly$ios)
system_data_frame <- rbind(system_data_frame,
    data.frame(System, Count))
System <- "Android"
Count <- sum(toutbay_weekly$android)
system_data_frame <- rbind(system_data_frame,
    data.frame(System, Count))
System <- "Other"
Count <- sum(toutbay_weekly$other_system)
system_data_frame <- rbind(system_data_frame,
    data.frame(System, Count))

pdf(file = "fig_toutbay_begins_user_systems_R.pdf", width = 11, height = 8.5)
system_bar_plot <- ggplot(data = system_data_frame,
    aes(x = System, y = Count)) +
    geom_bar(stat = "identity", width = 0.75,
        colour = "black", fill = "darkblue") +
    ylim(0, max(system_data_frame$Count + 15)) +
    theme(axis.title.y = element_text(size = 15, colour = "black")) +
    theme(axis.title.x = element_blank()) +
    theme(axis.text.x = element_text(size = 15, colour = "black")) +
    annotate("text", x = 1, y = system_data_frame$Count[1] + 5,
        label = paste(as.character(round(100 * system_data_frame$Count[1]/
            sum(system_data_frame$Count), digits = 0)), "%", sep = "")) +
    annotate("text", x = 2, y = system_data_frame$Count[2] + 5,
        label = paste(as.character(round(100 * system_data_frame$Count[2]/
            sum(system_data_frame$Count), digits = 0)), "%", sep = "")) +
    annotate("text", x = 3, y = system_data_frame$Count[3] + 5,
        label = paste(as.character(round(100 * system_data_frame$Count[3]/
            sum(system_data_frame$Count), digits = 0)), "%", sep = "")) +
    annotate("text", x = 4, y = system_data_frame$Count[4] + 5,
        label = paste(as.character(round(100 * system_data_frame$Count[4]/
            sum(system_data_frame$Count), digits = 0)), "%", sep = "")) +
    annotate("text", x = 5, y = system_data_frame$Count[5] + 5,
```

21
22

```
        label = paste(as.character(round(100 * system_data_frame$Count[5]/
            sum(system_data_frame$Count), digits = 0)), "%", sep = ""))
print(system_bar_plot)
dev.off()

# plot multiple time series for sessions, pageviews, and session duration
pdf(file = "fig_toutbay_begins_site_stats_R.pdf", width = 8.5, height = 11)
sessions_plot <- ggplot(data = toutbay_weekly,
    aes(x = week, y = sessions)) + geom_line()  +
    ylab("Sessions") +
    theme(axis.title.x = element_blank()) +
    annotate("rect", xmin = 11.75, xmax = 13.25,
        ymin = 0, ymax = max(toutbay_weekly$sessions),
        fill = "blue", alpha = 0.4)
pageviews_plot <- ggplot(data = toutbay_weekly,
    aes(x = week, y = pageviews)) + geom_line() +
    ylab("Page Views") +
    theme(axis.title.x = element_blank()) +
    annotate("rect", xmin = 11.75, xmax = 13.25,
        ymin = 0, ymax = max(toutbay_weekly$pageviews),
        fill = "blue", alpha = 0.4)

duration_plot <- ggplot(data = toutbay_weekly,
    aes(x = week, y = ave_session_seconds)) + geom_line() +
    xlab("Week of Operation") +
    ylab("Seconds") +
    theme(axis.title.x = element_text(size = 15, colour = "black")) +
    annotate("rect", xmin = 11.75, xmax = 13.25,
        ymin = 0, ymax = max(toutbay_weekly$ave_session_seconds),
        fill = "blue", alpha = 0.4) +
    annotate("text", x = 12.5, y = 20, size = 4, colour = "white",
        label = "UseR!")

mts_plot <- grid.arrange(sessions_plot, pageviews_plot,
    duration_plot, ncol = 1, nrow = 3)
print(mts_plot)
dev.off()

# construct Sankey diagram for home page scrolling
pdf(file = "fig_toutbay_begins_sankey_R.pdf", width = 8.5, height = 11)
nodes <- data.frame(ID = c("A","B","C","D","E","F","G"),
    x = c(1, 2, 3, 4, 5, 6, 6),
    y = c(7, 7.5, 8, 8.5, 9, 9.5, 6),
    labels = c("Top",
               "Video",
               "What?",
               "How?",
               "FAQ",
               "News",
               "Exit"),
               stringsAsFactors = FALSE,
               row.names = c("A","B","C","D","E","F","G"))
edges <- data.frame(N1 = c("A","B","C","D","E",
                           "A","B","C","D","E"),
                    N2 = c("G","G","G","G","G",
                           "B","C","D","E","F"),
    Value = c(
    sum(toutbay_weekly$sessions)- sum(toutbay_weekly$scroll_videopromo),
    sum(toutbay_weekly$scroll_videopromo) -
        sum(toutbay_weekly$scroll_whatstoutbay),
    sum(toutbay_weekly$scroll_whatstoutbay) -
        sum(toutbay_weekly$scroll_howitworks),
    sum(toutbay_weekly$scroll_howitworks) - sum(toutbay_weekly$scroll_faq),
    sum(toutbay_weekly$scroll_faq) - sum(toutbay_weekly$scroll_latestfeeds),
    sum(toutbay_weekly$scroll_videopromo),
```

23

```
    sum(toutbay_weekly$scroll_whatstoutbay),
    sum(toutbay_weekly$scroll_howitworks),
    sum(toutbay_weekly$scroll_faq),
    sum(toutbay_weekly$scroll_latestfeeds)), row.names = NULL)

selected_pallet <- brewer.pal(9, "Blues")
river_object <- makeRiver(nodes, edges,
    node_styles =
    list(A = list(col = selected_pallet[9], textcol = "white"),
    B = list(col = selected_pallet[8], textcol = "white"),
    C = list(col = selected_pallet[7], textcol = "white"),
    D = list(col = selected_pallet[6], textcol = "white"),
    E = list(col = selected_pallet[5], textcol = "white"),
    F = list(col = selected_pallet[5], textcol = "white"),
    G = list(col = "darkred", textcol = "white")))
plot(river_object, nodewidth = 4, srt = TRUE)
dev.off()
```

24

第 3 章
Web 爬行与抓取

Danny：“你必须先走再爬。”

Rusty：“请把顺序颠倒过来。”

Danny Ocean（George Clooney 饰）、Rusty Ryan（Brad Pitt 饰）

电影《十一罗汉》（2001 年）

也许是由于年纪的原因，也许是由于我急切地想减肥而养成的锻炼习惯，也许是由于我拒绝服用任何带有刺激素的药物，包括咖啡或茶。不管是什么原因，我很容易打瞌睡。

对我来说，在白天当中躺下来睡觉，2、3 个小时后醒来，然后疑惑时间都去了哪里，这已经是司空见惯的事情。我怀疑我打瞌睡的行为不正常，然而又无法确定这是不是不健康的信号。我需要关注打瞌睡的毛病吗？我不知道。我认为我需要了解一下我这个年纪的男人的睡眠习惯。

我可以花几小时的时间上网去寻找关于瞌睡行为方面的信息。我可以去浏览卫生和健美方面的网站。我可以去寻求医疗方面的咨询及查找医疗方面的研究成果。然而，要想系统地学习瞌睡方面的知识（在打瞌睡时也可以那么做），我会去做 Web 爬行与抓取。然后，我会对收集到的数据进行分析，使最终的结果很容易解读。

25

爬行比网上冲浪的速度要快，因为爬行是自动进行的。一个爬行软件或蜘蛛软件所做的远远不只是收集一个 Web 页面上的数据。像蜘蛛网里的蜘蛛那样，万维网中的爬行软件或蜘蛛软件可以很容易地穿过已知链接，并顺着这些链接从一个网页跳转到另一个网页，收集来源于多个数据源的信息。

Google、微软、雅虎（Yahoo!）这一类的公司都会对整个互联网进行爬行，并为它们各自的搜索引擎构建索引。我们所使用的爬行软件则只做局限爬行，我们称其为聚焦式爬行。

聚焦式爬行从一个或多个点开始，通常由一组相关的 Web 地址构成。我们在爬行之前会先进行浏览，目的是找到这些起始页面。表 3-1 中罗列了使用一个聚焦式爬行软件获得有关瞌睡行为方面的信息所选择的 10 个网站的地址。

表 3-1 聚焦式爬行使用的 Web 地址

简　称	描述及 Web 地址
AMA	美国医学协会 美国医学协会杂志（JAMA, Sleep） http://jama.jamanetwork.com/solr/searchresults.aspx?q=sleep&fd_JournalID=67&f_JournalDisplayName=JAMA&SearchSourceType=3
HARVARD1	哈佛大学医学院（新闻发布） http://www.health.harvard.edu/press_releases/snoozing-without-guilt--a-daytime-nap-can-be-good-for-health/
HARVARD2	哈佛大学医学院（健康睡眠） http://healthysleep.med.harvard.edu/healthy/science/variations/changes-in-sleep-with-age
MAYO1	梅奥医学教育与研究基金会 http://www.mayoclinic.org/
MAYO2	梅奥医学教育与研究基金会（瞌睡） http://www.mayoclinic.org/napping/ART-20048319?p=1
NIH1	国立卫生研究院 http://nih.gov/
SLEEP	国立睡眠基金会 http://sleepfoundation.org/sleep-topics/napping
HHS	美国卫生与人事服务部（睡眠与瞌睡） http://healthfinder.gov/search/?q=sleep+and+napping
WEBMD	WebMD（老龄化与睡眠） http://www.webmd.com/sleep-disorders/guide/aging-affects-sleep
WIKIPNAP	Wikipedia（瞌睡） http://en.wikipedia.org/wiki/Nap

聚焦式爬行也会终止。我们将爬行限制在指定领域的范围内。我们可以在爬行软件下载了指定领域中一定数量的页面后将其停止。我们可以设定下载页面的有关规则。另外，我们也可以限定每一次特定爬行下载页面的总数。

每一个聚焦式爬行都有一个特定的目标。因此，我们会保留满足条件的网页。针对当前的任务，即收集关于瞌睡与健康方面的信息，我们使用关键词对网页进行选择："睡眠""瞌睡"及"年龄"。

有了很好的网页爬行，下一项任务就是从网页中识别或抽取出我们想要寻找的特定信息。每一个网页都包含 HTML 标记，用于定义网页节点的层次结构（或称为树形结构）。我们将如此构建的树称作文档对象模型（Document Object Model，DOM）。Python 和 R 语言包都支持的 XPath 定义了一种专门的语法，用于对 DOM 中的节点及相关属性进行遍历，以抽取相关的数据。我们从节点中获取数据，例如从具有段落标记的节点中获取文本数据。

在从网页中获取了文本数据后，我们可以使用正则表达式对这些数据进行分析，定义网页格式的代码、无用的空格、标点符号等可以从文档中删除。任何一次成功的爬行在结束时

26～27

都应该获取了一个完整的文档集，以用于文本分析。

图3-1是一个Web自动数据获取框架，与Hoffman et al.（2014）开发的Python Web爬行与抓取软件一致。我们首先爬行，然后识别，最后再分析。这是基于Web的二次研究需要的工作。下一步，我们为文本选择一个合适的包装以及相对应的元数据，也许就是JSON或XML。我们为随后要进行的分析构建一个文档仓库或文本集合。可扩展的企业级数据库，如PostgreSQL或MongoDB，以及基于Hadoop分布式文件系统所构建的数据仓库都可以用于实现这个目标。

28

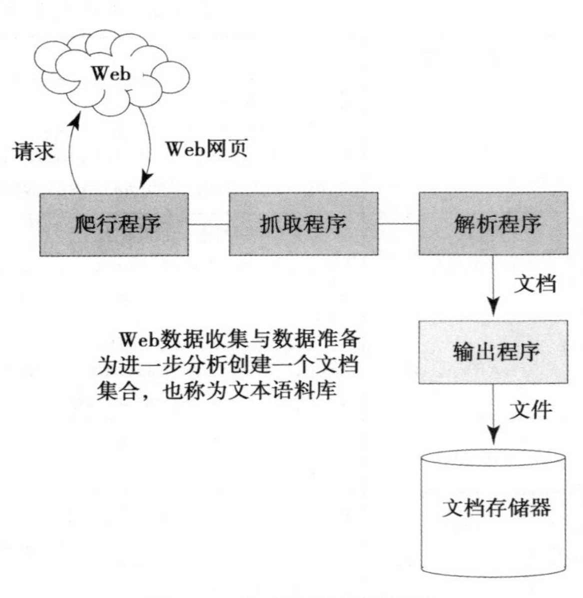

图3-1 自动数据获取框架

爬行与抓取是Web与网络数据科学的基础。建模与分析都是从数据开始，而Web是一个巨大的数据仓库。学习如何通过使用有效的方法获取有用的数据是数据科学的一项核心技能。

对于学习XML以及使用XPath对HTML网页进行识别来说，已经有大量的优秀参考资料（Tennison 2001; Simpson 2002; Kay 2008; Nolan and Lang 2014）。Friedl（2006）包含了文本分析中的正则表达式方面的内容。关于更多聚焦式Web爬行及数据挖掘方面的讨论，可以参考Chakrabarti（2003）和Liu（2011）。

代码3-1是一个从ToutBay网站的首页中抽取出HTML数据的Python程序。该程序通过使用XPath语法对HTML进行解析，从具有段落标记的节点中抽取出文本数据。这个程序来源于Behnel（2014），Reitz（2014b）和Richardson（2014）开发的Python应用软件包。代码3-2是相对应的R语言程序，来源于Lang（2014a, 2014b）开发的应用软件包。

代码3-1 抽取及识别网站数据（Python）

```
# Extracting and Scraping Web Site Data (Python)

# prepare for Python version 3x features and functions
from __future__ import division, print_function

# import packages for web scraping/parsing
import requests  # functions for interacting with web pages
import lxml   # functions for parsing HTML
from bs4 import BeautifulSoup  # DOM html manipulation

# -----------------------------------------------
# demo using the requests and lxml packages
# -----------------------------------------------
# test requests package on the home page for ToutBay
web_page = requests.get('http://www.toutbay.com/', auth=('user', 'pass'))
# obtain the entire HTML text for the page of interest
```

```
# show the status of the page... should be 200 (no error)
web_page.status_code
# show the encoding of the page... should be utf8
web_page.encoding

# show the text including all of the HTML tags... lots of tags
web_page_text = web_page.text
print(web_page_text)

# parse the web text using html functions from lxml package
# store the text with HTML tree structure
web_page_html = lxml.html.fromstring(web_page_text)

# extract the text within paragraph tags using an lxml XPath query
# XPath // selects nodes anywhere in the document  p for paragraph tags
web_page_content = web_page_html.xpath('//p/text()')
# show the resulting text string object
print(web_page_content)  # has a few all-blank strings
print(len(web_page_content))
print(type(web_page_content))  # a list of character strings

# ------------------------------------------------------------
# demo of scraping HTML with beautiful soup instead of lxml
# ------------------------------------------------------------
my_soup = BeautifulSoup(web_page_text)
# note that my_soup is a BeautifulSoup object
print(type(my_soup))

# remove JavaScript code from Beautiful Soup page object
# using a comprehension approach
[x.extract() for x in my_soup.find_all('script')]
# gather all the text from the paragraph tags within the object
# using another list comprehension
soup_content = [x.text for x in my_soup.find_all('p')]
# show the resulting text string object
print(soup_content)  # note absence of all-blank strings
print(len(soup_content))
print(type(soup_content))  # a list of character strings

# there are carriage return, line feed characters, and spaces
# to delete from the text... but we have extracted the essential
# content of the toutbay.com home page for further analysis
```

29 ~ 31

代码 3-2 抽取及识别网站数据（R 语言）

```
# Extracting and Scraping Web Site Data (R)

# install required packages

# bring packages into the workspace
library(RCurl)  # functions for gathering data from the web
library(XML)  # XML and HTML parsing

# gather home page for ToutBay using RCurl package
web_page_text <- getURLContent('http://www.toutbay.com/')

# show the class of the R object and encoding
print(attributes(web_page_text))

# show the text including all of the HTML tags... lots of tags
print(web_page_text)

# scrape the HTML DOM into an internal C data structure for XPath processing
web_page_tree <- htmlTreeParse(web_page_text, useInternalNodes = TRUE,
```

```
    asText = TRUE, isHTML = TRUE)
print(attributes(web_page_tree))

# extract the text within paragraph tags using an XPath query
# XPath // selects nodes anywhere in the document  p for paragraph tags
web_page_content <- xpathSApply(web_page_tree, "//p/text()")
print(attributes(web_page_content))
print(head(web_page_content))
print(tail(web_page_content))

# send content to external text file for review
sink("text_file_for_review.txt")
print(web_page_content)
sink()

# there are node numbers, line feed characters, and spaces
# to delete from the text... but we have extracted the essential
# content of the toutbay.com home page for further analysis
```

32

代码3-3展示的是如何使用Python对一个单独的网页进行识别，其中用到Hoffman et al. (2014) 提出的Python框架以及Lefkowitz et al. (2014) 开发的网络应用软件库。

代码 3-3　简单的单页Web识别软件（Python）

```
# Simple One-Page Web Scraper (Python)
#
# prepare for Python version 3x features and functions
from __future__ import division, print_function

# scrapy documentation at http://doc.scrapy.org/

# workspace directory set to outer folder/directory wnds_chapter_3b
# the operating system commands in this example are Mac OS X

import scrapy  # object-oriented framework for crawling and scraping
import os  # operating system commands

# function for walking and printing directory structure
def list_all(current_directory):
    for root, dirs, files in os.walk(current_directory):
        level = root.replace(current_directory, '').count(os.sep)
        indent = ' ' * 4 * (level)
        print('{}{}/'.format(indent, os.path.basename(root)))
        subindent = ' ' * 4 * (level + 1)
        for f in files:
            print('{}{}'.format(subindent, f))


# initial directory should have this form (except for items beginning with .):
#    wnds_chapter_3b
#        run_one_page_scraper.py
#        scrapy.cfg
#        scrapy_application/
#            __init__.py
#            items.py
#            pipelines.py
#            settings.py
#            spiders
#                __init__.py
#                one_page_scraper.py

# examine the directory structure
```

```
current_directory = os.getcwd()
list_all(current_directory)

# list the avaliable spiders, showing names to be used for crawling
os.system('scrapy list')

# decide upon the desired format for exporting output: csv, JSON, or XML
# run the scraper exporting results as a comma-delimited text file items.csv
os.system('scrapy crawl TOUTBAY -o items.csv')
# run the scraper exporting results as a JSON text file items.json
os.system('scrapy crawl TOUTBAY -o items.json')
# run the scraper exporting results as a dictionary XML text file items.xml
os.system('scrapy crawl TOUTBAY -o items.xml')
```

33

```
# ----------------------------
# MyItem class defined by
# items.py
# ----------------------------

# location in directory structure:
# wnds_chapter_3b/scrapy_application/items.py

# establishes data fields for scraped items

import scrapy  # object-oriented framework for crawling and scraping

class MyItem(scrapy.item.Item):
    # define the data fields for the item (just one field used here)
    paragraph = scrapy.item.Field()  # paragraph content


# ----------------------------
# MyPipeline class defined by
# pipelines.py
# ----------------------------

# location in directory structure:
# wnds_chapter_3b/scrapy_application/pipelines.py

class MyPipeline(object):
    def process_item(self, item, spider):
        return item


# ----------------------------
# settings for scrapy.cfg
# settings.py
# ----------------------------

# location in directory structure:
# wnds_chapter_3b/scrapy_application/settings.py

BOT_NAME = 'MyBot'
BOT_VERSION = '1.0'

SPIDER_MODULES = ['scrapy_application.spiders']
NEWSPIDER_MODULE = 'scrapy_application.spiders'
USER_AGENT = '%s/%s' % (BOT_NAME, BOT_VERSION)

COOKIES_ENABLED = False
DOWNLOAD_DELAY = 2
RETRY_ENABLED = False
DOWNLOAD_TIMEOUT = 15
REDIRECT_ENABLED = False
DEPTH_LIMIT = 50
```

34

```
# ----------------------------
# spider class defined by
# script one_page_scraper.py
# ----------------------------
# location in directory structure:
# wnds_chapter_3b/scrapy_application/spiders/one_page_scraper.py

# prepare for Python version 3x features and functions
from __future__ import division, print_function

# each spider class gives code for crawing and scraping

import scrapy  # object-oriented framework for crawling and scraping
from scrapy_application.items import MyItem  # item class

# spider subclass inherits from BaseSpider
# this spider is designed to crawl just one page of one website
class MySpider(scrapy.spider.BaseSpider):
    name = "TOUTBAY"  # unique identifier for the spider
    allowed_domains = ['toutbay.com']  # limits the crawl to this domain list
    start_urls = ['http://www.toutbay.com']  # first url to crawl in domain

    # define the scraping method for the spider
    # note that this function is called "parse," but we are actually scraping
    def parse(self, response):
        html_scraper = scrapy.selector.HtmlXPathSelector(response)
        divs = html_scraper.select('//div')  # identify all <div> nodes
        # XPath syntax to grab all the text in paragraphs in the <div> nodes
        results = []  # initialize list
        this_item = MyItem()  # use this item class
        this_item['paragraph'] = divs.select('.//p').extract()
        results.append(this_item)  # add to the results list
        return results

# Suggestions for the student: Use scrapy to scrape another web page,
# extracting additional DOM content, such as <a> link text and links.
# Utilize new links to move from a one-page scraper to a more complete
# crawler that goes from one page to the next within a web domain.
# Use regular expressions to parse the text from this focused crawl.
```

35

在Python框架之上，我们来展示如何构建一个聚焦式爬行软件（亲切地称呼为“十头蜘蛛”），用于从我们之前已经找到的网站上收集关于睡眠、瞌睡和年龄方面的信息。代码3-4是完整的用于爬行、识别及分析的代码。

代码3-4 瞌睡时进行爬行及识别（Python）

```
# Crawling and Scraping while Napping (Python)
#
# Focused Crawl with a Ten-Headed Spider Using the Scrapy Framework
#
# prepare for Python version 3x features and functions
from __future__ import division, print_function

# scrapy documentation at http://doc.scrapy.org/

# workspace directory set to outer folder/directory wnds_chapter_3b
# the operating system commands in this example are Mac OS X

import scrapy  # object-oriented framework for crawling and scraping
import os  # operating system commands

# function for walking and printing directory structure
```

```
def list_all(current_directory):
    for root, dirs, files in os.walk(current_directory):
        level = root.replace(current_directory, '').count(os.sep)
        indent = ' ' * 4 * (level)
        print('{}{}/'.format(indent, os.path.basename(root)))
        subindent = ' ' * 4 * (level + 1)
        for f in files:
            print('{}{}'.format(subindent, f))

# initial directory should have this form (except for items beginning with .):
#     wnds_chapter_3c
#         run_ten_headed_spider.py
#         scrapy.cfg
#         scrapy_application/
#             __init__.py
#             items.py
#             pipelines.py
#             settings.py
#             spiders
#                 __init__.py
#                 ten_headed_spider.py

# examine the directory structure
current_directory = os.getcwd()
list_all(current_directory)

# list the avaliable spiders, showing names to be used for crawling
os.system('scrapy list')

# decide upon the desired format for exporting output: csv, JSON, or XML

# here we employ JSON for each of the ten sites being crawled
# we run each spider subclass separately so that stored results
# may be identified with the website being crawled
# Test crawl
os.system('scrapy crawl TEST -o results_TEST.json')

# American Medical Association
os.system('scrapy crawl AMA -o results_AMA.json')

# Harvard Medical School (Press Releases)
os.system('scrapy crawl HARVARD1 -o results_HARVARD1.json')
# Harvard Medical School (Healthy Sleep)
os.system('scrapy crawl HARVARD2 -o results_HARVARD2.json')

# Mayo Foundation for Medical Education and Research
os.system('scrapy crawl MAYO1 -o results_MAYO1.json')
# Mayo Foundation for Medical Education and Research (Napping)
os.system('scrapy crawl MAYO2 -o results_MAYO2.json')

# National Institutes of Health
os.system('scrapy crawl NIH -o results_NIH.json')
# National Sleep Foundation
os.system('scrapy crawl SLEEP -o results_SLEEP.json')
# U.S. Department of Health and Human Services (Sleep and Napping)
os.system('scrapy crawl HHS -o results_HHS.json')
# WebMD (Aging and Sleep)
os.system('scrapy crawl WEBMD -o results_WEBMD.json')
# Wikipedia (Nap)
os.system('scrapy crawl WIKINAP -o results_WIKINAP.json')

# ----------------------------
# MyItem class defined by
# items.py
# ----------------------------
```

36

```
# location in directory structure:
# wnds_chapter_3c/scrapy_application/items.py

# establishes data fields for scraped items

import scrapy  # object-oriented framework for crawling and scraping

class MyItem(scrapy.item.Item):
    # define the data fields for the item (just one field used here)
    paragraph = scrapy.item.Field()  # paragraph content

# ----------------------------
# MyPipeline class defined by
# pipelines.py
# ----------------------------
# location in directory structure:
# wnds_chapter_3c/scrapy_application/pipelines.py

class MyPipeline(object):
    def process_item(self, item, spider):
        return item
# ----------------------------
# settings for scrapy.cfg
# settings.py
# ----------------------------
# location in directory structure:
# wnds_chapter_3c/scrapy_application/settings.py

BOT_NAME = 'MyBot'
BOT_VERSION = '1.0'

SPIDER_MODULES = ['scrapy_application.spiders']
NEWSPIDER_MODULE = 'scrapy_application.spiders'
USER_AGENT = '%s/%s' % (BOT_NAME, BOT_VERSION)

COOKIES_ENABLED = False
DOWNLOAD_DELAY = 2
RETRY_ENABLED = False
DOWNLOAD_TIMEOUT = 15
REDIRECT_ENABLED = False
DEPTH_LIMIT = 50

# ----------------------------
# spider class defined by
# script ten_headed_spider.py
# ----------------------------
# location in directory structure:
# wnds_chapter_3c/scrapy_application/spiders/ten_headed_spider.py

# prepare for Python version 3x features and functions
from __future__ import division, print_function

# each spider class gives code for crawing and scraping
import scrapy  # object-oriented framework for crawling and scraping
from scrapy_application.items import MyItem  # item class

# each spider subclass inherits from BaseSpider
# each spider subclass is designed to crawl one website
# each spider can have its own parsing logic based on the
# DOM of the website being crawled snd scraped...
class MySpiderTEST(scrapy.spider.BaseSpider):
    name = "TEST"  # unique identifier for the spider
    allowed_domains = ['toutbay.com']  # limits the crawl to this domain list
    start_urls = ['http://www.toutbay.com']  # first url to crawl in domain
    # define the parsing method for the spider
```

37

```
    def parse(self, response):
        html_scraper = scrapy.selector.HtmlXPathSelector(response)
        divs = html_scraper.select('//div')  # identify all <div> nodes
        # XPath syntax to grab all the text in paragraphs in the <div> nodes
        results = []  # initialize list
        this_item = MyItem()  # use this item class
        this_item['paragraph'] = divs.select('.//p').extract()
        results.append(this_item)  # add to the results list
        return results
```

38

```
# American Medical Association
class MySpiderAMA(scrapy.spider.BaseSpider):
    name = "AMA"  # unique identifier for the spider
    # limit the crawl to this domain list
    allowed_domains = ['ama-assn.org']
    # first url to crawl in domain
    start_urls = ['http://jama.jamanetwork.com/solr/searchresults.aspx?\
        q=sleep&fd_JournalID=67&f_JournalDisplayName=JAMA&SearchSourceType=3']
    # define the parsing method for the spider
    def parse(self, response):
        html_scraper = scrapy.selector.HtmlXPathSelector(response)
        divs = html_scraper.select('//div')  # identify all <div> nodes
        # XPath syntax to grab all the text in paragraphs in the <div> nodes
        results = []  # initialize list
        this_item = MyItem()  # use this item class
        this_item['paragraph'] = divs.select('.//p').extract()
        results.append(this_item)  # add to the results list
        return results


# Harvard Medical School (Press Releases)
class MySpiderHARVARD1(scrapy.spider.BaseSpider):
    name = "HARVARD1"  # unique identifier for the spider
    # limits the crawl to this domain list
    allowed_domains = ['health.harvard.edu']
    # first url to crawl in domain
    start_urls = ['http://www.health.harvard.edu/press_releases/\
        snoozing-without-guilt--a-daytime-nap-can-be-good-for-health']
    # define the parsing method for the spider
    def parse(self, response):
        html_scraper = scrapy.selector.HtmlXPathSelector(response)
        divs = html_scraper.select('//div')  # identify all <div> nodes
        # XPath syntax to grab all the text in paragraphs in the <div> nodes
        results = []  # initialize list
        this_item = MyItem()  # use this item class
        this_item['paragraph'] = divs.select('.//p').extract()
        results.append(this_item)  # add to the results list
        return results


# Harvard Medical School (Healthy Sleep)
class MySpiderHARVARD2(scrapy.spider.BaseSpider):
    name = "HARVARD2"  # unique identifier for the spider
    allowed_domains = ['med.harvard.edu']  # limits the crawl to this domain list
    # first url to crawl in domain
    start_urls = ['http://healthysleep.med.harvard.edu/healthy/science/\
        variations/changes-in-sleep-with-age']
    # define the parsing method for the spider
    def parse(self, response):
        html_scraper = scrapy.selector.HtmlXPathSelector(response)
        divs = html_scraper.select('//div')  # identify all <div> nodes
        # XPath syntax to grab all the text in paragraphs in the <div> nodes
        results = []  # initialize list
        this_item = MyItem()  # use this item class
        this_item['paragraph'] = divs.select('.//p').extract()
```

39

```
        results.append(this_item)  # add to the results list
        return results

# Mayo Foundation for Medical Education and Research
class MySpiderMAYO1(scrapy.spider.BaseSpider):
    name = "MAYO1"  # unique identifier for the spider
    # limit the crawl to this domain list
    allowed_domains = ['mayoclinic.org']
    # first url to crawl in domain
    start_urls = ['http://www.mayoclinic.org']
    # define the parsing method for the spider
    def parse(self, response):
        html_scraper = scrapy.selector.HtmlXPathSelector(response)
        divs = html_scraper.select('//div')  # identify all <div> nodes
        # XPath syntax to grab all the text in paragraphs in the <div> nodes
        results = []  # initialize list
        this_item = MyItem()  # use this item class
        this_item['paragraph'] = divs.select('.//p').extract()
        results.append(this_item)  # add to the results list
        return results

# Mayo Foundation for Medical Education and Research (Napping)
class MySpiderMAYO2(scrapy.spider.BaseSpider):
    name = "MAYO2"  # unique identifier for the spider
    # limit the crawl to this domain list
    allowed_domains = ['mayoclinic.org']
    # first url to crawl in domain
    start_urls = ['http://www.mayoclinic.org/napping/ART-20048319?p=1']
    # define the parsing method for the spider
    def parse(self, response):
        html_scraper = scrapy.selector.HtmlXPathSelector(response)
        divs = html_scraper.select('//div')  # identify all <div> nodes
        # XPath syntax to grab all the text in paragraphs in the <div> nodes
        results = []  # initialize list
        this_item = MyItem()  # use this item class
        this_item['paragraph'] = divs.select('.//p').extract()
        results.append(this_item)  # add to the results list
        return results

# National Institutes of Health
class MySpiderNIH(scrapy.spider.BaseSpider):
    name = "NIH"  # unique identifier for the spider
    allowed_domains = ['nih.gov']  # limits the crawl to this domain list
    start_urls = ['http://nih.gov']  # first url to crawl in domain
    # define the parsing method for the spider
    def parse(self, response):
        html_scraper = scrapy.selector.HtmlXPathSelector(response)
        divs = html_scraper.select('//div')  # identify all <div> nodes
        # XPath syntax to grab all the text in paragraphs in the <div> nodes
        results = []  # initialize list
        this_item = MyItem()  # use this item class
        this_item['paragraph'] = divs.select('.//p').extract()
        results.append(this_item)  # add to the results list
        return results

# National Sleep Foundation
class MySpiderSLEEP(scrapy.spider.BaseSpider):
    name = "SLEEP"  # unique identifier for the spider
    # limit the crawl to this domain list
    allowed_domains = ['sleepfoundation.org']
    # first url to crawl in domain
    start_urls = ['http://sleepfoundation.org/sleep-topics/napping']
    # define the parsing method for the spider
    def parse(self, response):
        html_scraper = scrapy.selector.HtmlXPathSelector(response)
```

40

```
        divs = html_scraper.select('//div')  # identify all <div> nodes
        # XPath syntax to grab all the text in paragraphs in the <div> nodes
        results = []  # initialize list
        this_item = MyItem()  # use this item class
        this_item['paragraph'] = divs.select('.//p').extract()
        results.append(this_item)  # add to the results list
        return results

# U.S. Department of Health and Human Services (Sleep and Napping)
class MySpiderHHS(scrapy.spider.BaseSpider):
    name = "HHS"  # unique identifier for the spider
    # limit the crawl to this domain list
    allowed_domains = ['healthfinder.gov']
    # first url to crawl in domain
    start_urls = ['http://healthfinder.gov/search/?q=sleep+and+napping']
    # define the parsing method for the spider
    def parse(self, response):
        html_scraper = scrapy.selector.HtmlXPathSelector(response)
        divs = html_scraper.select('//div')  # identify all <div> nodes
        # XPath syntax to grab all the text in paragraphs in the <div> nodes
        results = []  # initialize list
        this_item = MyItem()  # use this item class
        this_item['paragraph'] = divs.select('.//p').extract()
        results.append(this_item)  # add to the results list
        return results

# WebMD (Aging and Sleep)
class MySpiderWEBMD(scrapy.spider.BaseSpider):
    name = "WEBMD"  # unique identifier for the spider
    # limit the crawl to this domain list
    allowed_domains = ['webmd.com']
    # first url to crawl in domain
    start_urls =\
        ['http://www.webmd.com/sleep-disorders/guide/aging-affects-sleep']
    # define the parsing method for the spider
    def parse(self, response):
        html_scraper = scrapy.selector.HtmlXPathSelector(response)
        divs = html_scraper.select('//div')  # identify all <div> nodes
        # XPath syntax to grab all the text in paragraphs in the <div> nodes
        results = []  # initialize list
        this_item = MyItem()  # use this item class
        this_item['paragraph'] = divs.select('.//p').extract()
        results.append(this_item)  # add to the results list
        return results


# Wikipedia (Nap)
class MySpiderWIKINAP(scrapy.spider.BaseSpider):
    name = "WIKINAP"  # unique identifier for the spider
    allowed_domains = ['wikipedia.org']  # limits the crawl to this domain list
    # first url to crawl in domain
    start_urls = ['http://en.wikipedia.org/wiki/Nap']
    # define the parsing method for the spider
    def parse(self, response):
        html_scraper = scrapy.selector.HtmlXPathSelector(response)
        divs = html_scraper.select('//div')  # identify all <div> nodes
        # XPath syntax to grab all the text in paragraphs in the <div> nodes
        results = []  # initialize list
        this_item = MyItem()  # use this item class
        this_item['paragraph'] = divs.select('.//p').extract()
        results.append(this_item)  # add to the results list
        return results

# Suggestions for the student: Use source code and element inspection
```

41

```
# utilities provided in modern browsers such as Firefox and Chrome
# to examine the DOM of each website being crawled/scraped. Modify the
# scraping logic of the spider for that website as defined in the
# def parse functions for the spider class for that website.
# Add URLs to the spider classes by adding to the start.urls attribute
# of each spider class. Ensure that each website is crawled thoroughly
# using <a> links to additional pages within the website.
# Note that some of the start.urls used in the current spiders merely
# provide links to other sources. We need to drill down into those
# sources to find journal or web articles with titles relating
# to the problem at hand: sleep, napping, and age.
# Run the focused crawl with these def parse enhancements and
# examine the results. Repeat this process as needed.
# Use regular expressions to parse text from the final focused crawl.
# Build a text corpus for subsequent text analysis.
```

那么围绕着十头蜘蛛的故事是什么呢？瞌睡是常见的事情，也并不是不健康的信号。事实上，从国立睡眠基金会得到的数据表明很多人都跟我一样。许多名人，如温斯顿·丘吉尔、约翰·肯尼迪、阿尔伯特·爱因斯坦和托马斯·爱迪生，都是出了名的频繁瞌睡者。

对于瞌睡有利的数据已经非常多，如果愿意，我还可以收集更多的数据。我们所需要做的就是爬行、识别与解析。然而，先去做更重要的事情吧。又到了瞌睡的时间啦。

42

第 4 章

测试链接、外观与体验

Debbie：“哇！好漂亮的褶卷。你知道，我特别喜欢家具上褶卷的感觉。”

Terry：“是吗？”

Debbie：“对呀。”

Terry：“那么就进来吧，我让你摸一下。我的意思是，如果你愿意，你可以来摸一下，嗯，我会让你摸一下这些家具。”

Debbie Dunham（Candy Clark 饰）、Terry“The Toad”
Fields（Charles Martin Smith 饰）
电影《美国风情画》（1973 年）

我父亲曾讲过他学习游泳的故事。他的邻居，一群来自于宾夕法尼亚州 Schwenksville 乡邻里年长的男孩子把他扔进珀基奥门溪后朝他喊道：“要么沉下去，要么游上来。”这真的是一个脱胎换骨式的学习方法。

虽然我儿子丹尼尔与我父亲的名字相同，在威斯康星州麦迪逊市的 Swim West 学习游泳时却有着截然不同的经历。他通过参加从初级到高级十个层次的游泳课学习如何游泳。我亲眼见到这两种极端的教育方法的价值。当我们在学习新的语言或者新的做事方法时，逐步学习的方法很好。但当对这个过程不太了解时，我认为脱胎换骨式的学习会更好。

极端性在很多努力奋斗中都会产生不错的结果，包括在网上做生意。有些公司没有计划好就开始进行网络营销。另外一些公司则咨询了搜索引擎专家，并制定了详细的计划，以实现他们开展网络营销的目标。无论采用哪种途径，数据会告诉我们最终的结果是成功还是失败，数据也会指导我们对网站进行改进。

以下是现任雅虎公司总经理兼总裁 Marissa Mayer 就如何运用数据进行决策的一个故事。在谷歌任职时，Marissa Mayer 曾经因为下令测试 40 个深浅不同的蓝色来确定哪一个蓝色能够获得最高点击率而知名。这个实验解决了信息技术家和图形艺术家之间的纠纷。如今，实验和数据主宰天下，而不是网站设计者的观点和判断。

对网站设计和维护的方方面面进行测试非常关键。现在，让我们从链接、外观和体验三个方面来讨论对网站进行测试的问题。要对网站链接（从其他网站能够到达某个网站的链接的数量和质量）进行测试，我们通过搜索来评价性能。要对网站外观进行测试，我们会对拥有不同内容和设计风格的网页进行评估，看哪一个能够获得最高的点击率或最长的会话持续时间。要对网站体验进行测试，我们可以参考网站的可用性测试以及网站的响应能力，即网站的性能。

Web领域的咨询师可能会声称他们知道答案。但是，在Web快速发展的今天，咨询师的答案可能已经过时。要了解事实而不是建议，并且要及时跟上每一天都会出现的新情况。我们必须度量，而且要持续度量。我们必须做实验，必须测试，事实都存在于数据中。

我们已经讨论了从服务器端到客户端的网站性能，通常每次只专注于一个网站。然而，当讨论搜索时，我们对网站性能的评判则必须与其他网站进行比较。

各类机构都想了解其网站的存在状况，即网站在Web世界的知名度。设计到位的搜索查询会提供非常有用的竞争信息。假如一个研究人员想要评估雅虎网站在梦幻体育方面的知名度，那么这个研究人员对雅虎作为一个知名品牌、雅虎网站本身或者雅虎财经都没有兴趣。一般性的网站统计结果对于雅虎梦幻体育来说没有任何关联性。

44

将雅虎与其他梦幻体育提供商进行比较的一种方法是，针对梦幻体育，构建一个具有代表性的查询。我们可以通过搜索引擎执行这些查询，并观察搜索结果。搜索结果可以分为两大类：有机搜索结果与广告（即付费搜索）。当对链接进行测试时，重点关注的是有机搜索结果，这些才是搜索引擎根据用户的查询匹配出的相关度最高的自然搜索结果。到目前为止，相关性的一个最重要方面就是从其他网页链接到某个页面的链接数。

截至2013年12月，互联网用户总数为28亿人，约占全球总人口的39%。北美洲的互联网用户约为3亿人，占其人口总数的84.9%，欧洲的互联网用户约为5.66亿人，占其人口总数的68.6%（Miniwatts Marketing Group 2014）。以上这些数字转化为每天数十亿的网页搜索，此外还有一些是访问社交媒体以及网络游戏。

通过基于Web的应用获得知名度就像在店铺面前摆放告示牌一样。企业希望被发现，希望用户对企业本身以及所开展的业务予以高度认可。在Web环境中求生存，获得认可是实现销售的第一步。

基于有机搜索形成的网页排名反映了网站成功吸引访客的程度。更高的排名意味着更高的关注度。更多的人找到一个网站意味着更多的人能够看到网站传达的信息。对于电子商务销售商们来说，可以期望的是人们找到网站就意味着销售量。

在浏览器地址栏输入网站地址是直接访问网站的一个途径。但更多的用户是通过搜索引擎访问网站。代替在地址栏中输入www.toutbay.com，用户会在搜索框中输入“toutbay”，然后由搜索引擎来完成后续的工作。用户键入的是搜索查询，得到是搜索响应。

搜索结果通常只提供对链接的简短描述，包括付费链接和有机链接。用户如何使用搜索结果则由很多因素决定，包括搜索结果列表中每个搜索条目的相对排名以及具有更加重要作

用的与查询条件的相关程度。突出的关键字和标题都会对用户的选择产生影响。 45

通用目的的搜索引擎会持续地抓取可寻址到的网页。对于每一个可建立索引的页面，搜索引擎会提取关键字并将网页纳入相应的索引中。关键字构成超文本标记中的文本。关键字通常出现在纯文本域中，或者出现在用于描述音频、视频、数据库和JavaScript代码的源文本中。对于网站提供者来说，被索引是被发现的先决条件。

有些搜索引擎提供者会将用户的查询条件与用于索引的关键字列表中的关键字进行匹配，这些列表还包含与相关网页的链接，这些网页又包含与其他网页的链接。如果要出现在有机搜索的结果中，网页的关键字必须与用户查询条件中的关键字相匹配。这样才能确保搜索结果与用户查询的相关性。

是什么确定了一个网页在有机搜索结果中的排名？在其他条件相同的情况下，关键词密度高的网页排名要高于关键词密度低的网页，而关键词密度是指网页匹配的搜索查询条目数与网页中所有条目数之比。然而，搜索中所有条件都相同的情况很难做到。在很大程度上，一个网页在搜索结果中的排名取决于其他Web页面链接到这个网页的链接数。

搜索算法是得到严密保护的商业机密，其保密程度随不同的搜索引擎提供者以及时间的不同而不尽相同。我们可以从有关搜索的文献中了解很多算法，我们知道这些算法都充分利用了网络的结构，我们也知道PageRank（Brin and Page, 1998）是很多算法的核心，在某些网页中的链接也比在其他一些网页中的链接更加重要。

正如我们评价一个人在社交网络中的重要性或地位是通过这个人知晓他人或与别人的关系一样，我们评价一个网页的重要性也是看它与其他网页之间的链接。想象一下每一个页面都具有某种程度的重要性或与用户查询的相关性。当一个网页链接到其他网页，这些网页也就具有了某种程度的重要性或与用户查询的相关性。PageRank算法将所有这些重要性和相关性都加以考虑。在有机搜索结果中排名最高的网页自然也就是最重要或与用户查询最相关的网页。

当咨询者谈到“搜索引擎优化（SEO）”这个术语时，他们并不是在讨论如何提高搜索引擎的性能，因为这是搜索引擎提供者的责任；他们也不是在讨论我们通常所理解的数据科学或运筹学中的“优化”。搜索引擎优化是关于机构网站存在的问题，表面上看似在讲“搜索引擎优化”，其实真正目的是对链接进行测试，即在搜索环境中测试网站呈现出的效果。 46

要想网站的存在和呈现有很好的效果，需要针对其他网站和社交媒体做大量细致的工作，也要对不同媒体和硬件终端做很多正确的工作。为了抓住大好机会，网站管理员必须耐心地做大量重复性的工作。

作为一个知名的搜索性能培训与咨询公司，SEOMoz公司在2013年对100多名搜索咨询师做了一个问卷调查，以得到他们关于影响搜索性能因素的建议。在所有反馈的10类因素中，链接权威因素（即连接到一个网页的链接数量）被认为是最重要的，排名第二的是社交媒体的推荐，跟随其后的是网页中的核心内容以及关键词。调查结果见表4-1。

表 4-1　影响搜索引擎结果排名的因素

（排名）因素	说　明
（1）网页链接权威特征	对网页的链接进行度量的指标，如连接到某个网页的链接数量（PageRank）
（2）网页层面社交	对社交媒体的推荐进行度量的指标，如 Facebook、Twitter 和 Google+
（3）网页层面的压阵文本	对单独网页中的文本进行度量的指标，包括压阵文本和 URL（关键词位置）
（4）网页层面的关键词使用率	关键词所处的网页部分，包括标题（关键词密度）
（5）网页层面的待定关键词	对网页中非关键词的使用以及非链接进行度量的指标，如页面的长度和加载速度
（6）域链接权威特征	对网页所处的根域中的链接进行度量的指标
（7）域层面的压阵文本	对网页所处的根域中的压阵文本进行度量的指标
（8）域层面的关键词使用	关键词在根域或其子域中如何得到使用
（9）域中的待定关键词	与根域相关的非关键词特征，如域名的长度
（10）域层面的品牌度量指标	具有品牌效应的根域的特征（信誉）

来源：SEOMoz, Inc. (2013) at `http://moz.com/search-ranking-factors`

假设我们现在服务于洛杉矶道奇队，球队的拥有者想要增加观看比赛的观众人数，以提高收入。道奇队是洛杉矶地区两支职业棒球大联盟比赛队中的其中一支，而棒球也只是众多运动项目中的一个，运动也只是众多消遣娱乐活动中的一种。我们如何才能让道奇队的网页在洛杉矶脱颖而出呢？

设想以下这个极端的搜索查询方式：一般查询，如“找出洛杉矶有趣的事情”，相对于具体查询，如“购买道奇队的门票”。如果要使道奇队的主页面出现在搜索结果中，我们可以期望一般查询的有机搜索性能会大大低于具体查询。

在一般和具体之间，还存在很多其他的查询方式。对如何构建搜索查询进行的引导可能会基于对谷歌关键字数据的分析，就像附录 C 中介绍的那样。

47 ～ 48

设想要对搜索 Dodgers（道奇队）和 Angels（天使队）进行对比，我们构建的搜索字符串会指明是棒球比赛，但不会指明是 Dodgers 或者 Angels，也就是说，查询中的词不会偏向 Dodgers 或者 Angels。我们可以对测试实现自动化，使用一系列的搜索请求，然后获得相应的搜索结果。我们可以对这些搜索结果中的字符串进行剔除或分析，依照不同的指标估算性能。

我们可以使用通过 Google AdWords 关键字规划工具生成的数据识别出与棒球门票销售相关的关键词，这是评估 Dodgers 和 Angels 网站的必要条件。Google AdWords 关键字规划工具是 Google 向广告商发布定价模型的一种方式，提供与用户查询进行编码后的文字相对应的关键词，而这些关键词都来源于 Google 的关键字索引。Google 会为每个关键词赋予一个反映竞价和另外一个反映出价的值。竞价值的取值范围为 [0–1]，具体取值取决于愿意为此关键词出价的广告商数量。出价值是 Google 所建议的此关键词每一次被点击的价格（这个价格是很有可能在竞争性出价过程胜出的价格）。关键字规划工具还会对访问量进行预测，然而这些数值并不是可以直接观察到的，而是 Google 提供给广告商的一些假想数值，用以指导他们对过程进行规划。

当关键字规划无法告诉我们用户的搜索行为时，我们可以使用Google提供的用于对网站搜索性能进行测试的关键字。如果通过预测访问量来对关键字进行排名，我们可以确定哪些是用户最可能使用的关键词，而这些都由Google来进行识别、编码及建立索引。假定我们选择与Dodgers和Angels比赛门票搜索关联度最高的30个关键字，然后在搜索引擎的测试脚本中使用这些关键字。

有机搜索排名是评估网站的一项指标，能表明它在在线市场中所处的位置。为了构建一个适合评估的实例，我们分别考虑了使用台式电脑和笔记本电脑的在线用户，并将搜索作为网上的活动。将以上这个设备类型和在线活动作为前提，我们可以构造出一系列的测试实例，每一个测试实例都由发出的请求和搜索引擎所做的响应构成。

搜索性能是实现在网上得到认可这个大目标的一个组成部分。任何机构都想要知道其网站的知名度，而评估网站的知名度不是一个很容易通过度量而解决的问题。除了搜索引擎之外，用户也使用很多其他的应用。

49

在从前的大众媒体时代，设备能够监控观众在电视上看到的所有一切。而在今天，如果要做同样的事情，就需要在不同的设备中对用户的行为进行监控，如电视机、台式电脑、笔记本电脑、平板电脑、手机以及其他穿戴式设备。我们预计用户的在线行为会因为使用设备的不同而有所不同。

我们也同样预计在线行为会因为时间、使用地点以及应用程序的不同而有所不同。用户通过Email、短信和社交媒体与他人进行交流，他们也会听音乐、下载书籍、看视频、打游戏。由于用户现在可以在线完成大部分事情，对一个网站做一个完整的评估需要对用户的日常活动进行行为采样。这是对在线行为进行大规模研究的起步，但也许这就是对一个机构的网站进行充分评估所需要做的。

要想获得有关网站分析的更多信息，请参阅Kaushik (2010)和Clifton (2012)。要想获得设计高搜索性能网站方面的建议，请参阅Moran and Hunt (2009)和Dover (2011)。Croll and Power (2009)对网站监控与性能度量进行了概要介绍。

在网络上得到认可是传播信息和进行销售的第一步（如果你是一个在线零售商）。接下来就是网站设计，这就涉及网站的视觉（审美）和体验（易用性）。关于审美和图形设计，可以参考的资料包括Samara (2007)和Golombisky and Hagen (2013)。关于易用性，可以参考经典著作Steve Krug (2014)：*Don't Make Me Think!: A Common Sense Approach to Web Usability*。⊖

50

代码4-1 面向搜索中性能测试的关键字识别（R语言）

```
# Identifying Keywords for Testing Performance in Search (R)

# begin by installing necessary package RJSONIO

# load package into the workspace for this program
library(RJSONIO)  # JSON to/from R objects
```

⊖ 此书中文版已由机械工业出版社引进出版，书号是：978-7-111-48154-6。——编辑注

```
# read Angels keyword data from Google AdWords Keyword Planner
angels_1 <- read.csv("tickets_angels_arts_entertainment.csv",
    stringsAsFactors = FALSE)
angels_2 <- read.csv("tickets_angels_baseball.csv",
    stringsAsFactors = FALSE)
angels_3 <- read.csv("tickets_angels_sports_entertainment.csv",
    stringsAsFactors = FALSE)
angels_4 <- read.csv("tickets_angels_sports_events_ticketing.csv",
    stringsAsFactors = FALSE)
angels_5 <- read.csv("tickets_angels_sports_fitness.csv",
    stringsAsFactors = FALSE)
angels_6 <- read.csv("tickets_angels_sports.csv",
    stringsAsFactors = FALSE)

# read Dodgers keyword data from Google AdWords Keyword Planner
dodgers_1 <- read.csv("tickets_dodgers_arts_entertainment.csv",
    stringsAsFactors = FALSE)
dodgers_2 <- read.csv("tickets_dodgers_baseball.csv",
    stringsAsFactors = FALSE)
dodgers_3 <- read.csv("tickets_dodgers_sports_entertainment.csv",
    stringsAsFactors = FALSE)
dodgers_4 <- read.csv("tickets_dodgers_sports_events_ticketing.csv",
    stringsAsFactors = FALSE)
dodgers_5 <- read.csv("tickets_dodgers_sports_fitness.csv",
    stringsAsFactors = FALSE)
dodgers_6 <- read.csv("tickets_dodgers_sports.csv",
    stringsAsFactors = FALSE)

# check column names to ensure matches
names(angels_1) == names(angels_2)
names(angels_1) == names(angels_3)
names(angels_1) == names(angels_4)
names(angels_1) == names(angels_5)
names(angels_1) == names(angels_6)
names(angels_1) == names(dodgers_1)
names(angels_1) == names(dodgers_2)
names(angels_1) == names(dodgers_3)
names(angels_1) == names(dodgers_4)
names(angels_1) == names(dodgers_5)
names(angels_1) == names(dodgers_6)

# define simple column names prior to merging data frames
names(angels_1) <- names(angels_2) <- names(angels_3) <-
    names(angels_4) <- names(angels_5) <- names(angels_6) <-
    names(dodgers_1) <- names(dodgers_2) <- names(dodgers_3) <-
    names(dodgers_4) <- names(dodgers_5) <- names(dodgers_6) <-
    c("group", "keyword", "currency",
    "traffic", "october", "november", "december", "january",
    "february", "march", "april", "may", "june", "july",
    "august", "september", "competition", "cpcbid")

# add study category to each record of each data frame
angels_1$study <- rep("Arts and Entertainment", length = nrow(angels_1))
angels_2$study <- rep("Baseball", length = nrow(angels_2))
angels_3$study <- rep("Sports Entertainment", length = nrow(angels_3))
angels_4$study <- rep("Sports Events Ticketing", length = nrow(angels_4))
angels_5$study <- rep("Sports and Fitness", length = nrow(angels_5))
angels_6$study <- rep("Sports", length = nrow(angels_6))
dodgers_1$study <- rep("Arts and Entertainment", length = nrow(dodgers_1))
dodgers_2$study <- rep("Baseball", length = nrow(dodgers_2))
dodgers_3$study <- rep("Sports Entertainment", length = nrow(dodgers_3))
dodgers_4$study <- rep("Sports Events Ticketing", length = nrow(dodgers_4))
dodgers_5$study <- rep("Sports and Fitness", length = nrow(dodgers_5))
dodgers_6$study <- rep("Sports", length = nrow(dodgers_6))
```

51

```
# add team name to each record of each data frame
angels_1$team <- rep("Angels", length = nrow(angels_1))
angels_2$team <- rep("Angels", length = nrow(angels_2))
angels_3$team <- rep("Angels", length = nrow(angels_3))
angels_4$team <- rep("Angels", length = nrow(angels_4))
angels_5$team <- rep("Angels", length = nrow(angels_5))
angels_6$team <- rep("Angels", length = nrow(angels_6))
dodgers_1$team <- rep("Dodgers", length = nrow(dodgers_1))
dodgers_2$team <- rep("Dodgers", length = nrow(dodgers_2))
dodgers_3$team <- rep("Dodgers", length = nrow(dodgers_3))
dodgers_4$team <- rep("Dodgers", length = nrow(dodgers_4))
dodgers_5$team <- rep("Dodgers", length = nrow(dodgers_5))
dodgers_6$team <- rep("Dodgers", length = nrow(dodgers_6))

# combine the data frames
keyword_data_frame <- rbind(angels_1, angels_2, angels_3,
    angels_4, angels_5, angels_6, dodgers_1, dodgers_2,
    dodgers_3, dodgers_4, dodgers_5, dodgers_6)

# drop currency variable because everything is in US dollars
keyword_data_frame <- keyword_data_frame[, -3]   # currency is thrid column

# drop cases with missing values
keyword_data_frame <- na.omit(keyword_data_frame)

# examine the structure of the data frame
print(str(keyword_data_frame))

# select Sports category for both Dodgers and Angels
sports_data_frame <- subset(keyword_data_frame,
    subset = (study == "Sports"))
print(str(sports_data_frame))
# check on the keywords used for Sports (sports_data_frame)
with(sports_data_frame, print(table(keyword)))  # many not relevant

# distribution of cost-per-click bids
with(sports_data_frame, plot(density(cpcbid)))  # a few very high values

# relationship between traffic and cost-per-click bids
# weak positive relationship
with(sports_data_frame,
    cat("\n\nCorrelation between traffic and suggested CPC bid:",
        cor(traffic, cpcbid)))
with(sports_data_frame, plot(traffic, cpcbid))

# relationship between competition and cost-per-click bids
# moderate positive relationship
with(sports_data_frame,
    cat("\n\nCorrelation between competitors and CPC and suggested CPCbid:",
        cor(competition, cpcbid)))
with(sports_data_frame, plot(competition, cpcbid))

# select Baseball category for both Dodgers and Angels
baseball_data_frame <- subset(keyword_data_frame,
    subset = (study == "Baseball"))
print(str(baseball_data_frame))
# check on the keywords used for Baseball (baseball_data_frame)
with(baseball_data_frame, print(table(keyword)))  # many not relevant

# traffic estimates for keyword: "baseball tickets for sale"
# note identical values for Dodgers and Angels
baseball_tickets_for_sale_data_frame <- subset(baseball_data_frame,
        subset = (keyword == "baseball tickets for sale"))
print(baseball_tickets_for_sale_data_frame)
```

52

```
# traffic estimates for keyword: "dodgers tickets"
# note identical values for Dodgers and Angels
dodgers_tickets_data_frame <- subset(baseball_data_frame,
        subset = (keyword == "dodgers tickets"))
print(dodgers_tickets_data_frame)

# traffic estimates for keyword: "angels tickets"
# note identical values for Dodgers and Angels
# interesting that "angels tickets" has lower traffic
# estimates than "dodgers tickets" but a higher CPC
angels_tickets_data_frame <- subset(baseball_data_frame,
        subset = (keyword == "angels tickets"))
print(angels_tickets_data_frame)

# what about "baseball tickets" across all the studies
# this occurs in three of the categories for both Dodgers and Angels
# note the expected seasonal pattern in traffic
# also note identical traffic estimates for March, April, and May
# and identical values for June and July... a clear indication
# that these are not actual data... nor are they likely to have
# come from a data-based predictive model... there is too much
# regularity across the time series of monthly traffic estimates
baseball_tickets_data_frame <- subset(keyword_data_frame,
        subset = (keyword == "baseball tickets"))
print(baseball_tickets_data_frame)

# select Sports Entertainment for keyword search
working_data_frame <- subset(keyword_data_frame,
    subset = (study == "Sports Entertainment"))
print(str(working_data_frame))

# rank keyword records by traffic estimate (highest first)
sorted_data_frame <-
    working_data_frame[sort.list(working_data_frame$traffic,
        decreasing = TRUE),]

# consider only unique keywords in the sorted list
preliminary_keyword_list <- unique(sorted_data_frame$keyword)
cat("\n\n", length(preliminary_keyword_list),
  "keywords in preliminary list\n")

# output the list for review with the intention of selecting
# a subset of keywords relevant to both the Dodgers and Angels
# we use a JSON file for this purpose
json_string <- toJSON(preliminary_keyword_list)
# remove backslashes from string
sink("preliminary_json.txt")
json_string
sink()
```

53

54

第 5 章
关注竞争对手

“接近你的朋友，但更要接近你的敌人。”

Michael Corleone（Al Pacino 饰）
电影《教父Ⅱ》(1974 年)

最近，我乘坐从洛杉矶飞往芝加哥的 Spirit 航空公司的航班。作为知名的廉价航空公司，Spirit 收取的是最基本的票价，然而会对提供的便利设施再进行收费。我的首选并且是唯一的选择就是一个座位。我往返都选择了 10F 座位，这样只需要支付：2×14=28 美元。我可以选择购买单程为 25 美元的座位，但这似乎有点太多。网上订票很省时间，除了屏幕上会接二连三地弹出广告之外。航空公司询问我是否想要多付些钱以便可以提前登机时，我拒绝了，同时我也没有接受支付携带行李的费用。

到达机场后，我花 20 分钟重新整理了随身携带的物品，把两个帆布包中的书和衣物都塞进到一个包里。这样对我很有利，因为根据登机检票员的说明，如果随身携带两件行李，那我将需要再多支付 100 美元。我穿上夹克外套，并将鼠标、备用眼镜、Python 袖珍指南等物品都塞进口袋里。

飞机起飞后不久，有一个婴儿开始哭闹，并持续了整个航程 3 小时 20 分钟的大部分时间。这并不是 Spirit 公司的错，然而我有些期望空乘员会向我们宣布：“为了您能有一个愉快的旅程，我们将为您提供付费 20 美元的耳塞及 100 美元的消噪耳机”。

我们安全并且准时抵达芝加哥，两天后，回程也一样顺利，但没有婴儿的哭闹。然而，不得不为购买座位及随身携带的行李付费而造成的不便让我对再次乘坐 Spirit 航空公司的航班产生疑虑。有那么多家航空公司飞行在洛杉矶至芝加哥这条航线上，我怀疑 Spirit 航空公司能有多强的竞争力呢。

就好像是根深蒂固的网络爬虫一样，现在我们做研究都是从 Web 开始。然而，让我们从那么想当然的举动中暂且后退一步，来思考一下如何开展具有竞争性的情报收集工作。

大多数商业领域中的竞争都是显而易见的，无论我们是用经济领域中的术语完全性竞争、垄断性竞争或一些领先公司之间的竞争（垄断）来对竞争进行描述。无论竞争采用什么方式，大部分的管理者都认为了解竞争对手可以取得决定性的优势。

相较于其他类型的研究，竞争性情报还不为人们所了解，或者会遭到误解，经常还会被贬低，有时也会让人们产生畏惧。不道德的商业行为、商业间谍和偷窃都是会引起高度关注的媒体新闻（Penenberg and Barry, 2000），但真实的竞争性情报工作却是很平凡的。

回顾一下 Yogi Berra 提出的一个很不寻常的观点：你只需要观看就可以知晓很多。这就是竞争性情报的本质。我们研究，我们观察，我们学习。

在不了解竞争对手的情况下开展一个竞争性情报项目是完全可以的。事实上，没有任何先入之见可能也是一种优势，因为可以敞开双眼进入这个领域。

竞争性情报具有前瞻性。光是了解竞争对手过去都做过什么是不够的，我们必须预见未来。我们要预测到他们将来的决策、产品、服务以及价格。如果预测准确，竞争性情报将会引导制定商业策略。

商业发展的脚步非常快，企业必须对竞争格局、政治和监管环境以及技术的发展变化做出快速的反应。越来越多的企业面临全球性的竞争，所以我们必须对全世界进行研究、观察和学习。

56

很多博弈论模型都假设所有博弈方都具有同样的知识，并可获取到同样的信息。事实上，信息不对称却是常态。当一个参与者在竞争性博弈中获取的信息少于其他参与者时，这个参与者在竞争中就会处于不利地位。

竞争信息的来源有很多，其中很大一部分都来自于网络。我们通常相互交换商业出版物并从中寻找相关的会员列表，我们也与业内人士进行交谈。我们惊讶地发现，很多信息都可以在公共领域中找到，任何人都可以通过 Web 浏览器和好的搜索工具访问这些信息。很多公司都在它们各自的网站上对外披露出比应该披露出的更多的信息。很多信息都是有关当前和将来要推出的新产品、公司的组织结构、公司核心成员的简介以及公司的使命宣言。

Woodward 和 Bernstein 能"跟着钱走"，我们也可以。商业交易会披露很多信息。当一个企业的竞争对手决定建立一个新的制造工厂时，这个竞争对手会涉入大量的商业交易中。很多人都会牵扯其中，包括公司经理、房地产商、承包商、房地产经纪人、律师和银行家。这样就会产生很多公共记录，包括当地政府和联邦政府的许可与备案、银行文件、环境评估报告等，也可能会有很多新闻报道和在当地传播的小道消息。通过公共资源获得的信息可以推断出工厂的确切位置与规模。通过环境评估报告，也可以推断将要生产的产品类型以及产量。

已经登记在案的有关业务流程的专利都是公开的记录，能够提供当前和未来产品的信息。联邦政府会提供对专利信息进行的访问，很多企业的网站也会提供这样的访问。

针对政府项目的投标接受公众的监督，由此可以获得关于产品和定价方面的信息。有关竞争对手的其他信息可以通过其他渠道收集，如对外公布的价格指南、新闻稿、刊登在报纸和杂志上的文章、会议论文集、白皮书、技术报告、年度报告和财务分析报告。

新闻媒体会提供大量有关竞争对手的信息，广播和电视节目都会录制对企业核心员工以及行业咨询师进行的采访。收集竞争性情报的专业人员知道除了要在全国性的媒体上搜索信息外，还要在当地小的新闻媒体上搜索信息。

57

其他的信息来源包括只允许订阅用户访问的企业联合会数据源，通过这个途径还可以对大量的法律和商业数据库进行查询。这些收费的信息服务对收集竞争性情报通常非常有用。

在竞争性情报中，专业人员会区分主要来源和辅助来源（不要混淆对主要来源和辅助来源的搜寻）。主要来源是信息的直接来源，能够完全代表被研究的竞争对象。辅助来源是从主要来源或者其他辅助来源获得信息后转手再提供的信息，这个转手的过程可能会造成信息有意或无意的改变。站在新闻和法律的立场上来说，通常认为主要来源信息比辅助来源信息更加可靠和准确。

除了公开的信息之外，竞争性情报专业人员还喜欢采用另外一种他们都钟爱的方式，即与人们交谈。雇员、前雇员、销售人员、顾客，这些人都是潜在的竞争性情报信息的来源。

就像我们做所有的查询那样，竞争性情报工作也都遵循法律和道德规范。道德规范问题常常出现在竞争性情报工作中。只要研究者采用合理的方式对自己进行了介绍，提出问题就完全合乎法律和道德的要求。这就相当于研究者使用记者具有的技能，只是她解释清楚了她是为某一个公司工作。竞争性情报建立在商业专家之间自由和开放交换信息的基础上。贸易展览和专业会议都是找到知情人士的好地方，同时我们也可以在网络世界中找到这样的人。

到底能够获得多少竞争性信息？窃听、受贿和欺骗行为有违法律和道德准则，然而很多在灰色地带中采取的行为虽然会受到质疑，但并不违反法律或道德准则。竞争性情报专业人员必须小心谨慎，避免发生任何可能不当的行为。

当公司将信息发布在网络上，就相当于传播给了整个世界。为了保护具有竞争价值的信息，如专有信息和商业机密，公司就应该要当心网络。当然，在当今这个世界，说起来容易，要做到却非常难。

58

我们所了解的大部分有关通用的竞争性情报知识同样适用于网络世界中的竞争性情报。有一些关于在线搜索的书主要关注竞争性情报的来源（Miller 2000a; Vine 2000; Campbell and Swigart 2014）。一些学者也对竞争性情报爬虫进行过讨论（Chau and Chen 2003; Hemenway and Calishain 2004），大多数我们所了解的聚焦式爬虫也都在竞争性情报方面得到应用。Thompson（2000）以及 Houston、Bruzzese 和 Weinberg（2002）为撰写商业报告的记者提供了建议。这些建议对于竞争性情报专业人员来说同样适用。战略性与竞争性情报专业人员协会（Strategic and Competitive Intelligence Professionals, SCIP）以及很多大学通过图书馆与信息系统课程都提供竞争性情报调查方法方面的培训。

我们通过 Spirit 航空公司这个实例来展示一下竞争性情报的整个过程。以下是讨论的场景，我们是一家与 Spirit 公司在洛杉矶至芝加哥这条航线上竞争的航空公司。我们对 Spirit 的业绩增长非常关注，怀疑它的低票价策略可能会吸引很多乘客。我们想尽快对 Spirit 公司进行尽可能多的了解。我们会像往常一样上网查找信息[⊖]，表 5-1 总结了我们通过交互方式获得的结果。

⊖　此处所使用的数据的采集时间是 2014 年 11 月 4 日。

表 5-1 Spirit 航空公司的竞争性情报来源

来　源	目标（方法）	网址（查询字符串）
维基百科	概况	http://en.wikipedia.org/wiki/Spirit_Airlines
LinkedIn	机构	https://www.linkedin.com/company/spirit-airlines
indeed.com	招聘启事 （高级职位搜索） （高级职位搜索） （高级职位搜索） （高级职位搜索）	http://www.indeed.com/ FindJobs:what=company:(SpiritAirlineswhere=blank) FindJobs:what=company:(SpiritAirlineswhere=Miami,FL) FindJobs:what=company:(SpiritAirlineswhere=Detriot,MI) FindJobs:what=company:(SpiritAirlineswhere=Dallas,TX)
Glassdoor	招聘启事 （搜索） （搜索）	http://www.glassdoor.com/index.htm Jobs.Company:SpiritAirlines Salaries
Compete.com	网站活动 （搜索）	https://www.compete.com/ Website:http://www.spirit.com/
雅虎财经 （Yahoo! Finance）	财务数据 （搜索）	http://finance.yahoo.com/ SearchFinance:SAVE
谷歌财经 （Google Finance）	财务数据 （搜索）	https://www.google.com/finance Search:NASDAQ:SAVE
彭博资讯 （Bloomberg）	财经新闻 （搜索）	http://www.bloomberg.com/ http://www.bloomberg.com/quote/SAVE:US
Quandl	财务数据 （财务报表）	https://www.quandl.com/ https://www.quandl.com/c/stocks/save
Spirit 航空公司	公司网站 （投资者关系）	http://www.spirit.com/ http://ir.spirit.com/

我们希望能够设计出一个尽可能高效的过程对 Spirit 公司进行了解。我们首先通过维基百科对其概况进行了解：成立的时间、公司高管、总部所在地等。我们也要了解在该公司航班网络中排名前十的机场以及这些机场每天的航班数量。

接下来，我们研究公司的组织结构。我们已经知道公司董事长兼总经理是 Ben Baldanza，因此，我们可以从了解他开始，沿着与他有关的链接来建立一个以他为中心的网络。然而，假设我们走到了一个类似 LinkedIn 这样的面向专业人士的链接，找出了该网站中已有的 Spirit 公司员工的人数。我们可以通过点击“查看所有”（See all）来查看这些员工以及他们所处的位置：迈阿密 / 劳德代尔堡（656 人）、大底特律地区（123 人）、达拉斯 / 沃斯堡地区（81 人）等。这些数据可以帮助我们估算该航空公司在每一个地区员工的比例，为我们提供该航空公司在所服务区域内的大致情况。

59～60

职位招聘启事是一个可以很好了解公司发展计划的地方。我们在 Spirit 公司所有职位中进行高级职位搜索开始，然后再添加该航空公司服务的主要城市。因此，我们通过访问 indeed.com 网站来搜索该航空公司服务的主要城市。我们会发现共列出 50 个招聘岗位，其中包括在迈阿密地区的 27 个、底特律地区的 3 个和达拉斯地区的 3 个。

另一个招聘网站 Glassdoor 在我们搜索的这一天列出了 Spirit 公司发布的共 272 个招聘岗位。点击“薪酬”链接，我们可以看到正式员工和兼职员工的薪酬标准。该公司的网站开发人员每年的薪酬是 8.6 万美元，但是一个客服人员的工资仅仅只有每小时 10 美元。我们可以继续下去，将该航空公司的薪酬水平与其他航空公司进行比较。

网站统计数据通常会反映出顾客对公司所提供的服务感兴趣的程度。不过这可能不太适用于 Spirit 航空公司，因为很大一部分的销售是通过旅游和票务服务提供商来实现的。我们访问 Compete.com 去寻找有关网站方面的数据，注意到 2014 年 9 月有超过 180 万用户访问了 Spirit 航空公司的网站。更重要的是，我们看到在过去一年里相关的活动呈现出的是一个平坦的趋势（没有任何增长）。我们感兴趣的并不是网站本身，而是它的活跃情况，这是顾客对航空公司所提供的服务感兴趣程度的一个指标。

我们同样可以去一些诸如 Twitter、Facebook、Google+ 这样的社交网站，看看人们对该航空公司都发表些什么评论。这一点可以通过使用为每一个这样的社交网站所设计的应用程序编程接口（API）来更好地实现（Russell 2014）。

有关 Spirit 公司（纳斯达克股票代码为：SAVE）财务方面的数据可以从很多数据源获得，包括雅虎财经、Google 财经和彭博资讯。更棒的是，我们可以从 Quandl 收集更多关于该公司的数据。使用 R 语言，我们可以高速下载这些数据、显示时间序列，并构建财务预测。

当然，我们的竞争性情报工作少不了要去关注 Spirit 公司的网站。我们可以在公司网站上找到很多关于产品和服务方面的信息，可以通过公司的股东年度报告知道公司的营业收入以及员工总人数，可以阅读新闻发布稿，比如以下这条 2014 年 10 月 28 日发布的新闻："Spirit 航空公司宣布，2014 年第三季度的业绩创了新纪录：2014 年第三季度调整后的净利润增加了 27.6%，达到 7390 万美元。"

61

也许其他航空公司有理由关注 Spirit 航空公司以及在航空旅行方面的廉价 / 便利设施定价策略。图 5-1 展示的是 Spirit 公司过去四年的股价变化。

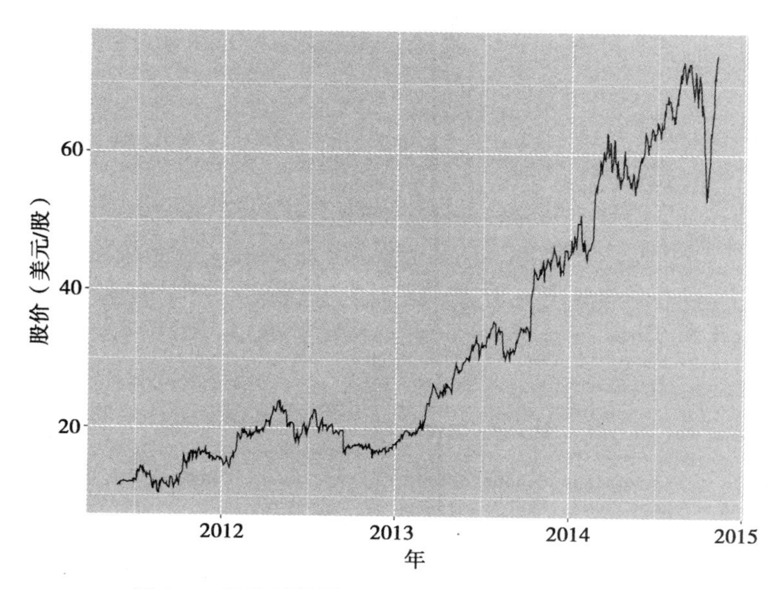

图 5-1　竞争性情报：Spirit 航空公司股价飞上天

Cascade Insights 的 Sean Campbell 和 Scott Swigart (2014) 为在线竞争性情报工作的方法做了一个很好的综述，就像我们通过 Spirit 航空公司这个实例所展示出的。选择和点击很容易做，我们通过交互所做的任何事情都可以通过编程来自动从网上获取数据。竞争性情报本身就是一种不会有哭闹婴儿的廉价旅行。

代码 5-1 是一段 R 语言程序，用于在网上收集 Spirit 航空公司的竞争性情报方面的数据，并将收集到的数据进行存储以供将来进行分析。航空业对经济周期特别敏感，这段代码包含了很多 R 语言程序包，用于对金融数据进行采集、处理、分析和展示（Grolemund and Wickham 2014; Lang 2014a; Lang 2014b; McTaggart and Daroczi 2014; Ryan 2014; Ryan and Ulrich 2014; Wickham and Chang 2014; Zeileis, Grothendieck and Ryan 2014）。

62

代码 5-1 竞争性情报：Spirit 航空公司财务档案（R 语言）

```
# Competitive Intelligence: Spirit Airlines Financial Dossier (R)

# install required packages

# bring packages into the workspace
library(RCurl)  # functions for gathering data from the web
library(XML)  # XML and HTML parsing
library(quantmod) # use for gathering and charting economic data
# online documentation for quantmod at <http://www.quantmod.com/>
library(Quandl)  # extensive financial data online
# online documentation for Quandl at <https://www.quandl.com/>
library(lubridate) # date functions
library(zoo)  # utilities for working with time series
library(xts)    # utilities for working with time series
library(ggplot2)  # data visualization

# -----------------------------------
# Text data acquisition and storage
# -----------------------------------
# get current working directory (commands for Mac OS or Linux)
cwd <- getwd()
# create directory for storing competitive intelligence data
ciwd <- paste(cwd, "/ci_data/", sep ="")
dir.create(ciwd)

# gather Wikipedia data using RCurl package
wikipedia_web_page <-
    getURLContent('http://en.wikipedia.org/wiki/Spirit_Airlines')
# store data in directory ciwd for future processing
sink(paste(ciwd, "/", "wikipedia_web_page", sep = ""))
wikipedia_web_page
sink()

# similar procedures may be used for all acquired data
# for the Spirit Airlines competitive intelligence study
# use distinct file names to identify the data sources

# -------------------------------------------------------------
# Yahoo! Finance for Spirit Airlines (NASDAQ stock symbol: SAVE)
# -------------------------------------------------------------
# stock symbols for companies can be obtained from Yahoo! Finance
# <http://finance.yahoo.com/lookup>

# get Spirit Airlines stock price data
getSymbols("SAVE", return.class = "xts", src = "yahoo")
print(str(SAVE)) # show the structure of this xts time series object
```

```
# plot the series stock price
chartSeries(SAVE,theme="white")
# examine the structure of the R data object
print(str(SAVE))
print(SAVE)
# convert character string row names to decimal date for the year
Year <- decimal_date(ymd(row.names(as.data.frame(SAVE))))
# Obtain the closing price of Spirit Airlines stock
Price <- as.numeric(SAVE$SAVE.Close)
# create data frame for Spirit Airlines Year and Price for future plots
SPIRIT_data_frame <- data.frame(Year, Price)

# similar procedures may be used for all airline competitors

# ---------------------------------------------------------------
# Google! Finance for Spirit Airlines (NASDAQ stock symbol: SAVE)
#   (basically the same data as from Yahoo! Finance)
# ---------------------------------------------------------------

# get Spirit Airlines stock price data
getSymbols("SAVE", return.class = "xts", src = "google")
print(str(SAVE)) # show the structure of this xts time series object
# plot the series stock price
chartSeries(SAVE,theme="white")
# examine the structure of the R data object
print(str(SAVE))
print(SAVE)
```

63

```
# -----------------------------------------
# ggplot2 time series plotting, closing
# price of Spirit Airlines common stock
# -----------------------------------------
# with a data frame object in hand... we can go on to use ggplot2
# and methods described in Chang (2013) R Graphics Cookbook

# use data frame defined from Yahoo! Finance
# the Spirit Airlines closing price per share SPIRIT_data_frame
# now we use that data structure to prepare a time series plot
# again, let's highlight the Great Recession on our plot for this time series
plotting_object <- ggplot(SPIRIT_data_frame, aes(x = Year, y = Price)) +
    geom_line() +
    ylab("Stock Price (dollars/share)") +
    ggtitle("Spirit Airlines Stock Price")
print(plotting_object)
# send the plot to an external file (sans title, larger axis labels)
pdf("fig_competitive_intelligence_spirit.pdf", width = 11, height = 8.5)
plotting_object <- ggplot(SPIRIT_data_frame, aes(x = Year, y = Price)) +
    geom_line() + ylab("Stock Price (dollars/share)") +
    theme(axis.title.y = element_text(size = 20, colour = "black")) +
    theme(axis.title.x = element_text(size = 20, colour = "black"))
print(plotting_object)
dev.off()

# ------------------------------------------
# FRED for acquiring general financial data
# ------------------------------------------
# general financial data may be useful in understanding what is
# happening with a company over time... here is how to get those data
```

64

```
# demonstration of R access to and display of financial data from FRED
# requires a connection to the Internet
# ecomonic research data from the Federal Reserve Bank of St. Louis
# see documentation of tags at http://research.stlouisfed.org/fred2/
# choose a particular series and click on it, a graph will be displayed
# in parentheses in the title of the graph will be the symbol
# for the financial series some time series are quarterly, some monthly
```

```
# ... others weekly... so make sure the time series match up in time
# see the documentation for quantmod at
# <http://cran.r-project.org/web/packages/quantmod/quantmod.pdf>

# here we show how to download the Consumer Price Index
# for All Urban Consumers: All Items, Not Seasonally Adjusted, Monthly
getSymbols("CPIAUCNS",src="FRED",return.class = "xts")
print(str(CPIAUCNS)) # show the structure of this xts time series object
# plot the series
chartSeries(CPIAUCNS,theme="white")

# Real Gross National Product in 2005 dollars
getSymbols("GNPC96",src="FRED",return.class = "xts")
print(str(GNPC96)) # show the structure of this xts time series object
# plot the series
chartSeries(GNPC96,theme="white")

# National Civilian Unemployment Rate,
#    not seasonally adjusted (monthly, percentage)
getSymbols("UNRATENSA",src="FRED",return.class = "xts")
print(str(UNRATENSA)) # show the structure of this xts time series object
# plot the series
chartSeries(UNRATENSA,theme="white")

# University of Michigan: Consumer Sentiment,
#    not seasonally adjusted (monthly, 1966 = 100)
getSymbols("UMCSENT",src="FRED",return.class = "xts")
print(str(UMCSENT)) # show the structure of this xts time series object
# plot the series
chartSeries(UMCSENT,theme="white")

# New Homes Sold in the US, not seasonally adjusted (monthly, thousands)
getSymbols("HSN1FNSA",src="FRED",return.class = "xts")
print(str(HSN1FNSA)) # show the structure of this xts time series object
# plot the series
chartSeries(HSN1FNSA,theme="white")

# ----------------------------------------
# Multiple time series plots
# ----------------------------------------
# let's try putting consumer sentiment and new home sales on the same plot

# University of Michigan Index of Consumer Sentiment (1Q 1966 = 100)
getSymbols("UMCSENT", src="FRED", return.class = "xts")
ICS <- UMCSENT # use simple name for xts object
dimnames(ICS)[2] <- "ICS" # use simple name for index
chartSeries(ICS, theme="white")
ICS_data_frame <- as.data.frame(ICS)
ICS_data_frame$date <- ymd(rownames(ICS_data_frame))
ICS_time_series <- ts(ICS_data_frame$ICS,
  start = c(year(min(ICS_data_frame$date)), month(min(ICS_data_frame$date))),
  end = c(year(max(ICS_data_frame$date)),month(max(ICS_data_frame$date))),
  frequency=12)

# New Homes Sold in the US, not seasonally adjusted (monthly, millions)
getSymbols("HSN1FNSA",src="FRED",return.class = "xts")
NHS <- HSN1FNSA
dimnames(NHS)[2] <- "NHS" # use simple name for index
chartSeries(NHS, theme="white")
NHS_data_frame <- as.data.frame(NHS)
NHS_data_frame$date <- ymd(rownames(NHS_data_frame))
NHS_time_series <- ts(NHS_data_frame$NHS,
  start = c(year(min(NHS_data_frame$date)),month(min(NHS_data_frame$date))),
  end = c(year(max(NHS_data_frame$date)),month(max(NHS_data_frame$date))),
  frequency=12)
```

65

```
# define multiple time series object
economic_mts <- cbind(ICS_time_series,
  NHS_time_series)
  dimnames(economic_mts)[[2]] <- c("ICS","NHS") # keep simple names
modeling_mts <- na.omit(economic_mts) # keep overlapping time intervals only

# examine the structure of the multiple time series object
# note that this is not a data frame object
print(str(modeling_mts))

# ----------------------------------------
# Prepare data frame for ggplot2 work
# ----------------------------------------
# for zoo examples see vignette at
# <http://cran.r-project.org/web/packages/zoo/vignettes/zoo-quickref.pdf>
modeling_data_frame <- as.data.frame(modeling_mts)
modeling_data_frame$Year <- as.numeric(time(modeling_mts))

# examine the structure of the data frame object
# notice an intentional shift to underline in the data frame name
# this is just to make sure we keep our object names distinct
# also you will note that programming practice for database work
# and for work with Python is to utilize underlines in variable names
# so it is a good idea to use underlines generally
print(str(modeling_data_frame))
print(head(modeling_data_frame))

# ---------------------------------------------
# ggplot2 time series plotting of economic data
# ---------------------------------------------
# according to the National Bureau of Economic Research the
# Great Recession extended from December 2007 to June 2009
# using our Year variable this would be from 2007.917 to 2009.417
# let's highlight the Great Recession on our plot
plotting_object <- ggplot(modeling_data_frame, aes(x = Year, y = ICS)) +
    geom_line() +
    annotate("rect", xmin = 2007.917, xmax = 2009.417,
        ymin = min(modeling_data_frame$ICS),
        ymax = max(modeling_data_frame$ICS),
        alpha = 0.3, fill = "red") +
    ylab("Index of Consumer Sentiment") +
    ggtitle("Great Recession and Consumer Sentiment")
print(plotting_object)

# --------------------------------------------------------------
# Quandl for Spirit Airlines (NASDAQ stock symbol: SAVE)
# obtain more extensive financial data for Spirit Airlines
# more documentation at http://blog.quandl.com/blog/using-quandl-in-r/
# --------------------------------------------------------------
Spirit_Price <- Quandl("GOOG/NASDAQ_SAVE", collapse="monthly", type="ts")
plot(stl(Spirit_Price[,4],s.window="periodic"),
    main = "Time Series Decomposition for Spirit Airlines Stock Price")

# Suggestions for the student: Employ search and crawling code to access
# all of the competitive intelligence reports cited in the chapter.
# Save these data in the directory, building a text corpus for further study.
# Scrape and parse the text documents using XPath and regular expressions.
# Obtain stock price series for all the major airlines and compare those
# series with Spirit's. See if there are identifiable patterns in these
# data and if those patterns in any way correspond to economic conditions.
# (Note that putting the economic/monthly and stock price/daily data
#   into the same analysis will require a periodicity change using the
#   xts/zoo data for Spirit Airlines. Refer to documentation at
#   <http://www.quantmod.com/Rmetrics2008/quantmod2008.pdf>
#   or from Quandl at <http://blog.quandl.com/blog/using-quandl-in-r/>.)
```

66

```
# Conduct a study of ticket prices on a round trips between two cities
# (non-stop flights between Los Angeles and Chicago, perhaps).
# Crawl and scrape the pricing data to create a pricing database across
# alternative dates into the future, taking note of the day and time
# of each flight and the airline supplying the service. Develop a
# competitive pricing model for airline travel between these cities.
# Select one of the competitive airlines as your client, and report
# the results of your competitive intelligence and competitive pricing
# research. Make strategic recommendations about what to do about
# the low-fare/amenities pricing approach of Spirit Airlines.
```

67

第 6 章

网络可视化

“好了，德米勒先生，我已经准备好拍摄我的特写镜头了。”

Norma Desmond（Gloria Swanson 饰）

电影《日落大道》（1950 年）

太平洋冲浪者号（Pacific Surfliner）从圣路易斯奥比斯波（San Luis Obispo）驶往圣地亚哥（San Diego）。我喜欢这趟火车。我从格兰岱尔（Glendale）上车，开始我的写作。驶往洛杉矶头十分钟的旅途很沉闷，因为沿途通过的都是电力变压器和废品回收公司。但从圣胡安（San Juan Capistrano）到欧申赛德（Oceanside）这一段就像其他旅程一样美好。

我喜欢知道我将要去哪里，已经到过哪里，以及这之间的地图或路线。但是伴随着映入眼帘的海洋和冲浪者，我更喜欢享受发生在眼前的一切。

对网络的研究做得越多，越能在我们的生活经历中感受到它的存在——我们去过的地方、经历的事情、阅读的文献、认识的人。各种事情都交织联系在一起。

网络以各类形态和不同大小的方式存在，每个网络都由网络节点和节点之间的通信连接构成。用数学术语来表达，网络是由点和边构成的一个图。网络的某一块能够让我们了解通过节点和连接而构成的网络结构。对于任意一个给定的规模足够大的网络，则可能划分出很多个块。我们尝试使用不同的可视化方式来发现如何更好地揭示网络的结构。

万维网就是一个网络，一个网页可以链接到另一个网页。一个软件系统就是将程序员编写的代码、对象和方法通过一个有向网络连接而成的。我们可以看到一个软件包依赖于另外一个软件包，功能模块之间也相互调用。

一个运输网络由多条连接构成，可以将连接的相交之处定义为虚拟节点。每条连接的长度和方向都具有含义。可视化的是一个地图，那么一个共同的目标就是找到从一个相交点到另外一个相交点之间的最短路径。

一个以出版物为代表的科学领域是以文章为节点，以引用为连接而形成的一个知识网。判断一篇文章的水平可以依据它所引用的参考文献的数量和质量，如果想要做到更加准确，则可以依据它被其他文章引用的次数。

在学术界、政界和商界，最关键的是连接关系的数量和质量。这些领域中的权力和影响力常常需要通过分析活跃的社交网络，从中找到具有影响力的成员，识别出重要的小团体或派系来确定。

一个更加方便的对网络进行描述的方法是基于这个网络的度分布。每一个无向图中的节点都有与其他节点相连的连接数量，称为度。度分布能够表明节点之间相互连接的程度。平均度或度中心性是对网络整体的一个统计度量。

为了计算一个网络的整体统计量，如度中心性，我们用矩阵来表达这个网络。我们用节点作为一个正方形矩阵的行和列，然后通过每对节点之间存在的特性得出矩阵单元中的值。例如，我们通过考虑节点之间的连接而得到一个邻接矩阵。如果节点 i 与节点 j 连接，则 $x_{ij}=1$，否则 $x_{ij}=0$。无向图产生的是对称矩阵，有向图则产生非对称矩阵。这些矩阵有时也被称作为社会矩阵。

简单的网络结构包括星形网络、环形网络、线形网络和树形网络。图 6-1 展示的是一个星形网络。度分布显示出实体（Amy）是这个无向网络的中心，毫无疑问地展示出这个实体在这个具有七人的图中比其他实体具有更好的连接关系。70

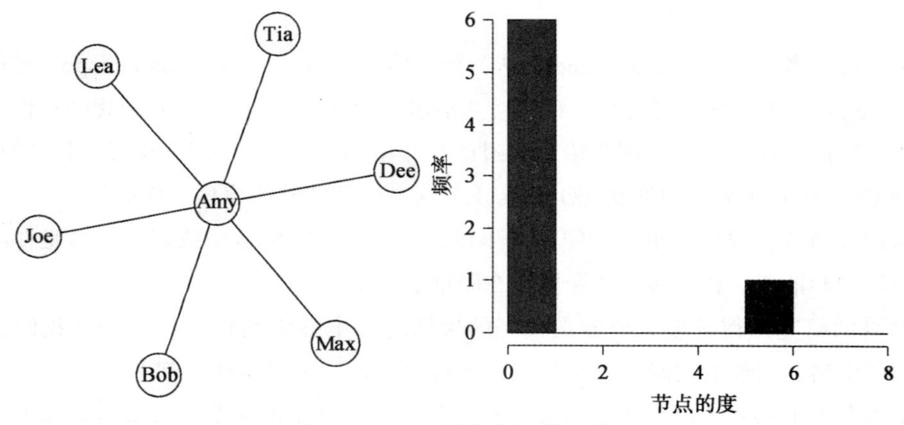

图 6-1　一个简单的星形网络

图 6-2 所展示的是一个环形网络。在环形网络中，每一个节点具有同样的重要性，每一个节点都连接到其他两个节点。这一点反映出它具有简单的度分布。

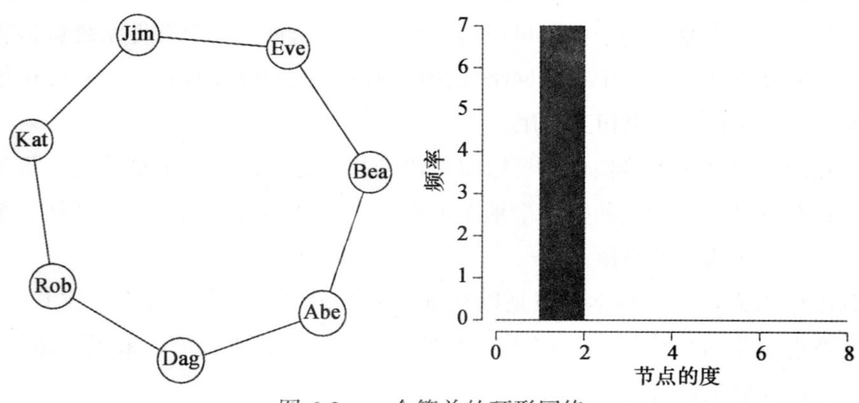

图 6-2　一个简单的环形网络

在图 6-3 所展示的线形或链形网络中，除了两个端节点外，每一个节点都与其他两个节点相连。值得注意的是，线形网络的拓扑结构与环形网络的拓扑结构相似。在这个例子中，在两个端节点（Roy 和 Zoe）之间增加一条边就将形成一个环。人们可以争辩说线形网络中处于中心的节点比其他节点更加重要，这是因为它处在网络中更多对节点所连接的路径上。这就是所谓的的介数中心性。

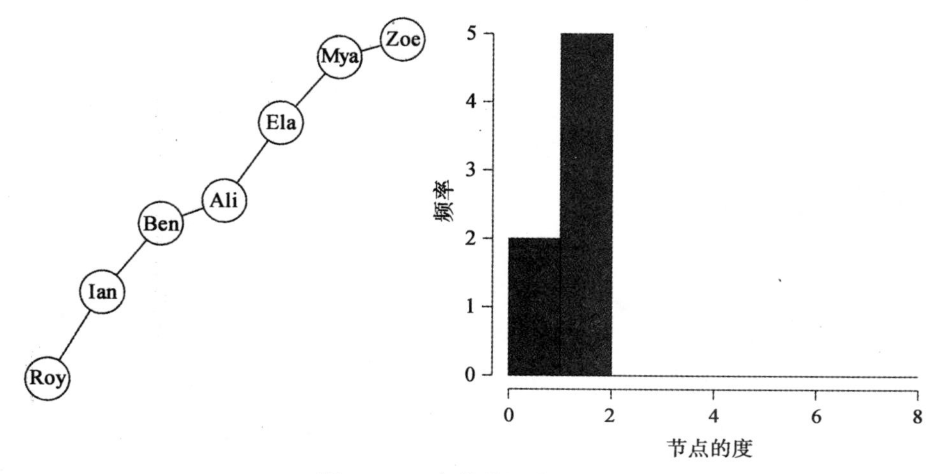

图 6-3　一个简单的线形网络

在一个具有 n 个节点的网络中，将所有节点都与至少其他一个节点相连接所需的最小连接数为 $(n-1)$，这一点可以从简单的星形网络和线形网络图中看出。

一个帮派（clique）或完全连接图是一个完全连接的网络，即每一个节点都与其他每一个节点相连接。一个具有 n 个节点的完全连接图将有 $\frac{n(n-1)}{2}$ 条连接。图 6-4 展示的是一个具有 7 个节点的完全连接图。在此图中，每一个节点都有 6 条连接（每一条连接都与另外一个节点相连接），因此，这个网络具有 $\frac{7(7-1)}{2}=21$ 条连接。完全连接图在社交网络分析中很重要，因为它可以帮助我们识别出一个社区中的核心成员。

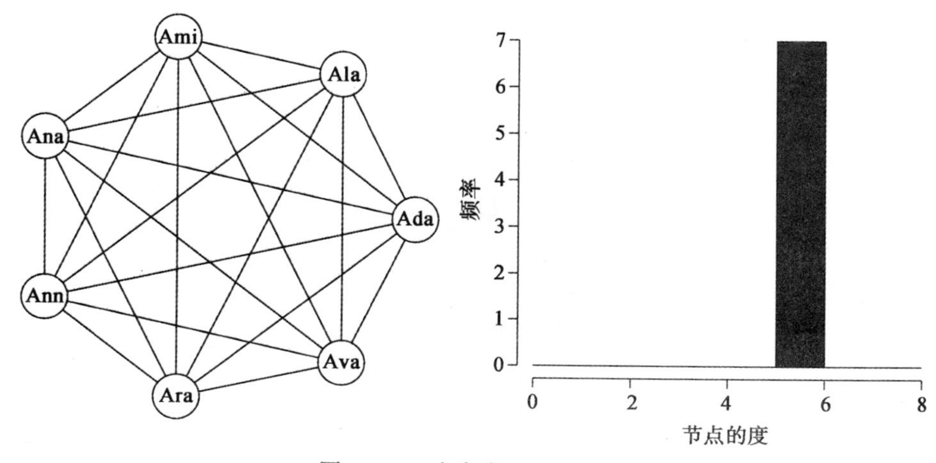

图 6-4　一个完全连接的网络

为一个网络进行正确布局也许需要一些运气。通常最好的方法是尝试使用不同的算法来看哪一个算法能够最好地展示网络的结构。图6-5让我们领略到从一个标准形式的网络可视化中发现一个树形或层次形结构是多么困难，即使这只是一个仅仅有8个节点的网络。因此，我们就有一个新的绕口令："只见树木不见森林"。在此，我们只见连接不见树。

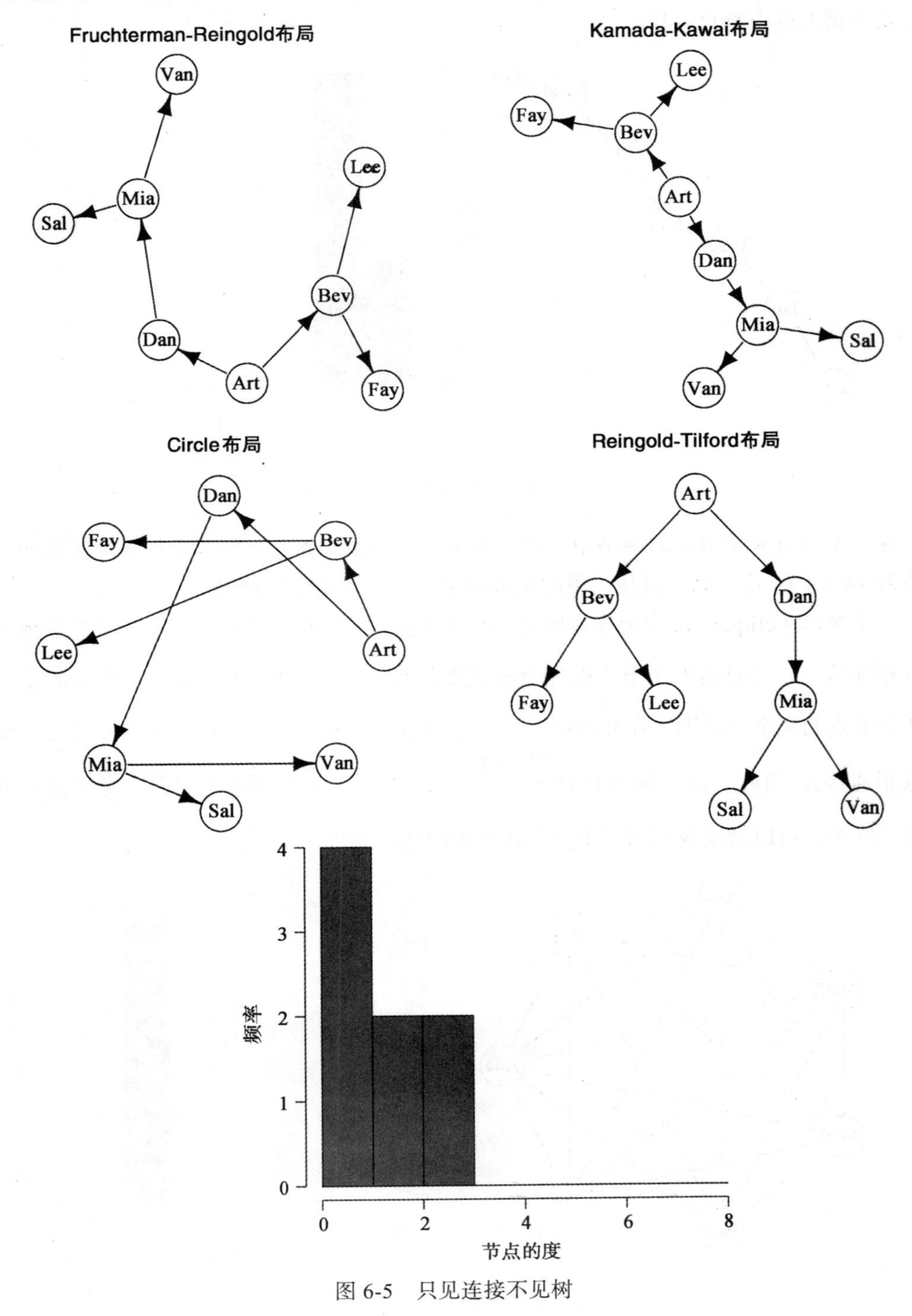

图6-5 只见连接不见树

随着节点数目的增多，用视觉来判断变得尤为困难。对于小型网络或者子网络，可视化则是非常有用的。

对于大型网络，我们可以寻找大量使用的或最强的连接。我们可以寻找那些中心性最高或者在网络中被视为最重要的节点。可视化可以为分析以自我为中心的（ego-centric）网络或者以特定兴趣节点开始的子网络提供帮助。我们可以在网上寻找节点集群，在社交网络中寻找社区或者派系。建模技术可以指导我们建立一个针对兴趣的子网络，然后再使用数据可视化对这个子网络做进一步的研究。

71
~
74

为了阐述可视化在社交网络分析中的应用，我们求助于附录 C 中的安然电子邮件档案。这是一个超过 36 000 个节点和 186 000 个连接的网络，因此我们通过分析不同的子网络（子图）来了解安然公司高管之间的交流情况。

方法之一是以一个关键节点为起点对这个自我为中心的网络进行分析，包括这个节点的直接邻居节点（即与这个节点有电子邮件往来的节点）以及这些节点之间存在的连接。图 6-6 是对以安然高管 1 为中心的网络进行可视化的结果。使用 Kamada-Kawai 算法很容易找到这样一组节点，在图中这些节点间都留有充足的空间以便于对它们进行标识。因此，我们可以使用该算法绘制出一张更大的带节点标识的图，如图 6-7 所示。

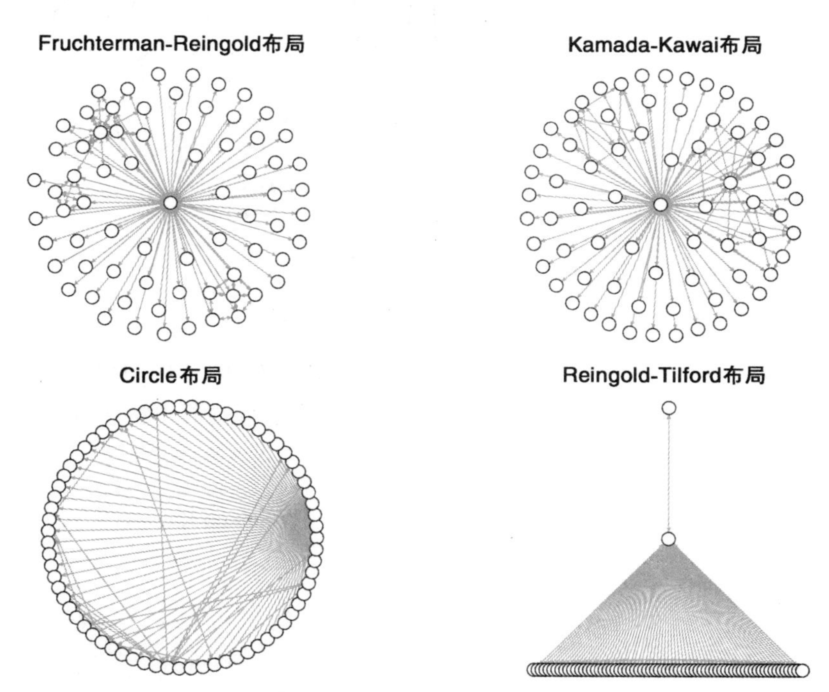

图 6-6　一个自我为中心网络的四种呈现方式

安然公司的电子邮件档案中包含安然公司高管之间往来的邮件，这就意味着非安然公司高管只能是接收终端或发送终端。如果我们使用大写字母 A 和 B 代表两个安然公司高管，用小写字母 a 和 b 代表两个非安然公司高管，那么我们就会看到以下的邮件来往记录：($A \Rightarrow B$)，($B \Rightarrow A$)，($A \Rightarrow a$)，($A \Rightarrow b$)，($a \Rightarrow A$)，($b \Rightarrow A$)，($a \Rightarrow B$) 和 ($b \Rightarrow B$)，但是绝对不会有

（$a \Rightarrow b$）或（$b \Rightarrow a$）这样的记录。也就是说，安然电子邮件档案以安然为中心，却具有不寻常的度分布。安然高管是这个网络中的活跃成员，但是在这个网络中还有成千上万个不是很重要的参与者节点。

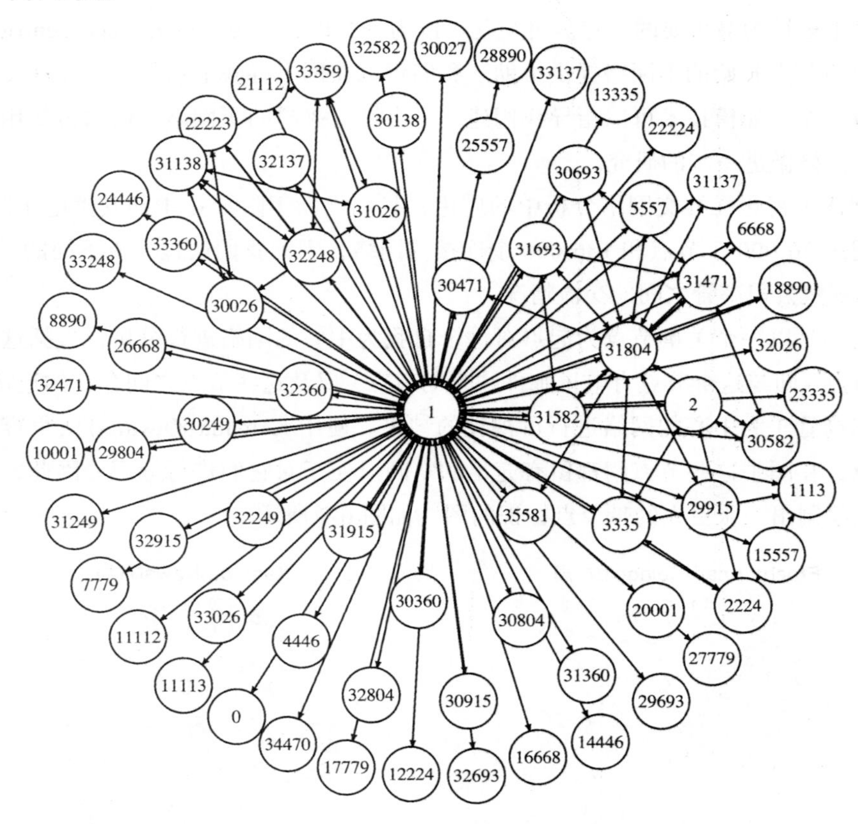

图 6-7 一个自我为中心的网络

假如我们要基于总连接数（度中心性）来找出排名前 50 的邮件用户，图 6-8 展示了使用不同方式来表达由这 50 个高管所构成的子图，图 6-9 则是对节点进行标识后的 Kamada-Kawai 布局图。

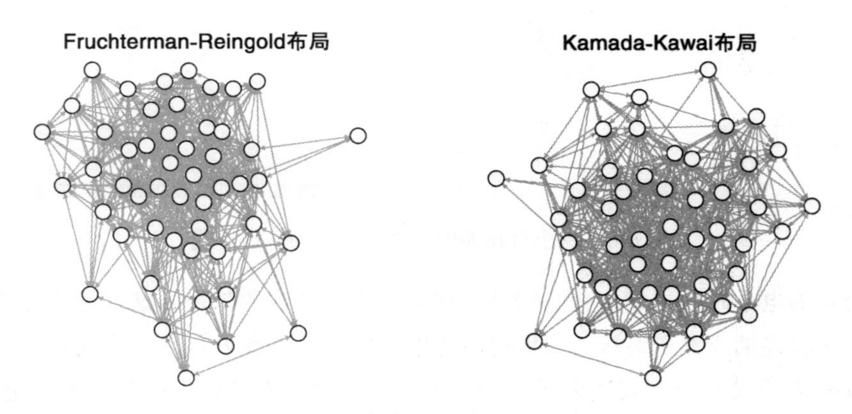

图 6-8 一个社区中最活跃成员的四种呈现方式

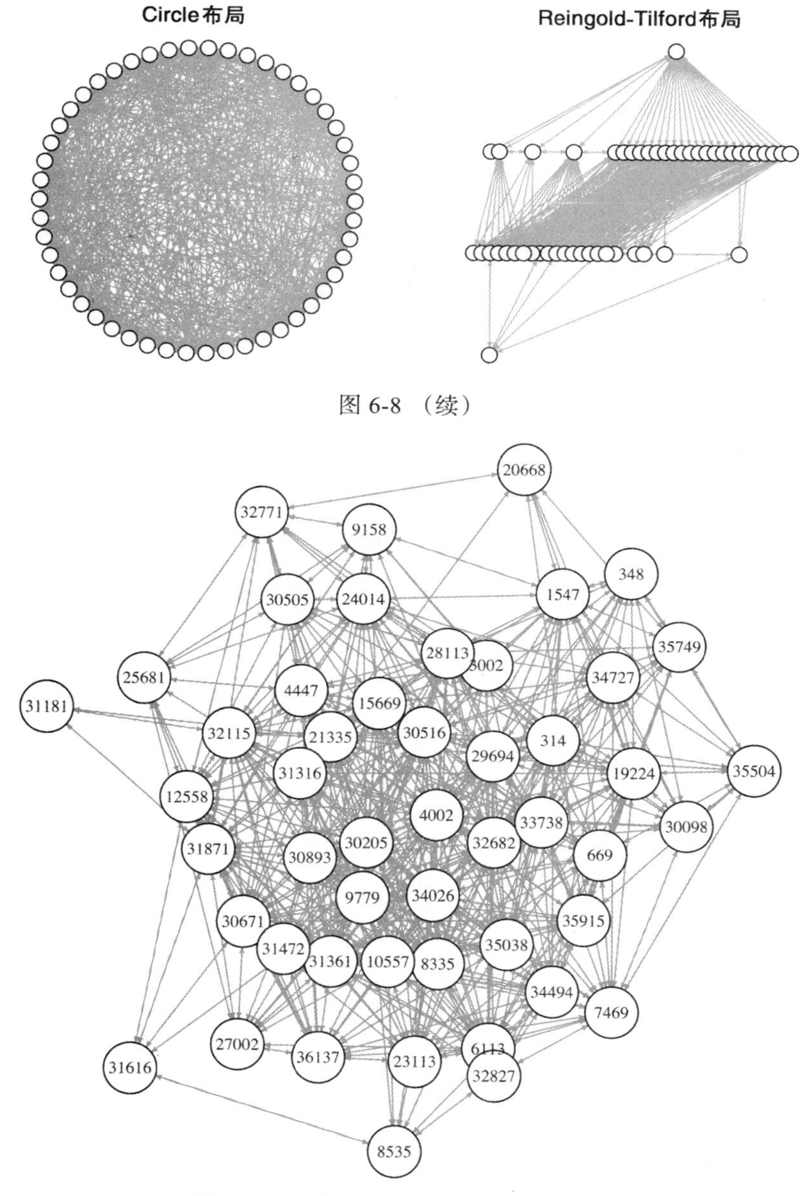

图 6-8 （续）

图 6-9　识别一个社区中最活跃的成员

接下来，我们用图论算法去深度探讨高管中的核心组织。发现这个核心的方法之一是找到具有完全连接关系的一组高管。在安然这个实例中，就是要找到满足在任意两个节点之间都存在发送或接收连接关系（邮件的发件箱或收件箱）的节点子集。图 6-10 显示排名前 50 的高管中存在的 14 个核心高管。

社交网络分析的其中一个目标是找到权利、影响力或重要性的根源。在社交网络分析中，这些概念反映在对中心性的度量上。

75 ~ 79

在安然公司这个实例中，我们可以针对这 50 个具有最高度中心性的节点中的每一个节点去度量其中心性。那么，网络图和对中心度进行度量是否可以帮助我们找到权利人物呢？

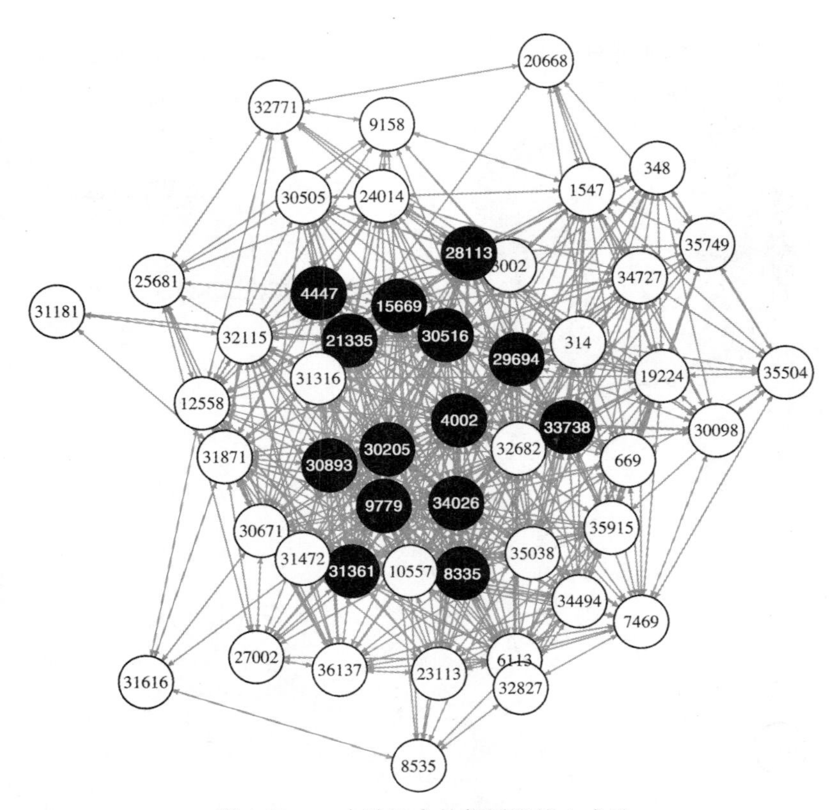

图 6-10 一个社区中的帮派及核心成员

或者说是否还存在不属于这个已经确认为核心组的高管呢？对安然电子邮件语料库本身进行进一步的研究可能有助于回答这些问题⊖。

我们对可视化所做的概述充分显示了这项技术对于了解社交网络以及机构结构方面的实用性，我们同时也了解到选择不同的可视化算法会获得完全不同的可视化效果。

我们希望绘制的图形很好看，也就是优美的视图。我们也希望绘制的图易读、易懂。可视化的目标也许就是要发现拓扑结构以及所属社区、小团体或完全连接图，或者是识别出偏离值或异常值。可视化区域的大小通常都有限制，因此我们希望绘制的图形能够填满可用于绘制的区域，同时使所有的节点和连接都能够很清楚地显示出来，使这些节点和连接容易看清楚及容易识别。连接之间的相互交叉以及对节点和标识的覆盖都会影响这个目标的实现。当然，使绘制所追求的目标在应用建模或数据科学的软件环境内得以实现是我们追求的另外一个目标。

有很多人对大型网络 / 图形的可视化持保留态度。Ben Fry (2008) 曾经说过："什么东西看上去都像是一张图，但几乎没有什么东西应该被绘制成为一张图。"然而，当与建模技术

⊖ 很多研究者都对安然邮件档案进行了进一步分析 (Berry and Browne 2005; Chapanond, Krishnamoorthy, and Yener 2005; Diesner, Frantz, and Carley 2005; Keila and Skillicorn 2005). 在对社交网络进行传统分析的基础上，随着时间的推移，我们也看到了很多应用于安然语料库中数据的网络模型、异常检测方法、波传播算法。具有代表性的研究包括 Priebe et al. (2005), Uddin, Hamra, and Hossain (2013), Wang et al. (2013), Savage et al. (2014), Uddin et al. (2014) 和 Ho and Xing (2014)。

结合起来使用，网络可视化可以提供丰富的信息。秘诀就是简化、识别出小团体和关键成员，并将注意力集中在大型网络中的子网上。

Di Battista et al. (1999) 以及 Kratochvíl (1999) 的贡献者对网络可视化所基于的数学和理论进行了回顾与总结。Kamada and Kawai (1989) 以及 Fruchterman and Reingold (1991) 提出了网络可视化的通用算法。Freeman (2005) 对网络可视化方法进行了概述，其中特别关注多变量方法，包括多维定标算法、一致性分析和主元素分析。

80
~
81

树形结构或层级结构是数据可视化中很有趣且具有挑战性的问题。已经有很多研究者讨论过专门针对树形结构的算法（Lengauer and Tarjan 1979; Reingold and Tilford 1981; Supowit and Reingold 1983; Rusu 2014）。例如，连接 / 边的绑定对于树形或者多层网络的可视化提供了一个很有效的思路，也可以用于发现网络中的子群组（Holden 2006; Crnovrsanin et al. 2014）。

除了可以使用 Python 语言和 R 语言软件包实现数据和网络的可视化外，还有很多其他绘图的程序（Junger and Mutzel 2004）。Gibson, Faith and Vickers (2013) 也对二维网络可视化技术进行了综述。

代码 6-1 是使用 Python 语言对简单网络进行定义和可视化的代码。这段代码的功能是绘制网络、决定网络的度分布，并将网络或图转换成为矩阵以便于进一步分析。这段代码来源于 Hagbert and Schult (2014) 所做的工作。

代码 6-1 对简单网络进行定义及可视化（Python）

```
# Defining and Visualizing Simple Networks (Python)

# prepare for Python version 3x features and functions
from __future__ import division, print_function

# load packages into the workspace for this program
import networkx as nx
import matplotlib.pyplot as plt  # 2D plotting
import numpy as np

# -------------------------
# star (undirected network)
# -------------------------
# define graph object for undirected star network
# adding one link at a time for pairs of nodes
star = nx.Graph()
star.add_edge('Amy', 'Bob')
star.add_edge('Amy', 'Dee')
star.add_edge('Amy', 'Joe')
star.add_edge('Amy', 'Lea')
star.add_edge('Amy', 'Max')
star.add_edge('Amy', 'Tia')

# examine the degree of each node
print(nx.degree(star))

# plot the star network and degree distribution
fig = plt.figure()
nx.draw(star, node_size = 2000, node_color = 'yellow')
plt.show()

fig = plt.figure()
plt.hist(nx.degree(star).values())
plt.axis([0, 8, 0, 8])
```

```
plt.xlabel('Node Degree')
plt.ylabel('Frequency')
plt.show()

# create an adjacency matrix object for the star network
# use nodelist argument to order the rows and columns
star_mat = nx.adjacency_matrix(star,\
    nodelist = ['Amy', 'Bob', 'Dee', 'Joe', 'Lea', 'Max', 'Tia'])

print(star_mat)  # undirected networks are symmetric

# determine the total number of links for the star network (n-1)
print(np.sum(star_mat)/2)
```

82
~
83

```
# ---------------------------
# circle (undirected network)
# ---------------------------

# define graph object for undirected circle network
# using a list of links for pairs of nodes
circle = nx.Graph()
circle.add_edges_from([('Abe', 'Bea'), ('Abe', 'Rob'), ('Bea', 'Dag'),\
    ('Dag', 'Eve'), ('Eve', 'Jim'), ('Jim', 'Kat'), ('Kat', 'Rob')])

# examine the degree of each node
print(nx.degree(circle))

# plot the circle network and degree distribution
fig = plt.figure()
nx.draw(circle, node_size = 2000, node_color = 'yellow')
plt.show()

fig = plt.figure()
plt.hist(nx.degree(circle).values())
plt.axis([0, 8, 0, 8])
plt.xlabel('Node Degree')
plt.ylabel('Frequency')
plt.show()

# create an adjacency matrix object for the circle network
# use nodelist argument to order the rows and columns
circle_mat = nx.adjacency_matrix(circle,\
    nodelist = ['Abe', 'Bea', 'Dag', 'Eve', 'Jim', 'Kat', 'Rob'])
print(circle_mat)  # undirected networks are symmetric

# determine the total number of links for the circle network
print(np.sum(circle_mat)/2)

# -------------------------
# line (undirected network)
# -------------------------
# define graph object for undirected line network
# using a list of links for pairs of nodes
line = nx.Graph()
line.add_edges_from([('Ali', 'Ben'), ('Ali', 'Ela'), ('Ben', 'Ian'),\
    ('Ela', 'Mya'), ('Ian', 'Roy'), ('Mya', 'Zoe')])

# examine the degree of each node
print(nx.degree(line))

# plot the line network and degree distribution
fig = plt.figure()
nx.draw(line, node_size = 2000, node_color = 'yellow')
plt.show()
```

```
fig = plt.figure()
plt.hist(nx.degree(line).values())
plt.axis([0, 8, 0, 8])
plt.xlabel('Node Degree')
plt.ylabel('Frequency')
plt.show()
```

84

```
# create an adjacency matrix object for the line network
# use nodelist argument to order the rows and columns
line_mat = nx.adjacency_matrix(line,\
    nodelist = ['Ali', 'Ben', 'Ela', 'Ian', 'Mya', 'Roy', 'Zoe'])
print(line_mat)  # undirected networks are symmetric

# determine the total number of links for the line network (n-1)
print(np.sum(line_mat)/2)

# ----------------------------
# clique (undirected network)
# ----------------------------
# define graph object for undirected clique
a_clique = nx.Graph()
a_clique.add_edges_from([('Ada', 'Ala'), ('Ada', 'Ami'), ('Ada', 'Ana'),\
('Ada', 'Ann'), ('Ada', 'Ara'), ('Ada', 'Ava'), ('Ala', 'Ami'),\
('Ala', 'Ana'), ('Ala', 'Ann'), ('Ala', 'Ara'), ('Ala', 'Ava'),\
('Ami', 'Ana'),('Ami', 'Ann'), ('Ami', 'Ara'), ('Ami', 'Ava'),\
('Ana', 'Ann'), ('Ana', 'Ara'), ('Ana', 'Ava'),\
('Ann', 'Ara'), ('Ann', 'Ava'), ('Ara', 'Ava')])

# examine the degree of each node
print(nx.degree(a_clique))

# plot the clique and degree distribution
fig = plt.figure()
nx.draw_circular(a_clique, node_size = 2000, node_color = 'yellow')
plt.show()

fig = plt.figure()
plt.hist(nx.degree(a_clique).values())
plt.axis([0, 8, 0, 8])
plt.xlabel('Node Degree')
plt.ylabel('Frequency')
plt.show()

# create an adjacency matrix object for the line network
# use nodelist argument to order the rows and columns
a_clique_mat = nx.adjacency_matrix(a_clique,\
    nodelist = ['Ada', 'Ala', 'Ami', 'Ana', 'Ann', 'Ara', 'Ava'])
print(a_clique_mat)  # undirected networks are symmetric

# determine the total number of links for the clique n(n-1)/2
print(np.sum(a_clique_mat)/2)

# -------------------------------
# tree (directed network/digraph)
# -------------------------------
# define graph object for undirected tree network
# using a list of links for pairs of from-to nodes
tree = nx.DiGraph()
tree.add_edges_from([('Art', 'Bev'), ('Art', 'Dan'), ('Bev', 'Fay'),\
    ('Bev', 'Lee'), ('Dan', 'Mia'), ('Mia', 'Sal'), ('Mia', 'Van')])
```

85

```
# examine the degree of each node
print(nx.degree(tree))

# create an adjacency matrix object for the line network
# use nodelist argument to order the rows and columns
```

```
tree_mat = nx.adjacency_matrix(tree,\
    nodelist = ['Art', 'Bev', 'Dan', 'Fay', 'Lee', 'Mia', 'Sal', 'Van'])
print(tree_mat)  # directed networks are not symmetric

# determine the total number of links for the tree
# upper triangle only has values
print(np.sum(tree_mat))

# plot the degree distribution
fig = plt.figure()
plt.hist(nx.degree(tree).values())
plt.axis([0, 8, 0, 8])
plt.xlabel('Node Degree')
plt.ylabel('Frequency')
plt.show()

# examine alternative layouts for plotting the tree
# plot the network/graph with default layout
fig = plt.figure()
nx.draw(tree, node_size = 2000, node_color = 'yellow')
plt.show()

# spring layout
fig = plt.figure()
nx.draw_spring(tree, node_size = 2000, node_color = 'yellow')
plt.show()

# circlular layout
fig = plt.figure()
nx.draw_circular(tree, node_size = 2000, node_color = 'yellow')
plt.show()

# concentric circles layout
fig = plt.figure()
nx.draw_shell(tree, node_size = 2000, node_color = 'yellow')
plt.show()

# note. plotting as tree may require pygraphviz

# Suggestions for the student: Define alternative network structures.
# Use matplotlib to create their plots. Create the corresponding
# adjacency matrices and compute network descriptive statistics,
# beginning with degree centrality. Plot the degree distribution
# for each network.  Read about pygraphviz and try to plot a tree.
```

86

代码6-2是使用R语言对简单网络进行定义和可视化的代码。代码6-3是一段用于分析安然邮件网络的R语言代码。这些代码来源于Bojanowski, Butts and Csardi (2014)所做的工作。

代码 6-2 对简单网络进行定义及可视化（R语言）

```
# Defining and Visualizing Simple Networks (R)

# begin by installing necessary package igraph

# load package into the workspace for this program
library(igraph)  # network/graph functions

# -------------------------
# star (undirected network)
# -------------------------
# define graph object for undirected star network
```

```
# using links for pairs of nodes by number
star <- graph.formula(1-2, 1-3, 1-4, 1-5, 1-6, 1-7)
# examine the graph object
print(str(star))

# name the nodes (vertices)
V(star)$name <- c("Amy", "Bob", "Dee", "Joe", "Lea", "Max", "Tia")

# examine the degree of each node
print(degree(star))

# determine the total number of links for the star (n-1)
print(sum(degree(star))/2)

# create an adjacency matrix object for the star network
star_mat <- get.adjacency(star)
print(star_mat)  # undirected networks are symmetric

# plot the network/graph to the console
plot(star, vertex.size = 20, vertex.color = "yellow")
# plot the network/graph and degree distribution to an external file
pdf(file = "fig_star_network.pdf", width = 5.5, height = 5.5)
plot(star, vertex.size = 25, vertex.color = "yellow", edge.color = "black")
hist(degree(star), col = "darkblue",
    xlab = "Node Degree", xlim = c(0,8), main = "",
    breaks = c(0,1,2,3,4,5,6,7,8))
dev.off()

# ---------------------------
# circle (undirected network)
# ---------------------------
# define graph object for undirected circle network
# using links for pairs of nodes by number
circle <- graph.formula(1-2, 1-7, 2-3, 3-4, 4-5, 5-6, 6-7)
# examine the graph object
print(str(circle))
# name the nodes (vertices)
V(circle)$name <- c("Abe", "Bea", "Dag", "Eve", "Jim", "Kat", "Rob")
# examine the degree of each node
print(degree(circle))
# determine the total number of links
print(sum(degree(circle))/2)

# create an adjacency matrix object for the circle network
circle_mat <- get.adjacency(circle)
print(circle_mat)  # undirected networks are symmetric

# plot the network/graph to the console
plot(circle, vertex.size = 20, vertex.color = "yellow")
# plot the network/graph and degree distribution to an external file
pdf(file = "fig_circle_network.pdf", width = 5.5, height = 5.5)
plot(circle, vertex.size = 25, vertex.color = "yellow", edge.color = "black")
hist(degree(circle), col = "darkblue",
    xlab = "Node Degree", xlim = c(0,8), main = "",
    breaks = c(0,1,2,3,4,5,6,7,8))
dev.off()

# -------------------------
# line (undirected network)
# -------------------------
# define graph object for undirected line network
# using links for pairs of nodes by number
line <- graph.formula(1-2, 1-3, 2-4, 3-5, 4-6, 5-7)
# examine the graph object
```

87

```
print(str(line))

# name the nodes (vertices)
V(line)$name <- c("Ali", "Ben", "Ela", "Ian", "Mya", "Roy", "Zoe")

# examine the degree of each node
print(degree(line))

# determine the total number of links (n-1)
print(sum(degree(line))/2)
# create an adjacency matrix object for the line network
line_mat <- get.adjacency(line)
print(line_mat)  # undirected networks are symmetric

# plot the network/graph to the console
plot(line, vertex.size = 20, vertex.color = "yellow")
# plot the network/graph and degree distribution to an external file
pdf(file = "fig_line_network.pdf", width = 5.5, height = 5.5)
plot(line, vertex.size = 25, vertex.color = "yellow", edge.color = "black")
hist(degree(line), col = "darkblue",
    xlab = "Node Degree", xlim = c(0,8), main = "",
    breaks = c(0,1,2,3,4,5,6,7,8))
dev.off()

# ----------------------------
# clique (undirected network)
# ----------------------------
# define graph object for undirected clique
# using links for pairs of nodes by number
a_clique <- graph.formula(1-2, 1-3, 1-4, 1-5, 1-6, 1-7,
                               2-3, 2-4, 2-5, 2-6, 2-7,
                                    3-4, 3-5, 3-6, 3-7,
                                         4-5, 4-6, 4-7,
                                              5-6, 5-7,
                                                   6-7)
# examine the graph object
print(str(a_clique))

# name the nodes (vertices)
V(a_clique)$name <- c("Ada", "Ala", "Ami", "Ana", "Ann", "Ara", "Ava")

# examine the degree of each node
print(degree(a_clique))

# determine the total number of links n(n-1)/2
print(sum(degree(a_clique))/2)

# create an adjacency matrix object for the clique
# this is a matrix of ones off-diagonal... fully connected
a_clique_mat <- get.adjacency(a_clique)
print(a_clique_mat)  # undirected networks are symmetric

# plot the network/graph to the console
plot(a_clique, vertex.size = 20, vertex.color = "yellow",
    edge.color = "black", layout = layout.circle)
# plot the network/graph and degree distribution to an external file
pdf(file = "fig_clique_network.pdf", width = 5.5, height = 5.5)
plot(a_clique, vertex.size = 25, vertex.color = "yellow",
    edge.color = "black", layout = layout.circle)
hist(degree(a_clique), col = "darkblue",
    xlab = "Node Degree", xlim = c(0,8), main = "",
    breaks = c(0,1,2,3,4,5,6,7,8))
dev.off()

# -------------------------------
```

88

```
# tree (directed network/digraph)
# --------------------------------
# define graph object for undirected tree network
# using links for pairs of nodes by number
tree <- graph.formula(1-+2, 1-+3, 2-+4, 2-+5, 3-+6, 6-+7, 6-+8)
# examine the graph object
print(str(tree))

# name the nodes (vertices)
V(tree)$name <- c("Art", "Bev", "Dan", "Fay", "Lee", "Mia", "Sal", "Van")

# examine the degree of each node
print(degree(tree))

# create an adjacency matrix object for the tree network
tree_mat <- get.adjacency(tree)
print(tree_mat)  # directed networks are not symmetric
# plot the network/graph to the console
plot(tree, vertex.size = 20, vertex.color = "yellow", edge.color = "black")

# plot the network/graph and degree distribution to an external file
# examine alternative layouts for plotting the tree
pdf(file = "fig_tree_network_four_ways.pdf", width = 5.5, height = 5.5)
par(mfrow = c(1,1))  # four plots on one page
plot(tree, vertex.size = 25, vertex.color = "yellow",
    layout = layout.fruchterman.reingold, edge.color = "black")
title("Fruchterman-Reingold Layout")
plot(tree, vertex.size = 25, vertex.color = "yellow",
    layout = layout.kamada.kawai, edge.color = "black")
title("Kamada-Kawai Layout")
plot(tree, vertex.size = 25, vertex.color = "yellow",
    layout = layout.circle, edge.color = "black")
title("Circle Layout")
plot(tree, vertex.size = 25, vertex.color = "yellow",
    layout = layout.reingold.tilford, edge.color = "black")
title("Reingold-Tilford Layout")
hist(degree(tree), col = "darkblue",
    xlab = "Node Degree", xlim = c(0,8), main = "",
    breaks = c(0,1,2,3,4,5,6,7,8))
dev.off()
```

89
90

代码 6-3 网络可视化——理解机构（R 语言）

```
# Visualizing Networks---Understanding Organizations (R)

# bring in packages we rely upon for work in predictive analytics
library(igraph)  # network/graph methods
library(network)  # network representations
library(intergraph)  # for exchanges between igraph and network

# ----------------------------------------------------------
# Preliminary note about data preparation
#
# The first two records of the original link file
# contain vertices referenced by a zero index.
# For R network algorithms we need to ensure that
# the from- and to-node identifiers are
# treated as character strings, not integers.

# ----------------------------------------------------------
# Read in list of links... (from-node, to-node) pairs
# ----------------------------------------------------------
all_enron_links <- read.table('enron_email_links.txt', header = FALSE)
```

```
cat("\n\nNumber of Links on Input: ", nrow(all_enron_links))
# check the structure of the input data data frame
print(str(all_enron_links))

# convert the V1 and V2 to character strings
all_enron_links$V1 <- as.character(all_enron_links$V1)
all_enron_links$V2 <- as.character(all_enron_links$V2)

# ensure that no e-mail links are from an executive to himself/herself
# i.e. eliminate any nodes that are self-referring
enron_links <- subset(all_enron_links, subset = (V1 != V2))
cat("\n\nNumber of Valid Links: ", nrow(enron_links))

# create network object from the links
# multiple = TRUE allows for multiplex links/edges
# because it is possible to have two or more links
# between the same two nodes (multiple e-mail messages
# between the same two people)
enron_net <- network(as.matrix(enron_links),
    matrix.type = "edgelist", directed = TRUE, multiple = TRUE)

# create graph object with intergraph function asIgraph()
enron_graph <- asIgraph(enron_net)

# name the nodes noting that the first identifer on input was "0"
node_index <- as.numeric(V(enron_graph))
V(enron_graph)$name <- node_name <- as.character(V(enron_graph) - 1)
# node name lookup table
node_reference_table <- data.frame(node_index, node_name, stringsAsFactors = FALSE)
print(str(node_reference_table))
print(head(node_reference_table))
# consider the subgraph of all people that node "1"
# communicates with by e-mail (mail in or out)
# node "1" corresponds to node index 2 in R
ego_1_mail <- induced.subgraph(enron_graph,
    neighborhood(enron_graph, order = 1, nodes = 2)[[1]])
# examine alternative layouts for plotting the ego_1_mail
pdf(file = "fig_ego_1_mail_network_four_ways.pdf", width = 5.5, height = 5.5)
par(mfrow = c(1,1))  # four plots on one page
set.seed(9999)  # for reproducible results
plot(ego_1_mail, vertex.size = 10, vertex.color = "yellow",
    vertex.label = NA, edge.arrow.size = 0.25,
    layout = layout.fruchterman.reingold)
title("Fruchterman-Reingold Layout")
set.seed(9999)  # for reproducible results
plot(ego_1_mail, vertex.size = 10, vertex.color = "yellow",
    vertex.label = NA, edge.arrow.size = 0.25,
    layout = layout.kamada.kawai)
title("Kamada-Kawai Layout")
set.seed(9999)  # for reproducible results
plot(ego_1_mail, vertex.size = 10, vertex.color = "yellow",
    vertex.label = NA, edge.arrow.size = 0.25,
    layout = layout.circle)
title("Circle Layout")
set.seed(9999)  # for reproducible results
plot(ego_1_mail, vertex.size = 10, vertex.color = "yellow",
    vertex.label = NA, edge.arrow.size = 0.25,
    layout = layout.reingold.tilford)
title("Reingold-Tilford Layout")
dev.off()

set.seed(9999)  # for reproducible results
pdf(file = "fig_ego_1_mail_network.pdf", width = 8.5, height = 11)
plot(ego_1_mail, vertex.size = 15, vertex.color = "yellow",
    vertex.label.cex = 0.9, edge.arrow.size = 0.25,
```

91

```
    edge.color = "black", layout = layout.kamada.kawai)
dev.off()

# examine the degree of each node in the complete Enron e-mail network
# and add this measure (degree centrality) to the node reference table
node_reference_table$node_degree <- degree(enron_graph)
print(str(node_reference_table))
print(head(node_reference_table))

# sort the node reference table by degree and identify the indices
# of the most active nodes (those with the most links)
sorted_node_reference_table <-
    node_reference_table[sort.list(node_reference_table$node_degree,
        decreasing = TRUE),]
# check on the sort
print(head(sorted_node_reference_table))
print(tail(sorted_node_reference_table))
# select the top K executives... set K
K <- 50
# identify a subset of K Enron executives based on e-mail-activity
top_node_indices <- sorted_node_reference_table$node_index[1:K]
top_node_names <- sorted_node_reference_table$node_name[1:K]
print(top_node_indices)
print(top_node_names)

# define a top nodes reference table as subset of complete reference table
top_node_reference_table <- subset(node_reference_table,
    subset = (node_name %in% top_node_names))
print(str(top_node_reference_table))
print(head(top_node_reference_table))

# construct the subgraph of the top K executives
top_enron_graph <- induced.subgraph(enron_graph, top_node_indices)
# examine alternative layouts for plotting the top_enron_graph
pdf(file = "fig_top_enron_graph_four_ways.pdf", width = 5.5, height = 5.5)
par(mfrow = c(1,1))  # four plots on one page
set.seed(9999)  # for reproducible results
plot(top_enron_graph, vertex.size = 10, vertex.color = "yellow",
    vertex.label = NA, edge.arrow.size = 0.25,
    layout = layout.fruchterman.reingold)
title("Fruchterman-Reingold Layout")
set.seed(9999)  # for reproducible results
plot(top_enron_graph, vertex.size = 10, vertex.color = "yellow",
    vertex.label = NA, edge.arrow.size = 0.25,
    layout = layout.kamada.kawai)
title("Kamada-Kawai Layout")
set.seed(9999)  # for reproducible results
plot(top_enron_graph, vertex.size = 10, vertex.color = "yellow",
    vertex.label = NA, edge.arrow.size = 0.25,
    layout = layout.circle)
title("Circle Layout")
set.seed(9999)  # for reproducible results
plot(top_enron_graph, vertex.size = 10, vertex.color = "yellow",
    vertex.label = NA, edge.arrow.size = 0.25,
    layout = layout.reingold.tilford)
title("Reingold-Tilford Layout")
dev.off()
# let's use the Kamada-Kawai layout for the labeled plot
set.seed(9999)  # for reproducible results
pdf(file = "fig_top_enron_graph.pdf", width = 8.5, height = 11)
plot(top_enron_graph, vertex.size = 15, vertex.color = "yellow",
    vertex.label.cex = 0.9, edge.arrow.size = 0.25,
    edge.color = "darkgray", layout = layout.kamada.kawai)
dev.off()
# a clique is a subset of nodes that are fully connected
```

92

```
# (links between all pairs of nodes in the subset)
# perform a census of cliques in the top_enron_graph
table(sapply(cliques(top_enron_graph), length))  # shows two large cliques
# the two largest cliques have thirteen nodes/executives
# let's identify those cliques
two_cliques <-
    cliques(top_enron_graph)[sapply(cliques(top_enron_graph), length) == 13]
# show the new index values for the top cliques... note the overlap
print(two_cliques)
# finding our way to the executive core of the company
# note index numbers are reset by the induced.subgraph() function
# form a new subgraph from the union of the top two cliques
core_node_indices_new <- unique(unlist(two_cliques))
non_core_node_indices_new <- setdiff(1:K, core_node_indices_new)
set_node_colors <- rep("white", length = K)
set_node_colors[core_node_indices_new] <- "darkblue"
set_label_colors <- rep("black", length = K)
set_label_colors[core_node_indices_new] <- "white"
# again use the Kamada-Kawai layout for the labeled plot
# but this time we use white for non-core and blue for core nodes
set.seed(9999)  # for reproducible results
pdf(file = "fig_top_enron_graph_with_core.pdf", width = 8.5, height = 11)
plot(top_enron_graph, vertex.size = 15,
    vertex.color = set_node_colors,
    vertex.label.color = set_label_colors,
    vertex.label.cex = 0.9, edge.arrow.size = 0.25,
    edge.color = "darkgray", layout = layout.kamada.kawai)
dev.off()
# check on the tree/hierarchy to search for the source of power
set.seed(9999)  # for reproducible results
plot(top_enron_graph, vertex.size = 15, vertex.color = "white",
    vertex.label.cex = 0.9, edge.arrow.size = 0.25,
    layout = layout.reingold.tilford)  # node name 314?
# compute centrality idices for the top executive nodes
# begin by expressing the adjacency matrix in standard matrix form
top_enron_graph_mat <- as.matrix(get.adjacency(top_enron_graph))
top_node_reference_table$betweenness <-
    betweenness(top_enron_graph)  # betweenness centrality
top_node_reference_table$evcent <-
    evcent(top_enron_graph)$vector  # eigenvalue centrality
print(str(top_node_reference_table))
print(top_node_reference_table)  # data for top executive nodes

# Suggestions for the student:  Experiment with techniques for identifying
# core executive groupings and sources of power in the organization.
# Could it be that node 314, a node not in the core group,
# is the true source of power in the organization?
# What do indices of centrality suggest about possible sources
# of power, influence, or importance in the Enron network?
# Try other network visualizations for the Enron e-mail network.
# Note that nowhere in our analysis so far have we looked at the
# number of e-mails sent from one player/node to another.
# All we have are binary links. If there is at least one e-mail between
# a pair of nodes, we have drawn a link between those nodes.
# Perhaps that is why everything looks like spaghetti or a ball of string.
# There is much that can be done with the Enron e-mail corpus.
# We could work with the original Enron e-mail case data, assigning
# executive names (not just numbers) to the nodes.  We could
# explore methods of text analytics using the e-mail message text.
```

93

94

第7章
了解社区

“其实她并不是我的妻子，我们只是有些孩子而已。不……不……，没有孩子。甚至不是孩子。不过，有时我们感觉好像有孩子。她并不漂亮，但是很容易相处。不，她并不容易相处，这就是为什么我不跟她住在一起。”

Thomas（David Hemmings 饰）
电影《放大》(1966 年)

我每天的一部分运动是早上五到六点之间晨走，除了经常遇见一位老者外，人行道上只有我一个人。据我所知，这位老者住在附近的一个公园，或者至少说那是他每天会度过大部分时间的地方。每当我跟他相遇时，我会向他挥手示意，他也会挥手回应我。我们之间的社会交往仅限于此。就像我们常说的“我们在城市里分享着这个空间，这样就很好。”

夫妻会结婚，我们会追溯家谱。契约将合作中的一方与合作中的另一方捆绑在一起，不论合作及合作方是什么。合并与收购、服务提供者与客户、学生与老师、朋友与恋人都是客观存在的。人们会成群结队地参与各种活动、集会或者会议。有些关系非常密切，有些关系则不是那么密切。

要想理解为什么人们会做他们所做的事情并不总是很容易，但是我们至少可以观察人们之间存在的关系。各种聚集的社区都由动机驱动，对此我们使用社交网络来表达。

什么是社交网络？社交网络就是相互连接的人的节点。那连接又是什么？那是我们所希望看到的任何可能形式。有时，就像我和那个老者那样，连接就是地理位置加上一个手势。

社交网络分析多年来一直都是心理学、社会学、人类学、政治科学等领域的研究热点。社交关系网图的发明以及社会结构的概念可以追溯到 80 多年前（Moreno 1934; Radcliffe-Brown 1940）。研究的主题包括隔离与流行、声望、权力与影响力、社会凝聚力、小团体与派系、在机构内部所处的地位与角色、平衡与互惠、市场关系以及中心性与连通性。

在过去的年代，收集社交网络数据是一个痛苦的过程。如今则不然，通过网络来实现的社交媒体重新燃起了人们对分析社交网络的兴趣。网络技术可以提供人们在做什么、来自何方、往哪里去以及跟什么人进行交流的记录，有时候还有他们都说过什么话的记录。现在有很多在线的朋友和粉丝、推特信息与转发、短消息、电子邮件和博客。数据非常之多，而我们面临的挑战就是要找到获取有用数据的途径，并使这些数据成为有意义的数据。

在意识到电子社交网络和移动智能设备应用的增长潜力后，企业看到了与朋友的朋友进行交流并向朋友的朋友进行销售的商机。以营利为目的的公司、非营利机构和政府组织都对研究社交媒体数据产生了兴趣。网络数据集虽然很大，但并不是不可以访问到，存在太多的可能性。

社交网络对于预测非常有用。我们认为，人们的态度和行为都会受他们所认识的人的影响。与过去一样，那些在“网络空间里”与我们最亲近的人可能对我们的态度和行为最具有影响力。

为了进一步对Web与网络数据科学领域进行探讨，我们寻求着关系中的模式或规律。关系意味着参与者之间的交互。社交网络中的参与者节点可以是人、机构、公司、市场中的买方和卖方、风险投资领域中的投资人和创业者，或者是国际商务活动中的不同国家。对参与者可以用其特征来进行描述，对他们之间的连接则可以用力度和方向来进行描述。有些参与者之间的连接是单向的，也有些是双向的，相对应的就是有向网络和无向网络。

96

各种对网络进行度量的方法都源自研究图论的数学家和研究社交网络的社会学家。网络度量用于描述网络节点之间的关系以及网络结构具有的特征。

无论是研究网络，还是研究一个内容领域或者一个社交团体，确定对哪些节点和连接进行研究可能都不是一件很容易的事。换言之，我们可能需要对网络划定一个界限，我们可能需要随意地定义网络从哪里开始，到哪里结束，即我们感兴趣的范围。

我们运用针对网络的数学模型来预测网络现象，研究网络度量之间的关系。目前有3个特别重要的网络模型：随机图模型、偏好依附模型和小世界网络模型。

第一个有趣的数学模型由Paul Erdös and Alfred Rényi (1959, 1960, 1961)首次提出，为以后提出的其他模型奠定了基础。随机图就是一组节点通过完全随机连接而形成的一个图。图7-1通过环形布局的方式展示了一个由50个节点和100个连接形成的随机图。这个网络不存在任何可以识别的模式。连接随机地与节点相关联，也就是说任意两个节点被连接起来的可能性都是相同的。

第二个面向网络的数学模型是偏好依附preferential allachment模型，该模型主要基于Barabási and Albert (1999)所做的研究。这个模型以一个随机图开始，然后添加新的连接，优先选择已经拥有大量连接的节点，即具有高度中心性的节点。

图7-2通过环形布局的方式展示了一个偏好依附网络，需要注意的是有些节点具有比其他节点更多的连接。在对这种偏好依附网络进行描述时，我们通常会想到“富者更富，穷者更穷。”

基于偏好依附形成的网络有时也被称为规模不受限（scale-free）网络，或称为长尾（long-tail）网络。在与网络科学相关的文献中，还有很多其他基于偏好依附的网络模型（Albert and Barabási 2002）。这些模型对于表达在线网络非常有用。

97

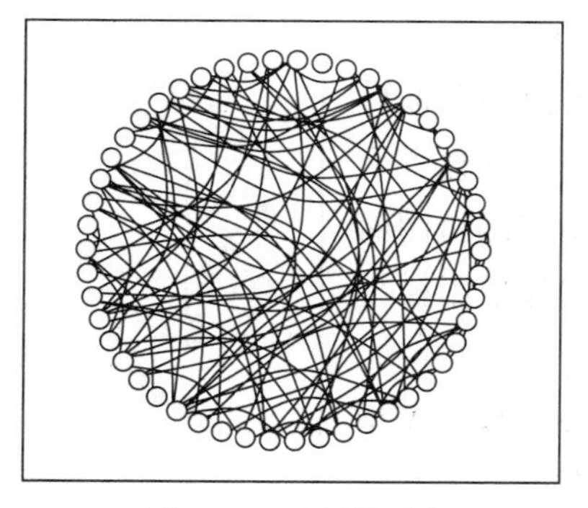

图 7-1　一个随机图

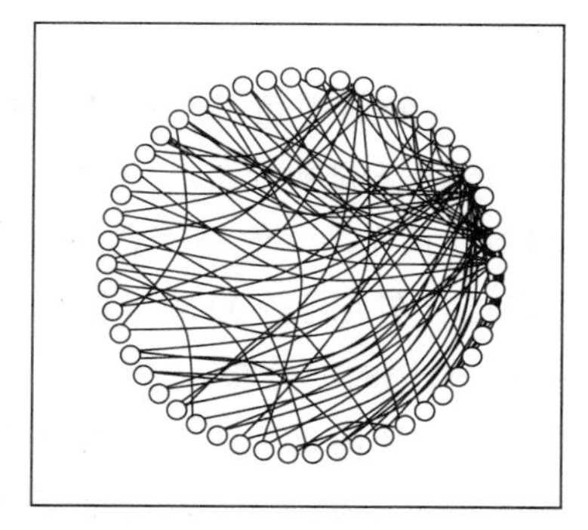

图 7-2　一个基于偏好依附形成的网络

第三个模型对于表示社交网络尤为重要，这就是小世界网络（Watts and Strogatz 1998）。在小世界模型定义的结构中，大多数节点或参与者都与处于它们附近的邻居节点相连接，当然也有一些与不在附近的节点相连接。小世界网络在社会心理学、社会学和网络科学中都得到了广泛的研究（Milgram 1967; Travers and Milgram 1969; Watts 1999; Schnettler 2009）。

要想生成一个无向的小世界网络，我们首先需要确定小世界网络中节点的个数以及这个小世界邻里的大小。然后，我们打乱原有的小世界，将所选的节点重新连接。就像邻里大小一样，随机依附或重新连接的概率也是定义此网络的一个参数。

图 7-3 所示是在对节点重新进行连接之前的一个基线小世界网络。使用圆形布局的方式，我们看到在左边的一个具有 10 个节点的网络中，节点只与相邻的节点相关联。每一个节点都有两个连接。随着网络区域逐步增大，从图中左边到右边，节点之间的连接个数在增加。在图中间的网络中，每一个节点都分别与其他四个节点相连接，而在图中右边的网络中，每一个节点都分别与其他 6 个节点相连接。小世界网络的一个特定属性是多数节点都与邻近的节点相关联。

98 ~ 99

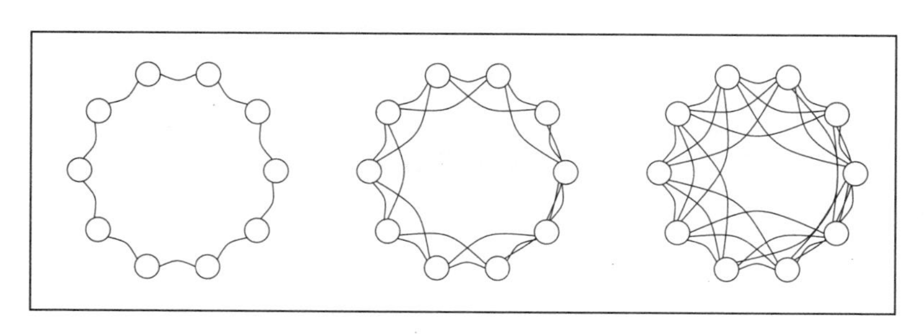

图 7-3　为小世界网络构建基线

从一个基线小世界网络到一个实际的小世界网络涉及重新布线。我们选择连接来重新进行布线，将连接与原来的节点分离，然后连接到另一个节点上。这样来选择连接都是随意的。

再一次使用环形布局方式，图 7-4 展示的是一个具有 50 个节点和 100 条连接的小世界网络。在这个网络中，有一半的连接都是在最邻近节点之间，另外一半连接与随机选择的节点相关联。

图 7-5 展示的是分别用随机图模型、小世界模型和偏好依附模型所产生的网络的度分布。图中的每个模型都包含大约 50 个节点和 100 条连接。值得注意的是，相对于随机图而言，小世界网络的优势在于节点的度较低，而偏好依附模型则具有高度倾斜或长尾的度分布。

在对网络进行统计仿真中，我们可以选择一个数学模型作为起点。它们对网络参数统计进行推断以及对网络未来状态进行预测也具有价值。

100

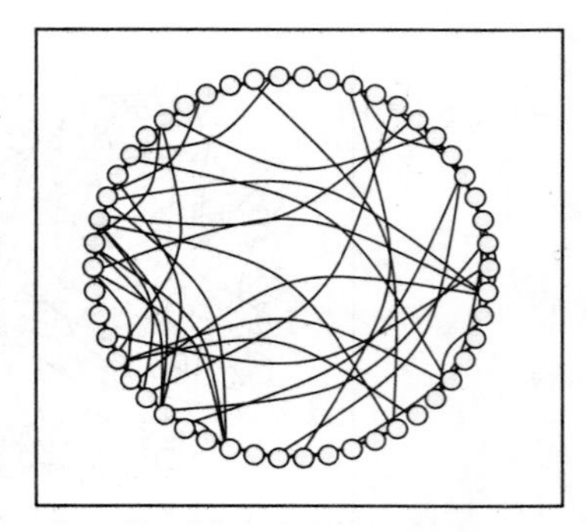

图 7-4　一个小世界网络

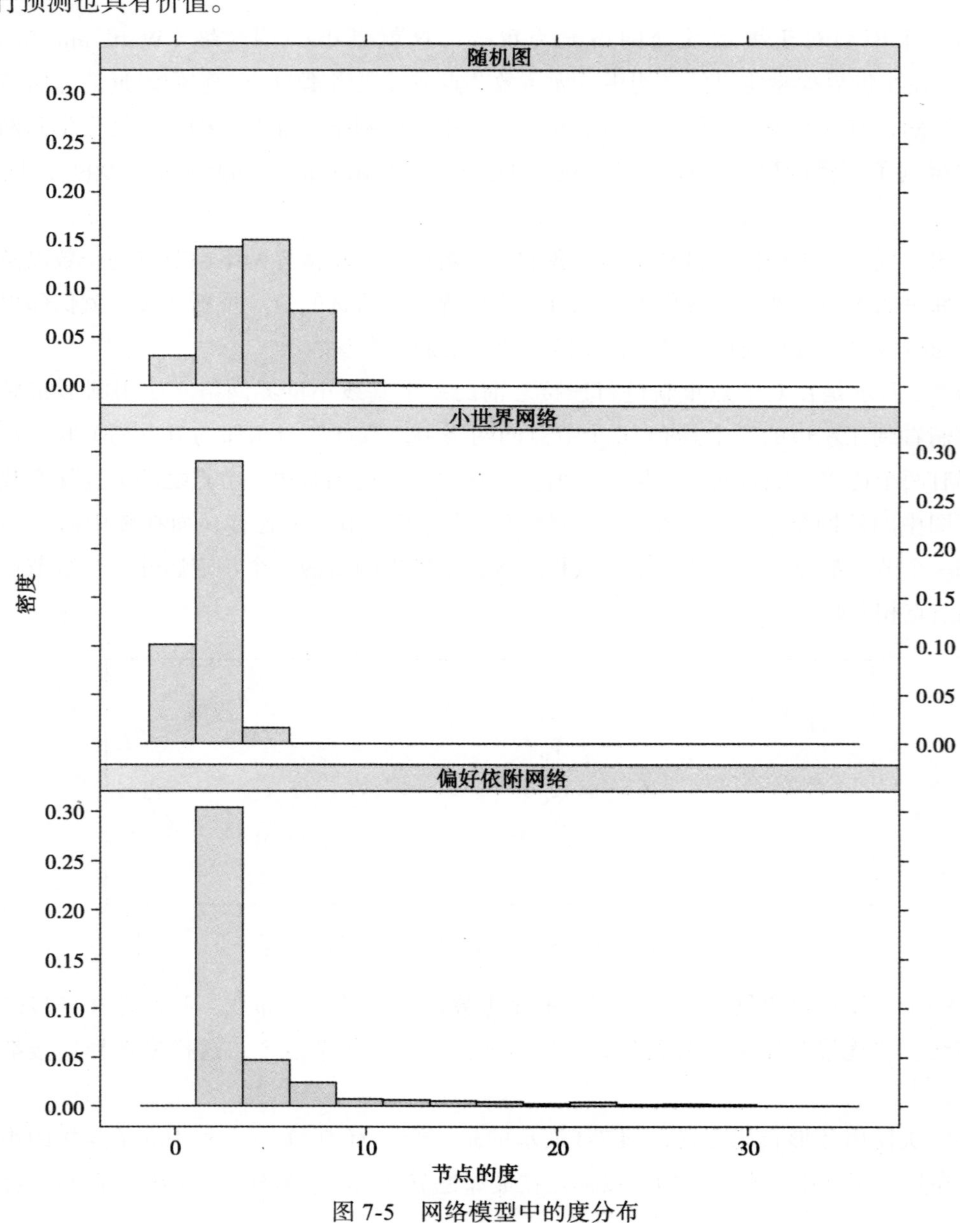

101

图 7-5　网络模型中的度分布

然而，在使用任何一个网络模型来进行推断和预测前，我们可能都要提出一个基本问题，即使用的模型是否适合所研究的问题。另外，信任通过随机图模型、偏好依附模型、小世界模型或其他数学模型所产生的样本数据是否合理？

度量节点的重要性对于找到权力、影响力或者重要性的根源非常有益。当将对中心性的度量作为政治权力的指标时，我们想要描绘出一个政治家对于他所属政党的重要性。我们也会考虑一个网页对于一个网站的重要性或者一篇杂志上的文章对于一个学科的影响程度。

我们可以从连接到一个节点的边开始。对于一个有向网络，我们可以确定一个节点出来和进去的边的个数，即我们对每一个节点的度进行度量，包括入度（in-degree）和出度（out-degree）。描述一个网络结构的方法就是基于网络的度分布，平均度或者度中心性是总结一个网络整体连通性的方式。

虽然一个节点的度中心性仅需要通过查看这个节点就可以计算出来，其他对中心性的度量则需要基于整个网络结构。接近中心性是一种对节点的度量，用来描述网络中某个节点与所有其他网络节点的紧密性，而紧密性是指节点之间的跳数或连接边的数量。度量接近中心性对于不连通的网络来说是一个问题，因此，相对于其他的度量方法，此度量方法不常用于对中心性的度量（Borgatti, Everett, and Johnson, 2013）。

在介数中心性（betweenness centrality）的度量中，我们考虑的是某个节点处于其他节点之间最短路径上的比例。介数中心性反映的是一个节点作为代理节点或中间节点的程度，它会影响网络流量或信息的流动。

如果我们认为，一个节点由于在距离上靠近一些重要的节点而成为一个重要的节点，而这些重要的节点又由于在距离上靠近其他一些重要的节点而成为重要的节点，如此循环下去。这种思路把我们引向特征向量中心性（eigenvector centrality）这个概念。就如用第一主要分量来描述一组变量的共同变化性那样，特征向量中心性可以用来描述网络中的某个节点对于构成网络的整组节点的中心性程度。

102

特征向量中心性是对整体连通性或重要性的一个度量方法。一个节点通过连接到其他具有很高连接程度的节点而得到更高的特征向量中心性。特征向量中心性取决于整个网络，是节点重要性的一个很好的整体度量。

不足为奇的是，对节点的重要性进行度量的不同方法具有很高的关联性，这一点可以通过以下步骤看出：先使用网络数学模型随机生成数据，计算出网络中每一个节点重要性或中心度的不同指数，然后将这些指数进行关联。

图7-6所展示的是一组由50个节点和100条连接构成的小世界网络的平均关联性的热点图。使用随机图模型和偏好依附模型进行仿真实验，可以看到对中心度的度量之间呈现出更高的正相关性。对使用3种数学模型（小世界、随机图和偏好依附模型）产生出的网络进行仿真实验结果表明，对于度中心性来说，与介数中心性和特征向量中心性相比较，接近中心性具有较低的相关性。

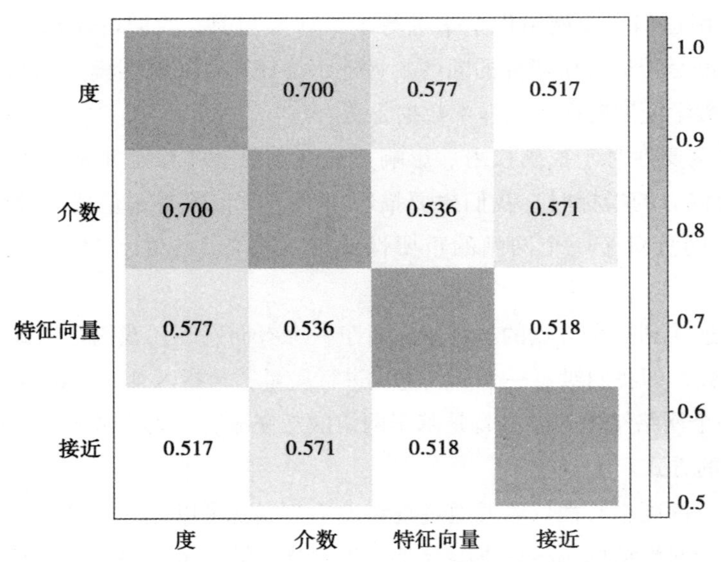

图 7-6 对中心度的不同度量方法都是相关联的

尽管 PageRank 算法的发布获得了许多赞誉（Brin and Page 1998），Google 对网站进行排名的方法基本上还是基于以下的想法：使用邻接矩阵的第一个特征向量来判断一个节点的重要性。与特征向量中心性一样，PageRank 算法是一种基于全网络的度量方法。从一个高重要性或高可信度的网页获得的链接要比从一个低重要性或低可信度的网页获得的链接更好。针对网站的 PageRank 算法和针对社交网络的特征向量中心性并不算是新的想法，它们均出自于 20 世纪初期数学家和统计学家的研究工作。⊖

一个网页的重要性可以通过链接到这个网页的其他网页的数量和质量来评估。一个人的重要性也可以通过他的同事、朋友或者亲属的重要性来判断。通过度量中心度，我们就有了找到可信度、权力或者影响力根源的方法。这些都与个体参与者（也就是社交网络中的节点）有关。那么网络中的连接或边又有什么作用呢？

当看着流过网络的信息或者穿过网络的关键路径时，我们就会关注到连接节点的边。我们通常使用介数中心度（betweenness）来对边进行区分。就像节点一样，边位于连接成对节点的路径上。而且，通过一条边进行通信的节点越多，这条边的介数中心度就越高。

各种对全网络进行度量的方法都会考虑节点和边，其中一个度量就是密度。一组完全不

⊖ Franceschet (2011) 指出，PageRank 的数学基础可以追溯到 1906 年提出的马尔科夫理论和 1912 年证明的 Perron-Frobenius 定理。特征值 / 特征向量分析也应用在排名和配对比较问题上以及对网站进行排名上。Keener (1993) 对排名方法所基于的数学进行过总结。配对比较已经应用于味觉测试实验、心理物理研究、优先尺度法、法官一致性的研究（考核者可靠性）以及对运动员和运动队进行排序、体育锦标赛的设计、体育锦标赛中种子队的确定 (Thompson 1975; Groeneveld 1990; Appleton 1995; Carlin 1996; West 2006)。在 Davidson and Farquhar (1976) 中罗列的有关配对比较与排名方法方面的文献来自于以下研究者所做的工作：Thurstone (1927), Guilford (1936), Kendall and Smith (1940), Wei (1952)。此外，Bradley and Terry (1952) 将传统的数理逻辑模型应用于分析配对比较研究上。配对比较方法综述由 Torgerson (1958), David (1963) 和 Bradley (1976) 提供。Langville and Meyer (2006, 2012) 对 PageRank 以及其他排名方法所基于的线性代数进行了综述。

连接或断开连接的节点没有任何边，而具有 n 个节点的完全连接图则有 $\frac{n(n-1)}{2}$ 条边。如果一个网络具有 l 条边，那么网络的密度通过公式 $\frac{l}{n(n-1)/2}$ 计算。

网络密度很容易计算，取值范围为 0～1。它也很容易解释为一个网络中实际存在的连接与所有可能存在的连接之比。但是，在描述不同大小的网络时，密度作为一个度量值的价值却非常有限。

社交网络关心的是人，而人却受到时间、距离和交流媒介的约束。人们可以维持的社交关系的数量是非常有限的。随着网络越来越大，能够观察到的网络密度呈下降趋势。

另外一个对整个网络的度量是*传递性*（transitivity），表达的是闭合*三元组*（triad）的比例。每一个三元组都由三个节点组成，而一个闭合三元组则是由三个节点以及两两节点之间的连接边组成。我们可以将每个闭合三元组看作是一个微型全连接图。与其他度量一样，传递性也可以通过邻接矩阵计算出来。

还有很多我们能够想到的对网络的度量。Iacobucci 在 Wasserman and Faust (1994) 的著作中对基于图和矩阵的度量进行了总体概述。

105

我们至今讨论过的所有度量都与网络拓扑或结构相关。这些度量描述了节点与节点、边与边，以及对于密度来说，边与可能的边之间的关系。这些度量可以帮助我们在大型网络中找出完全连接子图以及核心的社区。

除了单纯考虑拓扑或结构之外，我们还可以考虑元数据。节点具有与所扮演的角色、人口统计、态度和行为等相关的特征，边具有与宽度、带宽、长度 / 距离、遍历成本等相关的特征。想要基于社交网络做出预测时，我们通常就会基于元数据来进行思考。

消费者会购买某个产品吗？公民会投票给某个候选人吗？明年某人在体育赛事上会支出多少钱？了解由消费者、选民或者体育迷构成的社交网络可以帮助我们回答这些问题。

用于预测人们态度和行为的最近邻居方法不要求对基础流程或者统计分布做任何假设。这些方法都是数据驱动的，而不依赖理论，因此，这些方法可以适用于各种数据结构。

最近邻居可以是时间上的、地理上的、人口特征上的或心理上的。对于时间上的最近邻居来说，在时间上彼此更近的观察结果通常比时间上存在较大差距的观察结果具有更高程度的关联性。最近邻居可以用在面向时间序列的自回归与移动平均模型中。

在所有其他条件都相同的情况下，在空间上彼此更接近的观察结果比空间上存在较长距离的观察结果具有更高程度的关联性。地理上的相关性也是如此，居住在距离道奇体育馆 4 英里的人比住在佛罗里达坦帕市的人更可能是道奇队的球迷。

我们可以使用一系列有关人口特性或心理的变量集来计算多元距离的度量。例如，我们决定研究中的每一对个体之间的距离，然后用这些距离来识别出每个人最近的五个、十个或十五个邻居。谁会获得 Mary 的选票？那就要通过看居住在离 Mary 最近的邻居的投票结果来回答这个问题，看看他们都投了谁的票。

106

为了使用最近邻居法，我们需要得到时间、物理空间、人口统计、心理统计特征或网络拓扑上的距离。对于网络来说，我们找出每一个人所在的街坊，然后再找出在这个街坊里的人们。Joe会购买一个能够显示生理功能以及时间的多功能手表吗？谁在以Joe为中心的个人或以自我为中心的网络中与他的距离最近？他们是否也在购买这款多功能手表？

对邻近度的不同解释可能会对预测最近邻居产生完全不同的结果。我在晨练时遇到的那位老者与我在地理上很相近并且（我必须承认）在年龄上很相近，然而，我们在其他方面没有任何共同之处。我们不会经常出现在同样的社交圈。

关于更多的有关最近邻居法的讨论，可以参考以下文献：Duda, Hart, and Stork (2001), Shakhnarovich, Darrell, and Indyk (2006) 和 Hastie, Tibshirani, and Friedman (2009)。Kolaczyk (2009) 对最近邻居法以及其他预测模型方法做过综述，包括马尔可夫随机域与内核回归。

Kadushin (2012) 对社交网络进行了介绍。Ackland (2013) 将社交网的概念应用到网络和社交媒体网络中。Watts (2003), Lewis (2009) 和 Newman (2010) 对网络科学进行了综述。Jackson (2008) 和 Easley and Kleinberg (2010) 对将网络科学与经济领域中的博弈论进行集成做出了努力。

聚类、信息社区、核心群组和最大完全连接图的检测是社交网络分析中的一个重要研究领域。最近，Zubcsek, Chowdhury, and Katona (2014) 发表了一篇综述。

社交网络分析通常涉及多维度缩放、多层次聚类分析、对数线性建模以及其他各类很专业的方法。Kolaczyk (2009) 和 Kolaczyk and Csárdi (2014) 对网络科学中的统计方法进行过综述。Wasserman and Faust (1994) 和 Brandes and Erlebach (2005) 的贡献者们对社交网络分析的相关方法进行过综述。对数线性模型已被广泛地应用在这个领域的研究中。想要了解数线性模型的概况，请参考 Bishop, Fienberg, and Holland (1975), Christensen (1997) 和 Fienberg (2007) 这些文献。Wasserman and Iacobucci (1986) 和 Wasserman and Pattison (1996) 讨论了将对数线性模型和逻辑回归模型应用于社交网络。

107

随机块模型在对大型网络进行分析方面也非常有用。独立的节点以及它们之间的连接可以在一个网络中组合成不同的块，这些块也可以作为网络进行分析。要进一步了解块模型，可以参考 Holland et al. (1983) 和 Wasserman and Anderson (1987)。

千万不要低估在网络科学中分析员所做判断的重要性。分析员选择采样、建模和分析的方法。更重要的是，分析员对网络本身进行定义。

边将节点连接起来，但是分析员会确定这些连接的含义。回到安然公司邮件网络这个实例，是否高管之间的一封邮件信息就构成一条连接呢？或者是两封、三封或四封？分析员来设置构成一个连接所需的邮件 K 这个数值。有多少个可能的 K 值，就有多少种对原安然电子邮件网络的不同表现方式。总而言之，网络结构、边和节点的拓扑结构，就像网络的边界那样，都是分析员需要判断的事情。

网络科学和社交网络为大众媒体提供了有趣的素材，这些在以下的书籍中都进行了讨论：Buchanan (2002), Barabási (2003, 2010), Christakis and Fowler (2009) 和 Pentland (2014)。阅读这些书籍我们可以看到，世上物物相连，乱久必治，机遇也许根本就不是机遇，集体的智慧才是至高无上的。虽然集体的智慧这种提法很不寻常地被 Jung (1968) 提起，虽然网络科学可能会像宗教那样具有让人们无法抵抗的诱惑力，但是我们不能忘记，如果没有每一个独立个体的参与，就没有什么集体，也就根本没有网络，没有科学。

代码 7-1 展示的是基于随机图模型、小世界模型和偏好依附模型生成网络的 R 语言程序代码。这段代码来自于 Csardi (2014b), Kolaczyk and Csárdi (2014) 和 Sarkar (2014) 中的 R 语言软件包。Albert and Barabási (2002), Cami and Deo (2008), Goldenberg et al. (2009), Kolaczyk (2009) 和 Toivonen et al. (2009) 都对随机网络模型进行了综述。

代码 7-1　网络模型与度量（R 语言）

```
# Network Models and Measures (R)

# install necessary packages

# bring packages into the workspace
library(igraph)  # network/graph models and methods
library(lattice)  # statistical graphics

# load correlation heat map utility
load(file = "correlation_heat_map.RData")

# note. evcent() often produces warning about lack of convergence
# we will ignore these warnings in this demonstration program
options(warn = -1)
# number of iterations for the statistical simulations
NITER <- 100

# user-defined function to compute centrality index correlations
get_centrality_matrix <- function(graph_object) {
    adjacency_mat <- as.matrix(get.adjacency(graph_object))
    node_degree <- degree(graph_object)
    node_betweenness <- betweenness(graph_object)
    node_closeness <- closeness.estimate(graph_object,
        mode = "all", cutoff = 0)
    node_evcent <- evcent(graph_object)$vector
    centrality <- cbind(node_degree, node_betweenness,
    node_closeness, node_evcent)
        colnames(centrality) <-  c("Degree", "Betweenness",
            "Closeness", "Eigenvector")
    return(cor(centrality))
    }
# ---------------------------------
# Random Graphs
# ---------------------------------
# show the plot of the first random graph model
set.seed(1)
# generate random graph with 50 nodes and 100 links/edges
random_graph <- erdos.renyi.game(n = 50, type = "gnm", p.or.m = 100)
pdf(file = "fig_network_random_graph.pdf", width = 5.5, height = 5.5)
plot(random_graph, vertex.size = 10, vertex.color = "yellow",
    vertex.label = NA, edge.arrow.size = 0.25, edge.color = "black",
    layout = layout.circle, edge.curved = TRUE)
dev.off()

# express adjacency matrix in standard matrix form
random_graph_mat <- as.matrix(get.adjacency(random_graph))
# verify that the network has one hundred links/edges
print(sum(degree(random_graph))/2)

aggregate_degree <- NULL  # initialize collection of node degree values
correlation_array <- array(NA, dim = c(4, 4, NITER))  # initialize array
```

108 ~ 110

```
for (i in 1:NITER) {
    set.seed(i)
    random_graph <- erdos.renyi.game(n = 50, type = "gnm", p.or.m = 100)
    aggregate_degree <- c(aggregate_degree,
        degree(random_graph))
    correlation_array[,,i] <- get_centrality_matrix(random_graph)
    }
average_correlation <- matrix(NA, nrow = 4, ncol = 4,
    dimnames = list(c("Degree", "Betweenness", "Closeness", "Eigenvector"),
        c("Degree", "Betweenness", "Closeness", "Eigenvector")))
for (i in 1:4)
    for(j in 1:4)
        average_correlation[i, j] <- mean(correlation_array[i, j, ])

pdf(file = "fig_network_random_graph_heat_map.pdf", width = 11,
  height = 8.5)
correlation_heat_map(cormat = average_correlation)
dev.off()

# create data frame for node degree distribution
math_model <- rep("Random Graph", rep = length(aggregate_degree))
random_graph_degree_data_frame <- data.frame(math_model, aggregate_degree)

# ---------------------------------
# Small-World Networks
# ---------------------------------
# example of a small-world network (no random links)
set.seed(1)
# one-dimensional small-world model with 10 nodes,
# links to additional adjacent nodes in a lattice
# (nei = 1 implies degree = 2 for all nodes prior to rewiring)
# rewiring probability of 0.00... no rewiring
small_world_network_prelim <- watts.strogatz.game(dim = 1, size = 10,
    nei = 1, p = 0.00, loops = FALSE, multiple = FALSE)
# remove any multiple links/edges
small_world_network_prelim <- simplify(small_world_network_prelim)
# express adjacency matrix in standard matrix form
# show that each node has four links
print(degree(small_world_network_prelim))
# verify that the network has one hundred links/edges
print(sum(degree(small_world_network_prelim))/2)
pdf(file = "fig_network_small_world_nei_1.pdf", width = 5.5, height = 5.5)
plot(small_world_network_prelim, vertex.size = 25, vertex.color = "yellow",
    vertex.label = NA, edge.arrow.size = 0.25, edge.color = "black",
    layout = layout.circle, edge.curved = TRUE)
dev.off()

# another of a small-world network (no random links)
set.seed(1)
# one-dimensional small-world model with 10 nodes,
# links to additional adjacent nodes in a lattice
# (nei = 2 implies degree = 2 for all nodes prior to rewiring)
# rewiring probability of 0.00... no rewiring
small_world_network_prelim <- watts.strogatz.game(dim = 1, size = 10,
    nei = 2, p = 0.00, loops = FALSE, multiple = FALSE)
# remove any multiple links/edges
small_world_network_prelim <- simplify(small_world_network_prelim)
# express adjacency matrix in standard matrix form
# show that each node has four links
print(degree(small_world_network_prelim))
# verify that the network has one hundred links/edges
print(sum(degree(small_world_network_prelim))/2)

pdf(file = "fig_network_small_world_nei_2.pdf", width = 5.5, height = 5.5)
plot(small_world_network_prelim, vertex.size = 25, vertex.color = "yellow",
```

111

```
    vertex.label = NA, edge.arrow.size = 0.25, edge.color = "black",
    layout = layout.circle, edge.curved = TRUE)
dev.off()

# yet another of a small-world network (no random links)
set.seed(1)
# one-dimensional small-world model with 10 nodes,
# links to additional adjacent nodes in a lattice
# (nei = 3 implies degree = 6 for all nodes prior to rewiring)
# rewiring probability of 0.00... no rewiring
small_world_network_prelim <- watts.strogatz.game(dim = 1, size = 10,
    nei = 3, p = 0.00, loops = FALSE, multiple = FALSE)
# remove any multiple links/edges
small_world_network_prelim <- simplify(small_world_network_prelim)
# express adjacency matrix in standard matrix form
# show that each node has four links
print(degree(small_world_network_prelim))
# verify that the network has one hundred links/edges
print(sum(degree(small_world_network_prelim))/2)
pdf(file = "fig_network_small_world_nei_3.pdf", width = 5.5, height = 5.5)
plot(small_world_network_prelim, vertex.size = 25, vertex.color = "yellow",
    vertex.label = NA, edge.arrow.size = 0.25, edge.color = "black",
    layout = layout.circle, edge.curved = TRUE)
dev.off()

# rewire a selected proportion of the links to get small world model
set.seed(1)
small_world_network <- watts.strogatz.game(dim = 1, size = 50, nei = 2,
    p = 0.2, loops = FALSE, multiple = FALSE)
# remove any multiple links/edges
small_world_network <- simplify(small_world_network)
# express adjacency matrix in standard matrix form
# show that each node has four links
print(degree(small_world_network))
# verify that the network has one hundred links/edges
print(sum(degree(small_world_network))/2)
pdf(file = "fig_network_small_world.pdf", width = 5.5, height = 5.5)
plot(small_world_network, vertex.size = 10, vertex.color = "yellow",
    vertex.label = NA, edge.arrow.size = 0.25, edge.color = "black",
    layout = layout.circle, edge.curved = TRUE)
dev.off()
aggregate_degree <- NULL  # initialize collection of node degree values
correlation_array <- array(NA, dim = c(4, 4, NITER))  # initialize array
for (i in 1:NITER) {
    set.seed(i)
    small_world_network <- watts.strogatz.game(dim = 1, size = 50, nei = 1,
         p = 0.2, loops = FALSE, multiple = FALSE)
    aggregate_degree <- c(aggregate_degree,
        degree(small_world_network))
    correlation_array[,,i] <- get_centrality_matrix(small_world_network)
    }
average_correlation <- matrix(NA, nrow = 4, ncol = 4,
    dimnames = list(c("Degree", "Betweenness", "Closeness", "Eigenvector"),
        c("Degree", "Betweenness", "Closeness", "Eigenvector")))
for (i in 1:4)
    for(j in 1:4)
        average_correlation[i, j] <- mean(correlation_array[i, j, ])

pdf(file = "fig_network_small_world_correlation_heat_map.pdf", width = 11,
  height = 8.5)
correlation_heat_map(cormat = average_correlation)
dev.off()

# create data frame for node degree distribution
math_model <- rep("Small-World Network", rep = length(aggregate_degree))
```

112

```
small_world_degree_data_frame <- data.frame(math_model, aggregate_degree)

# ---------------------------------
# Scale-Free Networks
# ---------------------------------
# show the plot of the first scale-free network model
set.seed(1)
# directed = FALSE to generate an undirected graph
# fifty nodes to be consistent with the models above
scale_free_network <- barabasi.game(n = 50, m = 2, directed = FALSE)

# remove any multiple links/edges
scale_free_network <- simplify(scale_free_network)

pdf(file = "fig_network_scale_free.pdf", width = 5.5, height = 5.5)
plot(scale_free_network, vertex.size = 10, vertex.color = "yellow",
    vertex.label = NA, edge.arrow.size = 0.25, edge.color = "black",
    layout = layout.circle, edge.curved = TRUE)
dev.off()

# express adjacency matrix in standard matrix form
scale_free_network_mat <- as.matrix(get.adjacency(scale_free_network))

# note that this model yields a graph with almost 100 links/edges
print(sum(degree(scale_free_network))/2)

aggregate_degree <- NULL  # initialize collection of node degree values
correlation_array <- array(NA, dim = c(4, 4, NITER))  # initialize array
for (i in 1:NITER) {
    set.seed(i)
    scale_free_network <- barabasi.game(n = 50, m = 2, directed = FALSE)
    # remove any multiple links/edges
     scale_free_network <- simplify(scale_free_network)
    aggregate_degree <- c(aggregate_degree,
        degree(scale_free_network))
    correlation_array[,,i] <- get_centrality_matrix(scale_free_network)
    }
average_correlation <- matrix(NA, nrow = 4, ncol = 4,
    dimnames = list(c("Degree", "Betweenness", "Closeness", "Eigenvector"),
        c("Degree", "Betweenness", "Closeness", "Eigenvector")))
for (i in 1:4)
    for(j in 1:4)
        average_correlation[i, j] <- mean(correlation_array[i, j, ])

pdf(file = "fig_network_scale_free_correlation_heat_map.pdf", width = 11,
  height = 8.5)
correlation_heat_map(cormat = average_correlation)
dev.off()

# create data frame for node degree distribution
math_model <- rep("Preferential Attachment Network",
    rep = length(aggregate_degree))
scale_free_degree_data_frame <- data.frame(math_model, aggregate_degree)

# ---------------------------------
# Compare Degree Distributions
# ---------------------------------
plotting_data_frame <- rbind(scale_free_degree_data_frame,
    small_world_degree_data_frame,
    random_graph_degree_data_frame)

# use lattice graphics to compare degree distributions

pdf(file = "fig_network_model_degree_distributions.pdf", width = 8.5,
  height = 11)
```

113

```
lattice_object <- histogram(~aggregate_degree | math_model,
    plotting_data_frame, type = "density",
    xlab = "Node Degree", layout = c(1,3))
print(lattice_object)
dev.off()

# Suggestions for the student.
# Experiment with the three models, varying the numbers of nodes
# and methods for constructing links between nodes. Try additional
# measures of centrality to see how they relate to the four measures
# we explored in this program.  Explore summary network measures
# of centrality and connectedness to see how they relate across
# networks generated from random graph, preferential attachment,
# and small-world models.
```

114

当面对超大型的网络时，对运行时间和内存的要求都会很高。我们必须常常从大型网络中进行数据采样，以得到一个大小可控的网络来进行建模与分析。

数据采样带来了额外的挑战。我们是对节点，还是对边（成对的节点），还是对节点与边的组合进行采样？好的采样方法产生的样本所具有的中心度值和连通性值会与完整网络中的这些值相似。很多学者都对采样以及数据丢失问题进行过综述（Kossinets 2006; Leskovec and Faloutsos 2006; Handcock and Gile 2007; Huisman 2010; Smith and Moody 2013 ）。

代码 7-2 展示的是一个实现 Leskovec and Faloutsos (2006) 中所描述的采样方法的 R 语言程序。我们研究这些采样方法时使用了来自维基百科选票的数据。研究的问题是：“我们应该采用什么样的采样方法才能获取能够代表整个网络的样本”。为了评估示例程序中所使用到的不同采样方法的效果，我们使用通过计算完全连接三元组的比例而得到的传递性这个指标。传递性，也被称作平均聚类系数，是对网络连通性的一个度量。这个完整的维基百科投票网络的传递性为 0.1646。我们可以把这个数值作为一个全体参数的已知值，而我们想要找到一个能够有效估计此参数的网络采样方法。

代码 7-2　大型网络中的数据采样方法（R 语言）

```
# Methods of Sampling from Large Networks (R)

# install packages sna and network

# Background reference for sampling procedures:
# Leskovec, J. & Faloutsos, C. (2006). Sampling from large graphs.
# Proceedings of KDD '06. Available at
# <http://cs.stanford.edu/people/jure/pubs/sampling-kdd06.pdf>

# load packages into the workspace for this program
library("sna")  # social network analysis
library("network")  # network data methods

# user-defined function for computing network statistics
# this initial function computes network transitivity only
# additional code could be added to compute other network measures
# note that node-level measures may be converted to network measures
# by averaging across nodes
report.network.statistics <- function(selected.edges) {
    # selected.edges is a data frame of edges with two columns
    # corresponding to FromNodeId and ToNodeId for a directed graph
    # analysis of selected.edges can begin as follows
    # create a directed network/graph (digraph)
```

```
    selected.network <- network(as.matrix(selected.edges),
        matrix.type="edgelist",directed=TRUE)
    # convert to a selected.matrix/graph)
    selected.matrix <- as.matrix(selected.network)
    # transitivity (clustering coefficient) probability that
    # two nodes with a # common neighbor are also linked
    # (Consider three nodes A, B, and C.
    # If A is linked to C and B is linked to C,
    # what is the probability that A will be linked to B?)
    # If value above 0.50... indicates clustering of nodes.
    network.transitivity <-
        gtrans(selected.matrix, use.adjacency = FALSE)
    # report results for this run
    cat("\n\n","Network statistics for N = ",
      nrow(selected.edges),"\n    Transitivity = ",
          network.transitivity,sep="")
    }

# network data from the Wikipedia Votes case
# read in the data and set up a binary R data file
# wiki.edges = read.table("wiki_edges.txt",header=T)
# save(wiki.edges, file ="wiki_edges.Rdata")

# with the binary data file saved in your working directory
# you are ready to load it in and begin an analysis
load("wiki_edges.Rdata")  # brings in the data frame object wiki.edges
```

115

```
print(str(wiki.edges))  # shows 103,689 initial observations/edges
# check to see if there are any self-referring edges
self.referring.edges <- subset(wiki.edges,
   subset = (FromNodeId == ToNodeId))
print(nrow(self.referring.edges))  # shows that there are no such nodes

# this is a large network... so we will explore samples of the edges
# begin by identifying the number of unique nodes
unique.from.nodes <- unique(wiki.edges$FromNodeId)
print(length(unique.from.nodes))  # there are 6,110 unique from-nodes
unique.to.nodes <- unique(wiki.edges$ToNodeId)
print(length(unique.to.nodes))  # shows that there are 2,381 unique to-nodes
unique.nodes <- unique(c(unique.from.nodes,unique.to.nodes))
print(length(unique.nodes))  # there are 7,115 unique nodes total

# compute the transitivity for the complete network commented out here
# report.network.statistics(selected.edges = wiki.edges)
# RESULT: Transitivity of complete network: 0.1645504

# set the sample size N by setting a proportion of the edges to be sampled
P <- .1  # what proportion will work... .10, .20, .30, or higher?
N <- trunc(P * nrow(wiki.edges))

# specify the sampling method three methods are implemented here
# one is selected for each time the program is run
METHOD <- "RE"  # random edge sampling
# METHOD <- "RN"  # random node sampling
# METHOD <- "RNN"  # random node neighbor sampling

# seed the random number generator to obtain reporducible results
# if a loop is utilized to repeat sampling, ensure that the seed changes
# within each iteration of the loop
set.seed(9999)  # reporducible results are desired

# ------------------- random edge sampling -------------------
# Random edge (RE) sampling. Using the complete data frame with
# each record representing an edge or connection between nodes,
# we select N edges at random.
```

```
if(METHOD == "RE") { # begin if-block for random edge sampling
    selected.indices <- sample(nrow(wiki.edges),N)
    selected.edges <- wiki.edges[selected.indices,]
    }  # end if-block for random edge sampling

# ------------------- random node sampling -------------------
# Random node (RN) sampling. Using the list of unique node numbers,
# we select nodes at random without replacement. This sampling is
# carried out using the combined set of FromNodeId and ToNodeId
# node identifiers. The preliminary sample data frame consists
# of all edges containing these nodes. Then we sample exactly N
# edges or rows from the preliminary sample data frame.
if(METHOD == "RN") { # begin if-block for random node sampling
    selected.node.indices <-
        sample(length(unique.nodes),P*length(unique.nodes))
    selected.nodes <- unique.nodes[selected.node.indices]
    from.edge.samples <- subset(wiki.edges,
        subset = (FromNodeId %in% selected.nodes))
    to.edge.samples <- subset(wiki.edges,
        subset = (ToNodeId %in% selected.nodes))
    all.edge.samples <- rbind(from.edge.samples,to.edge.samples)
    selected.indices <- sample(nrow(all.edge.samples),N)
    selected.edges <- all.edge.samples[selected.indices,]
    } # end if-block for random node sampling

# ------------------- random node-neighbor sampling -------------------
Random node neighbor (RNN) sampling. We begin by selecting a node at
# random (referring to the FromNodeId values), together with all
# of its out-going neighbors (ToNodeId values). We continue selecting
# random FromNodeId values and their associated ToNodeId values until
# we have N edges in the sample.
if(METHOD == "RNN") { # begin if-block for random node neighbor sampling
    # obtain from-node values in permuted order
    permuted.from.nodes <- sample(unique.from.nodes)
    selected.edges <- NULL  # initialize the sample data frame
    number.of.selected.edges.so.far <- 0  # initialize edge count
    # initialize index of permuted.from.nodes
    index.of.permuted.from.node <- 0
    while(number.of.selected.edges.so.far < N) {
        index.of.permuted.from.node <- index.of.permuted.from.node + 1
        this.selected.set.of.edges <- subset(wiki.edges,
          subset = (FromNodeId ==
              permuted.from.nodes[index.of.permuted.from.node]))
        selected.edges <- rbind(selected.edges,this.selected.set.of.edges)
        number.of.selected.edges.so.far <- nrow(selected.edges)
        }
    # just use the first N of the selected edges
    selected.edges <- selected.edges[1:N,]
    } # end if-block for random node neighbor sampling

# report on network sampling results with the
# user-defined function report.network.statistics
cat("\n\n","Results for sampling method ", METHOD, sep="")
report.network.statistics(selected.edges)

# Suggestions for the student:
# Building on the example code, your job is to write and execute
# a program to answer a number of questions. How large of a sample
# is needed to obtain accurate estimates of transitivity?
# What type of sampling method (if any) provides the best
# estimate of transitivity for the complete network?
# Are there other network measures that might be more accurately
# estimated with network sampling?  Which network sampling method
# would you recommend for work in web and network data science?
# Go on to implement other network-wide measures, such as
```

116

```
# closeness, betweenness, and eigenvector centrality, and
# network density (number of observed links divided by
# the total possible links).
```

示例程序展示如何从维基百科投票网络中具有 N 条边的图获取样本的三种方法：随机边采样、随机节点采样和随机节点邻居采样。正如名字所表达的那样，随机边采样就是随机地从网络或者所有的边中随机选取边。使用整个边数据集，每一条数据记录都代表一条边或节点之间的一条连接，我们随机地选取其中 N 条边。对于随机节点采样，我们对每一个节点都进行唯一编号，然后不重复地随机选择节点。采样中同时使用了有向图中边的入节点标识编号和出节点标识编号。初步的样本数据集包含连接节点的所有边，然后，我们从初步样本数据集中取出正好 N 条边或者行。随机节点邻居采样会更复杂些。我们首先随机选取一个节点（称为出节点标识编号），然后再取出所有与这个节点有入连接边的邻居节点（使用入节点标识标号）。我们继续随机地选取有出标识编号的节点以及这些节点连接到的有入标识编号的邻居节点，直到样本中边的数量达到 N 条。[117]

第8章
度量情感

“我已经像地狱中疯狂的恶魔一样，我再也受不了了！”

Howard Beale（Peter Finch 饰）
电影《电视台风云》(1976年)

我已经作为一名非传统教育家而闻名，因为我拒绝对作业采用分级评鉴和按部就班指导的方式。如同听起来那么疯狂，我布置的大部分作业都是对迷你案例的研究：这是一个商务问题，这里有一些凌乱的数据，理解它们的含义并撰写一份报告给管理层。

我想强调的是我们在每天的工作和生活中都要面对这些问题。数据很凌乱，通常都是一些非结构化的文本。管理者们并不指出该如何去处理数据或者解决问题，因为这是数据科学家们的任务。

开放式的作业给了学生们一个勇于创新以及培养他们具备数据科学家应有的良好判断力的机会。尽管学生们在对同一个题目的理解上存在较大差距，但是他们仍然可能高产。成功取决于“能够容忍模糊不清”。

什么是对模糊不清能够容忍的程度？这是一种对于接受的态度。我们接受存在不确定性这一事实。虽然知道很多细节是一件好事，但不知道所有细节也无关紧要。我们今天不理解的事情也许以后会变得更加清晰。

情感分析是我们在数据科学中需要深入理解的事情之一。为情感分析（或者有时被称为为意见挖掘）而设计的工具如雨后春笋般地在网络中出现，然而几乎没有人理解它们的含义。

公众的情绪有什么特征？选民是否喜欢候选人？精心设计的信息是否可以改变对候选人、产品或者品牌的看法？这些问题我们都可能通过调查和深入研究得出答案。然而，在今天这个信息丰富的世界里，我们可以使用辅助数据源来回答这些问题。情感分析可以应用于从网络中搜集到的大部分数据，可以应用于新闻网站、博客和社交媒体。许多情感分析方面的研究都已经在推特中展开。

情感分析是一个我们还知之甚少的度量问题，提出的度量方法几乎还没有可靠有效的文献资料。数据科学家和建模者通常都没有良好的度量理论方面的基础。这对于如何正确进行情感分析来说是一个巨大的不确定性因素，甚至对于是否存在正确的度量方法也存在不确定性。采样方面的问题也比比皆是。

对模糊能够容忍的态度对高效学习是很有益的。但是，当我们的工作是要依据所做的度量去做出预测时，模糊就不是一件好事情了。我们希望度量具有一定的含义。我们希望这些度量尽可能可靠和有效。要正确地进行情感分析，我们需要设计可行的文本度量。这些度量本身也仅仅是一个开始。情感分析中真正的任务是证明这些度量可以被用于进行预测。

情感分析最重要的核心部分是文本特征选择以及文本度量（text measure）定义。文本度量就是对描述文本的属性或特征进行打分。按照最基本的理解，度量就是基于规则为属性打分。我们可以使用文本度量来评估个性、消费者的偏爱以及政治观点，就像我们可以使用调查工具那样。文本度量和调查工具的区别在于文本度量从由非结构化或半结构化的文本作为输入数据开始，而不是对问卷调查的回复。

我们使用IMDb影评数据库中的数据来说明一个完整的情感分析过程。除了IMDb数据之外，我自己也写过一些电影评论。我个人所做的影评展示在图8-1和图8-2中，我对每一部电影给予了从1到10不同的评价，1为“讨厌极了”，10为“棒极了。”

伽马射线效应（The Effect of Gamma Rays on Man-in-the-Moon Marigolds，1972）
基于Paul Zindel这个普利策奖得主的发挥，以及Paul Newman对Joanne Woodward进行的指导，这应该是你曾经看过的最令人振奋的影片之一。这部影片是对人类精神的致敬，崇尚超越个人当前处境以及克服逆境。这部影片同样也具有教育意义。如果有机会观看就别错过。
我的影评打分：10

银翼杀手（Blade Runner，1982）
甚至比Harrison Ford主演的另外一部电影《突袭者》(Raiders)更好看。这部影片保持了《突袭者》的许多优点，雷同之处也挺感人，这一点还挺绕人。视觉效果与细腻的摄影技术会吸引住你。就如同你身临其境一样。虽然按照我们的套路并不是一部非常好的影片，但仍然是一部很好的影片。
我的影评打分：9

我的表兄维尼（My Cousin Vinny，1992）
Joe Pesci和Marisa Tomei目前是一对很奇特的夫妻。这部电影建立在布鲁克林和阿拉巴马州的固有偏见上，很难去同情不幸的表弟以及他的男性朋友。这部电影也没有太多悬念，因为我们能够猜到最后的结局是什么。必须是那样的结果。Tomei在整个影片中的演技还行。如果不去关注她如何很自然地融入恋人的角色这个过程，以上所说全都是废话。
我的影评打分：4

火星人玩转地球（Mars Attacks，1996）
作为一个不需用心的娱乐还可以。我收藏了此片的DVD，每6个月左右就会为了笑一笑而观看一次。Nicholson在影片中扮演了两个角色：POTUS和房地产开发商。会爆炸的火星人脑袋非常棒。但是，这部影片不适合乡村音乐爱好者。
我的影评打分：7

图8-1 Tom所做的几个影视评论

搏击俱乐部（Fight Club，1998）
我在一家二手音像店找到了这部影片。我以为这是一部不错的电影。我的天哪！是我弄错了吧？我猜想可能是我对充满暴力的生活或娱乐方式不感兴趣。我无法同影片中的任何演员产生同感。如果有机会，我会用这部电影去以旧换新。
我的影评打分：2

选美俏卧底 2（Miss Congeniality 2，2006）
你一定是在开玩笑吧。第一部就是一部垃圾片，而这第二部更加糟糕。难道我们真的每隔 5 年就必须要忍受一次对我们智商的侮辱？我喜欢 Sandra Bullock，但不明白她怎么会去演这种烂片。
我的影评打分：1

判我有罪（Find Me Guilty，2006）
我并不真正是 Vin Diesel 的一个粉丝，但是我儿子建议我去看这部影片。我必须说，我对他们能够将一部围绕法庭场景的电影拍得这么好感到十分惊讶。 Vin Diesel 演得非常棒。也许是 Sidney Lumet 效应。很可惜我们已经失去了他。
我的影评打分：7

点球成金（Moneyball，2011）
你也许会认为我会喜欢这部电影，因为我喜爱运动和分析学，但是奥克兰运动队（Oakland Athletics）或者电影主角很难引起我的共鸣。并不又是因为 Brad Pitt。那个家伙肯定是闲不下来。我假定这部电影是由一个真实的故事改编而成。毕竟，你不能让运动队赢得世界职业棒球大赛，因为这件事没有发生过。然而，看完影片后给我留下的是一种空荡荡的感觉。我想这个故事从书呆子分析师的角度去诉说会更好。让 Jonah Hill 有更多上镜头的机会来看看他能做什么。也许你甚至在爱的驱动下去工作。我的一些学生问过我对这部电影的感受。我觉得还可以。
我的影评打分：4

图 8-2 Tom 所做的另外几个影视评论

得到情感分数仅仅只是一个开始。情感分析的真正工作是证明分数具有一定的含义，可以用于进行预测。在这个实例中，我们探讨各种如何形成面向情感的文本度量的方法，包括基于列表的度量、条目加权的度量以及文本分类模型。我们还要使用一个学习与测试方案对文本度量和模型的预测性能进行评估。

一种对情感进行分析的方法基于能够表达人类情绪或感受的正面和负面词汇集（词典，字典）。这些词汇集受到所使用语言以及应用场景的限制。为了说明如何基于列表进行情感度量，我们使用 Hu 和 Liu (2004) 开发的一个词汇集。这个词汇集的列表包含 2006 个正面和 4783 个负面意见及情感词汇。下面让我们来看看这些列表如何与 Mass et al. (2011)[⊖]中的一

⊖ 这些影评数据可以从 Mass et al. (2011) 的作者那里获得，他们是 Andrew L. Maas、Raymond E. Daly、Peter T. Pham、Dan Huang、Andrew Y. Ng 和 Christopher Potts，这些数据已经被以上几位好心的作者存放在公众都可以访问的区域。这些数据最初来源于 IMDb (The Internet Movie Database)。其中一个数据集里包含有 50 000 条没有打分的影评。还有与这个数据集差不多大小的学习与测试数据集，都包括有对电影的评分。在本章中，我们分别使用每一个数据集中一个很小的数据样本。这些影评也许可以从 http://ai.stanford.edu/~amaas/data/sentiment/ 网站上下载。

个影评全集相匹配。

Mass et al. (2011) 影评全集中提供了三种类型的数据集。第一种类型的数据集只有影评文本而没有分级。我们将使用这种数据集来识别出用于情感文本度量的词汇集。第二种类型的数据集包含影评以及分级。我们将根据这种数据集建立文本度量模型。与第二种一样，第三种类型的数据集也有分级。我们将用这种数据集对提出的文本度量进行评估。在整个过程结束前，我们将会按照在本章开始提出的分级对建立的文本度量及模型进行评估。

图 8-3 展示的是 Hu and Liu (2004) 所提供的词汇集中的 50 个词汇，应用于只有影评而没有分级的影评数据集。对这些正面和负面的词汇进行的分析告诉我们，正面的词可能来源于爱情片和喜剧、令人振奋或者“感觉良好”的电影，而负面的词汇则更倾向于与恐怖、暴力和悲剧相关[⊖]。

121~123

Positive Words（正面词）				
amazing（令人惊讶的）	beautiful（美丽的）	classic（经典的）	enjoy（享受）	enjoyed（享受过的）
entertaining（有娱乐趣味的）	excellent（杰出的）	fans（粉丝）	favorite（最钟爱的）	fine（好极了）
fun（有趣）	humor（幽默）	lead（领导）	liked（喜欢过）	love（爱）
loved（爱过）	modern（现代的）	nice（友好的）	perfect（完美的）	pretty（漂亮的）
recommend（推荐）	strong（强大的）	top（顶级）	wonderful（精彩的）	worth（值得）

Negative Words（负面词）				
bad（很差的）	boring（无聊的）	cheap（廉价的）	creepy（令人毛骨悚然的）	dark（黑暗的）
dead（呆板的）	death（死神）	evil（邪恶）	hard（困难的）	kill（致命）
killed（杀害了）	lack（缺少）	lost（迷失方向的）	miss（错过）	murder（谋杀）
mystery（神秘）	plot（阴谋）	poor（低劣的）	sad（悲伤的）	scary（吓人的）
slow（迟钝的）	terrible（可怕的）	waste（浪费）	worst（最差的）	wrong（错误的）

图 8-3　表达情感的 50 个词汇

值得注意的是，对正面词汇是好的或者对负面词汇是坏的没有任何固定标准。正像我们在日常交流中可能会使用任何词汇那样，上下文会给这些词汇赋予含义。事实上，我们很容易使用正面词汇列表里的词汇来评价一部烂电影，或者用负面词汇列表里的词汇来评价一部好电影。

在建立这些列表后，假使我们统计一下每篇影评中正面词汇和负面词汇出现的次数，从而计算出两个基于列表的文本度量。假如使用 POSITIVE（正面）表达影评中与正面词汇列表包含的 25 个词实现匹配的词汇比例，使用 NEGATIVE（负面）表达影评中与负面词汇列表包含的 25 个词实现匹配的词汇比例，则图 8-4 展示了 4 篇影评以及所对应的 POSITIVE（正面）和 NEGATIVE（负面）得分。

124

图 8-5 中的散点图展示的是将基于这样两个情感词汇列表的度量应用于 500 篇影评得出的结果。如果 POSITIV（正面）和 NEGATIVE（负面）处于一个维度的两端，我们期望

⊖ 需要注意的是，我们没有使用那些源于创建词汇集的词汇，就像我们会看到“享受”和“享受了”以及“相爱”和“相爱了”出现在正面词汇集中，而“杀掉”和“杀掉了”出现在负面词汇集中那样。归源涉及将词汇映射到一个根词汇，或者称作源。也许在今后针对这些数据的研究中会考虑归源的问题。

变种女狼2：释放（Ginger Snaps 2: Unleashed, 2004）

I liked it a lot, in fact even more than the first movie. I loved the character of Ghost and all the comic book shots and her third person lines. Good ending. One thing they could have done was make the identity of the werewolf clearer. Also when the sister appeared it was kind of forced.. it didn't seem like she was a delusion

（20个分析词，2个来自于正面词列表，0个来自于负面词列表）
基于列表的文本度量：正面词=10，负面词=0

强尼的“八头牛”妻子（Johnny Lingo, 1969）

Beyond the tremendous and true romantic love Johnny Lingo proves for his dear Mahana, he gives a tremendous object lesson in how to properly treat others, and bring out the very best in them. If all husbands would treat their wives the way Johnny treated Mahana, there could be no evil in the world.

（20个分析词，1个来自于正面词列表，1个来自于负面词列表）
基于列表的文本度量：正面词=5，负面词=5

永恒的明天（Tomorrow Is Forever, 1946）

The greatest and most poignant anguish we conscious beings experience is our recognition of the irretrievability of the past. All else we endure could be easily borne. The background strains of music as Orson Welles first recognizes Claudette Colbert haunts me still. She experienced a fragmentary nuance of remembrance that did not reach the level of her conscious recall.

（25个分析词汇，0个来自于正面词列表，0个来自于负面词列表）
基于列表的文本度量：正面词=0，负面词=0

世事艰难（Malas temporadas, 2005）

There are some exciting scenes in this movie but in general it is second-rate. The shoots are overextended, the characters are not life-like and some actors don't perform well either. I also didn't like multiple nationalist statements which have nothing to do with the plot. I guess the director intended to make his characters mysterious but instead they came out to be unnatural. We are supposed to see how different people successfully struggle with hard times in their lives. But two stories, the one of Carlos and that of Mikel, end up with nothing and the third, the story of Ana, makes a turn without any reason. The movie is very depressive but without any message that derives from it.

（40个分析词，0个来自于正面词列表，2个来自于负面词列表）
基于列表的文本度量：正面词=0，负面词=5

图 8-4 针对4个影评所做的基于列表的文本度量

这两个文本度量的关联性接近于 –1.0。事实上，当计算 POSITIVE（正面）和 NEGATIVE（负面）之间的关联性时，我们会发现这个关联性是个负值，但是却更接近于 0，而不是 –1.0。

将 POSITIVE（正面）和 NEGATIVE（负面）应用于本章开头的 8 篇影评会得出什么样的结果呢？表 8-1 表明这个结果并不是很理想。这些基于列表的度量应用于电影《伽马射线效应》（The Effect of Gamma Rays on Man-in-the-Moon Marigolds）和《判我有罪》（Find Me Guilty）时漏掉了一些好评，而在应用于电影《点球成金》（Moneyball）时则漏掉了一些差评。

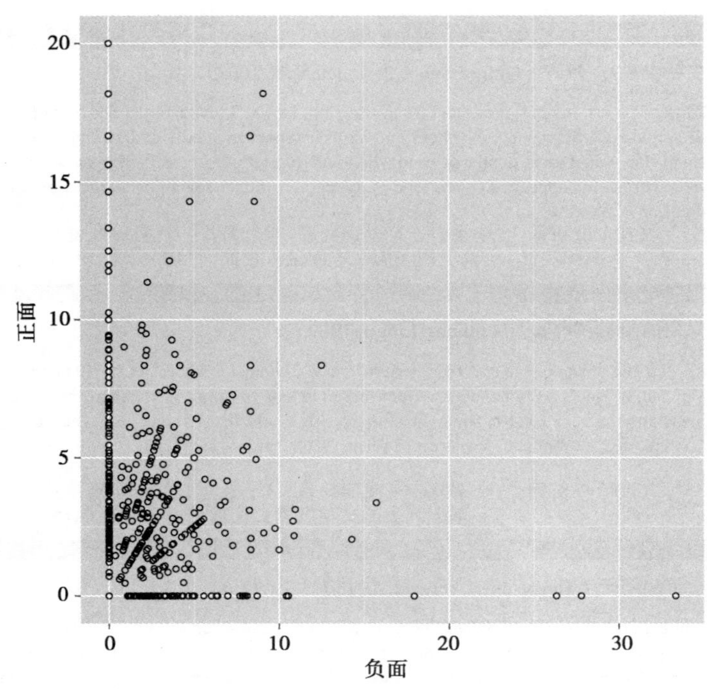

图 8-5 对正面、负面情感进行文本度量的散点图

表 8-1 针对 Tom 的影评所做的基于列表的情感度量

电影	词的总数	正面词	负面词	文本度量		打分	拇指向上 / 拇指朝下
				正面	负面		
伽马射线效应	26	0	1	0.00	3.85	10	拇指向上
银翼杀手	21	2	0	9.52	0.00	9	拇指向上
我的表兄维尼	29	1	2	3.45	6.90	4	拇指朝下
火星人玩转地球	20	1	0	5.00	0.00	7	拇指向上
搏击俱乐部	18	0	2	0.00	11.11	2	拇指朝下
选美俏卧底	10	0	1	0.00	10.00	1	拇指朝下
判我有罪	18	0	2	0.00	11.11	7	拇指向上
点球成金	36	2	1	5.56	2.78	4	拇指朝下

也许可以将 POSITIVE（正面）和 NEGATIVE（负面）通过某种方式相互结合，以对影评进行更加有效的预测。或者换一种方式，我们可以返回那个包含 50 个词汇的列表，然后对它们使用基于数据的模型来对影评进行预测。想要在文本度量的模型方面取得进一步的进展，让我们来尝试使用具有已知分级影评的训练集。

我们从正面影评训练集（获得 7～10 分的影评）和从负面影评训练集（获得 1～4 分的影评）中各选择 500 条记录。我们将它们混合在一起，形成一个具有 1 000 条影评的训练集。我们使用同样的过程创建一个具有 1 000 条影评的测试集。

125～127

使用不同的文本度量和预测建模技术，我们采用的是一个训练然后再测试的方法，即基于训练集来建立度量和模型，然后对测试集进行测试。我们的目标就是预测一篇影评是拇指向下（thumbs-up）的（评分高于 6 分），还是拇指朝下（thumbs-down）的（评分低于 5 分）。这是一个可能需要使用不同的技术才能得到解决的文本分类问题。下面是针对这个实例所使用的 6 种度量和建模技术。

- 简单差异。我们计算出分差值，即 POSITIVE（正面）分数减去 NEGATIVE（负面）分数，并且使用通过训练集而得出的一个分界值来预测拇指向上还是拇指朝下。
- 回归差异。我们用线性回归来确定 POSITIVE 的分数和 NEGATIVE 的分数的权重，最终形成一个线性预测。这里我们也使用一个分界值将预测出的分级确定为是拇指向上还是拇指朝下。
- 单词 / 字条分析。基于原有的 50 个词汇，我们使用训练数据来识别出具有正面倾向和具有负面倾向的词。然后再创建一个小的词汇集，给其中每一个拇指向上的词赋予权重 +1，给每一个拇指朝下的词赋予权重 –1。我们定义一个基于字条的文本度量，即所有字条分值的总和。预测拇指向上还是拇指朝下就取决于这个文本度量的结果（正或负）。这个过程与心理学中所用的传统的项目分析过程类似。相关内容可以参考 Nunnally (1967) 或 Lord and Novick (1968)。
- 逻辑回归。我们采用传统的统计建模方法来预测出一个二元值响应。特别需要注意的是，我们使用渐进逻辑回归从这个具有 50 个情感词汇的数据集中选择有用的词汇来进行预测。线性预测中的系数或者权重是通过最大相似法来决定的。逻辑回归是解决二元分类问题最常见的方法。目前已经有很多针对逻辑回归问题的讨论（Ryan 2008; Fox and Weisberg 2011; Hosmer, Lemeshow, and Sturdivant 2013）。
- 支持向量机。这种机器学习算法已经被证明是一种有效解决文本分类问题的技术，特别是解决有大量解释性变量的问题，就像我们在这里有 50 个情感词汇。Vladimir Vapnik 被公认为在此领域做出了很多贡献（Boser, Guyon, and Vapnik 1992; Vapnik 1998; Vapnik 2000）。更多关于支持向量机的讨论可以参考 Cristianini and Shawe-Taylor (2000), Izenman (2008) 和 Hastie, Tibshirani, and Friedman (2009)。Tong and Koller (2001) 讨论了支持向量机如何应用于对文本进行分类。 [128]
- 随机森林。这是一个被称为委员会或全体参与的方法，它使用上千个树形结构的分类器，最终得出一个预测结果。这些树形结构分类器使用 Breiman et al. (1984) 中描述的方法。Izenman (2008) 和 Hastie, Tibshirani, and Friedman (2009) 对树形结构的方法进行了综述。这是一个将训练集用递归的方式进行分解，最终得到预测拇指向上或拇指朝下的分类树。解释性变量集由这 50 个情感词汇构成。很多通过这样构成的分类树形成一个随机森林。与支持向量机一样，这个方法被证明在应对大量解释性变量时非常有效。这个方法基于自助重采样技术，由 Breiman (2001a) 首先提出，Izenman (2008) 和 Hastie, Tibshirani, and Friedman (2009) 对此方法做了更多介绍。

我们在此对各种度量与建模技术进行的测试构成基准实验的第一次迭代。我们要如何对研究得出的预测结果的精度进行评估呢？对于影评文本分类问题，我们使用一个很容易被管理者理解的指数：影评测试数据集中正确预测出拇指向上 / 拇指朝下结果的百分比。表 8-2

列出了基于测试数据集的统计结果以及正确预测的百分比。

表 8-2 对影评进行文本分类的精度（拇指向上或拇指朝下）

文本度量 / 模型	影评得到正确分类的百分比	
	训 练 集	测 试 集
简单差异	67.4	66.1
回归差异	67.3	66.4
词 / 条目分析	73.9	74.0
逻辑回归	75.2	72.6
支持向量机	79.0	71.6
随机森林	82.2	74.0

129

虽然随机森林的效果最好，基于传统心理学方法和随机森林方法的单词 / 字条分析对测试集进行文本分类可以取得同样好的效果。随机森林还有另外一个优点，即能够对解释性变量的重要性（在文本分类中单字的重要性）进行度量，如图 8-6 中的散点图所示。

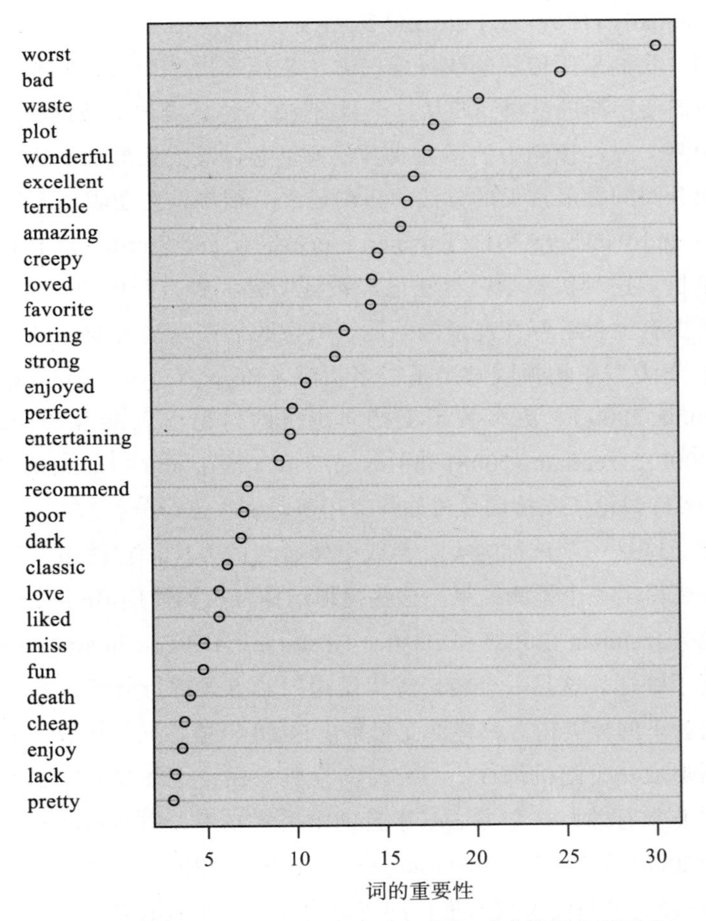

图 8-6 将影评区分为拇指向上或拇指朝下时单词的重要性

为了完成影评中的工作，基于我们的基准研究，我们选择出性能最好的度量模型，即随机森林，并应用这个模型对本章开头讨论的影评进行分类，结果列在表 8-3 中。可以看到，8 篇影评中的 5 篇影评（62.5%）获得了正确的分类。就像我们之前已经总结过的基于列表度量中的 NEGATIVE（负面）和 POSITIVE（正面）一样，随机森林方法在对《伽马射线效应》（The Effect of Gamma Rays on Man-in-the-Moon Marigolds）和《判我有罪》（Find Me Guilty）得到拇指向上的分类上和在对《点球成金》(Moneyball）得到拇指朝下的分类上均告失败。

表 8-3　随机森林文本度量模型在 Tom 影评中的应用

电　影	排名打分	实际的拇指向上 / 拇指朝下	预测的拇指向上 / 拇指朝下
伽马射线效应	10	拇指向上	拇指向下
银翼杀手	9	拇指向上	拇指向上
我的表兄维尼	4	拇指朝下	拇指朝下
火星人玩转地球	7	拇指向上	拇指向上
搏击俱乐部	2	拇指朝下	拇指朝下
选美俏卧底	1	拇指朝下	拇指朝下
判我有罪	7	拇指向上	拇指朝下
点球成金	4	拇指朝下	拇指向上

如果针对这些电影分级数据建立一个简单的树形分类器，它看起来就像图 8-7 所显示的那样。这个简单的树告诉我们，如果我们要基于一个单词，而且仅仅一个单词而已来进行分类，那么这个单词就应该是“最差（worst)”。人们倾向于使用“最差（worst)”这个单词来给出一个拇指朝下的评价，紧随其后的是单词“差（bad)”和“浪费（waste)”可以作为拇指朝下的预测标识。如果一篇影评中没有出现这三个单词，而是出现单词“令人惊讶的（amazing)”，这个简单的树形分类器就会做出拇指向上的预测。除此之外，我们还要看其他的单词，如“情节（plot)”“很喜欢（favorite)”“非常糟糕（terrible)”和“死神（death)”。

130～131

一个简单的树不一定是最好的预测器，然而它能够帮助我们解释模型的工作原理。在研究中，我们发现最好的文本分类器随机森林是由数百棵树所组成的，每一棵树都与另外一棵有所不同。

模型的价值在于用来进行预测的质量。简单树形分类器对本章开始讨论的那些影评预测的质量如何？答案是：不是很好。它和基于列表及基于随机森林的分类器存在同样的问题，此外还有一个问题，就是它把《搏击俱乐部》（Fight Club）分类为一个拇指向上的电影，但实际上这部电影应该分类为拇指朝下。想要了解为什么会发生这种情况，我们只需要往下看。电影《搏击俱乐部》（Fight Club）的影评不包含树中的任何单词，所以我们最终终止在右边数过来的第二个终端节点上，那是一个拇指向上的节点。

情感分析之前的研究是在内容分析、专题、语义及网络文本分析方面（Roberts 1997; Popping 2000; West 2001; Leetaru 2011; Krippendorff 2012）。这些方法在社会科学领域得到了广泛的应用，包括分析政治论述。一个早期在计算机上实现的内容分析出现在 General Inquirer 程序中（Stone et al. 1966; Stone 1997）。Buvac and Stone（2001）介绍了一个基于众多语义类型进行单词计数的文本度量程序。

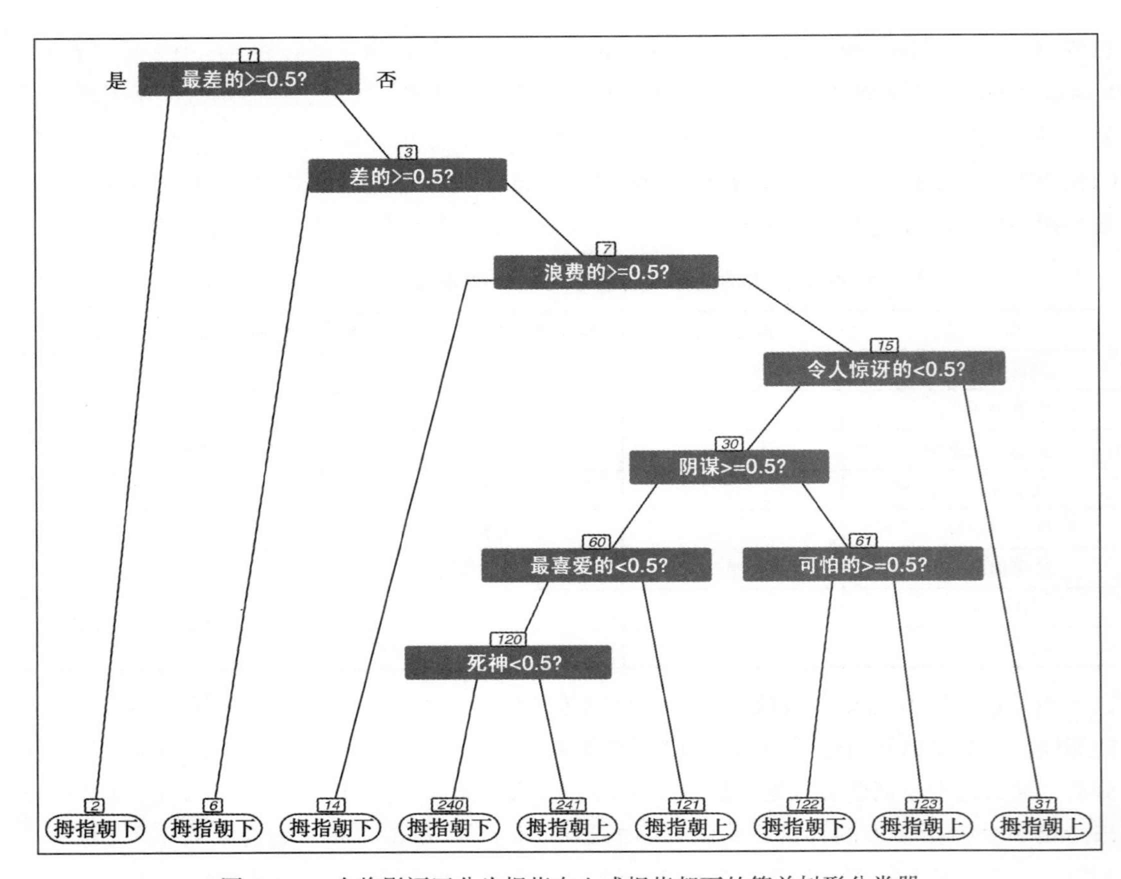

图 8-7　一个将影评区分为拇指向上或拇指朝下的简单树形分类器

文本度量起源于度量模型（得分算法）与一个词典，两者都需要由研究员或分析师来定义。这里所说的词典不是传统意义上的词典，不是按照字母顺序列出的单词以及对单词的释义。这个用于构建文本度量的词典是由一个单词列表仓库构成的，比如同义词和反义词、正面词和负面词，表达强烈和微弱词，双极形容词，演讲的部分内容，等等。这些列表来自专家对词义所做的判断。文本度量就是根据规则为文档赋予数字，这些规则是基于这些单词列表、得分算法以及预测分析学中使用的建模技术来确定的。

Charles Osgood 提出的语义差异（Osgood, Suci, and Tannenbaum 1957; Osgood 1962）是最流行的度量技术之一。具有代表性的双极维度包括正与负、强与弱、主动与被动。

文本度量作为一种了解消费者意见以及市场技术的方式非常有发展前景。就像政治研究员可以从公众、出版社和政客的言论中了解情况一样，商业研究员可以从消费者和竞争者的言论中了解情况。已经存在的有客服日志、通话录音和销售预测报告，再加上用户群、邮件列表和博客。此外，我们还有普适的社交媒体来构建用于文本和情感度量的文档库。

对于基于推特的文本度量，在电影到达全国各个影院上映之前就已经有预测电影是否会成功的各种尝试（Sharda and Delen 2006; Delen, Sharda, and Kumar 2007）。最有说服力的工作是由惠普实验室完成的利用推特中的聊天内容来预测电影的票房收入（Asur and Huberman 2010）。Bollen, Mao, and Zeng (2011) 运用推特情感分析来预测股市的涨跌。Taddy's (2013b,

132～133

2014) 所做的情感分析工作建立在 Cook (1998, 2007) 提出的逆回归方法之上。Taddy (2013a) 运用推特数据调查政治民意。

我们预计情感分析将成为未来很多年的一个非常活跃的研究领域。某些学者已经对非多维的情感度量表达出担忧。目前已经有了很多建立更加完善的情感词集以及多维度量方法方面的研究（Turney 2002; Asur and Huberman 2010）。最近几年在机器学习和量化语言学研究方面取得的进展也指出情感度量方法应该采用自然语言处理，而不仅仅是依赖于正面和负面词集（Socher et al. 2011）。

对于由 Geert Lovink(Lovink 2011) 所描述的“我们生活的谷歌化”而出现的社交媒体及互联网文化有很多批评者。《Mute Magazine》和《Proud to be Flesh》专辑就艺术与技术的相互作用提出了很有趣的观点（Slater et al. 2013）。对于社交媒体的更多批评来自于担忧社交媒体对个人隐私产生影响的人士（Rosen 2001; Turow 2013）。

代码 8-1 展示的是影评实例的 Python 程序。代码 8-2 则是相应的 R 语言程序。这个代码展示如何进行情感的文本度量以及建立基于影评文本对影评进行拇指向上或者拇指朝下预测的模型，其中应用了传统的方法和机器学习技术。这些代码来源于 Feinerer (2014), Wickham (2014b), Wickham and Chang (2014), Sarkar and Andrews (2014), Kuhn (2014), Liaw and Wiener (2014), Therneau, Atkinson, and Ripley (2014), Meyer et al. (2014) 和 Milborrow (2014) 开发的 R 语言软件包。 134

代码 8-1 情感分析与影视排名分类（Python）

```
# Sentiment Analysis Using the Movie Ratings Data (Python)

# Note that results from this program may differ from the results
# documented in the book because algorithms for text parsing
# and text classification vary between Python and R.
# The objectives of the analysis and steps in completing the analysis
# are consistent with those in the book. And results, although
# not identical between Python and R, should be very similar.

# prepare for Python version 3x features and functions
from __future__ import division, print_function

# import packages for text processing and machine learning
import os  # operating system commands
import re  # regular expressions
import nltk  # draw on the Python natural language toolkit
import pandas as pd  # DataFrame structure and operations
import numpy as np  # arrays and numerical processing
import matplotlib.pyplot as plt  # 2D plotting
import statsmodels.api as sm  # logistic regression
import statsmodels.formula.api as smf  # R-like model specification
import patsy  # translate model specification into design matrices
from sklearn import svm  # support vector machines
from sklearn.ensemble import RandomForestClassifier  # random forests
# import user-defined module
from python_utilities import evaluate_classifier, get_text_measures,\
    get_summative_scores

# list files in directory omitting hidden files
def listdir_no_hidden(path):
    start_list = os.listdir(path)
    end_list = []
```

```
    for file in start_list:
        if (not file.startswith('.')):
            end_list.append(file)
    return(end_list)
# define list of codes to be dropped from document

# carriage-returns, line-feeds, tabs
codelist = ['\r', '\n', '\t']

# there are certain words we will ignore in subsequent
# text processing... these are called stop-words
# and they consist of prepositions, pronouns, and
# conjunctions, interrogatives, ...
# we begin with the list from the natural language toolkit
# examine this initial list of stopwords
nltk.download('stopwords')
# let's look at that list
print(nltk.corpus.stopwords.words('english'))
# previous analysis of a list of top terms showed a number of words, along
# with contractions and other word strings to drop from further analysis, we add
# these to the usual English stopwords to be dropped from a document collection
more_stop_words = ['cant','didnt','doesnt','dont','goes','isnt','hes',\
    'shes','thats','theres','theyre','wont','youll','youre','youve', 'br'\
    've', 're', 'vs']

some_proper_nouns_to_remove = ['dick','ginger','hollywood','jack',\
    'jill','john','karloff','kudrow','orson','peter','tcm','tom',\
    'toni','welles','william','wolheim','nikita']

# start with the initial list and add to it for movie text work
stoplist = nltk.corpus.stopwords.words('english') + more_stop_words +\
    some_proper_nouns_to_remove

# text parsing function for creating text documents
# there is more we could do for data preparation
# stemming... looking for contractions... possessives...
# but we will work with what we have in this parsing function
# if we want to do stemming at a later time, we can use
#     porter = nltk.PorterStemmer()
# in a construction like this
#     words_stemmed =  [porter.stem(word) for word in initial_words]

def text_parse(string):
    # replace non-alphanumeric with space
    temp_string = re.sub('[^a-zA-Z]', '  ', string)
    # replace codes with space
    for i in range(len(codelist)):
        stopstring = ' ' + codelist[i] + '  '
        temp_string = re.sub(stopstring, '  ', temp_string)
    # replace single-character words with space
    temp_string = re.sub('\s.\s', ' ', temp_string)
    # convert uppercase to lowercase
    temp_string = temp_string.lower()
    # replace selected character strings/stop-words with space
    for i in range(len(stoplist)):
        stopstring = ' ' + str(stoplist[i]) + ' '
        temp_string = re.sub(stopstring, ' ', temp_string)
    # replace multiple blank characters with one blank character
    temp_string = re.sub('\s+', ' ', temp_string)
    return(temp_string)

# read in positive and negative word lists from Hu and Liu (2004)
with open('Hu_Liu_positive_word_list.txt','rt') as f:
    positive_word_list = f.read().split()
with open('Hu_Liu_negative_word_list.txt','rt') as f:
```

135

```
    negative_word_list = f.read().split()

# define counts of positive, negative, and total words in text document
def count_positive(text):
    positive = [w for w in text.split() if w in positive_word_list]
    return(len(positive))

# define text measure for negative score as percentage of negative words
def count_negative(text):
    negative = [w for w in text.split() if w in negative_word_list]
    return(len(negative))

# count number of words
def count_total(text):
    total = [w for w in text.split()]
    return(len(total))

# define text measure for positive score as percentage of positive words
def score_positive(text):
    positive = [w for w in text.split() if w in positive_word_list]
    total = [w for w in text.split()]
    return 100 * len(positive)/len(total)


# define text measure for negative score as percentage of negative words
def score_negative(text):
    negative = [w for w in text.split() if w in negative_word_list]
    total = [w for w in text.split()]
    return 100 * len(negative)/len(total)

def compute_scores(corpus):
    # use the complete word lists for POSITIVE and NEGATIVE measures
    # to score all documents in a corpus or list of documents
    positive = []
    negative = []
    for document in corpus:
        positive.append(score_positive(document))
        negative.append(score_negative(document))
    return(positive, negative)

# we use movie ratings data from Mass et al. (2011)
# available at http://ai.stanford.edu/~amaas/data/sentiment/
# we set up a directory under our working directory structure
# /reviews/train/unsup/ for the unsupervised reviews
# /reviews/train/neg/ training set negative reviews
# /reviews/train/pos/ training set positive reviews
# /reviews/test/neg/ text set negative reviews
# /reviews/test/pos/ test set positive reviews
# /reviews/test/tom/ eight movie reviews from Tom
# function for creating corpus and aggregate document
# input is directory path for documents
# document parsing accomplished by text_parse function
# directory of parsed files set up for manual inspection

def corpus_creator (input_directory_path, output_directory_path):
    # identify the file names in unsup directory
    file_names = listdir_no_hidden(path = input_directory_path)
    # create list structure for storing parsed documents
    document_collection = []
    # initialize aggregate document for all documents in set
    aggregate_document = ''
    # create a directory for parsed files
    parsed_file_directory = output_directory_path
    os.mkdir(parsed_file_directory)
    # parse each file and write to directory of parsed files
```

136
137

```
    for filename in file_names:
        with open(os.path.join(input_directory_path, filename), 'r') as infile:
            this_document = text_parse(infile.read())
            aggregate_document = aggregate_document + this_document
            document_collection.append(this_document)
            outfile = parsed_file_directory + filename
            with open(outfile, 'wt') as f:
                f.write(str(this_document))
    aggregate_words = [w for w in aggregate_document.split()]
    aggregate_corpus = nltk.Text(aggregate_words)
    return(file_names, document_collection, aggregate_corpus)

# function for extracting rating from file name
# for file names of the form 'x_y.txt' where y is the rating
def get_rating(string):
    return(int(string.partition('.')[0].partition('_')[2]))

# dictionary for mapping of ratings to thumbsupdown
map_to_thumbsupdown = {1:'DOWN', 2:'DOWN', 3:'DOWN', 4:'DOWN',
    6:'UP', 7:'UP', 8:'UP', 9:'UP', 10:'UP'}

# begin working with the unsup corpus
unsup_file_names, unsup_corpus, unsup_aggregate_corpus = \
    corpus_creator(input_directory_path = 'reviews/train/unsup/',\
        output_directory_path = 'reviews/train/unsup_parsed/')

# examine frequency distribution of words in unsup corpus
unsup_freq = nltk.FreqDist(unsup_aggregate_corpus)
print('\nNumber of Unique Words in unsup corpus',len(unsup_freq.keys()))
print('\nTop Fifty Words in unsup Corpus:',unsup_freq.keys()[0:50])

# identify the most frequent unsup words from the positive word list
# here we use set intersection to find a list of the top 25 positive words
length_test = 0  # initialize test length
nkeys = 0  # slicing index for frequency table extent
while (length_test < 25):
    length_test =\
        len(set(unsup_freq.keys()[:nkeys]) & set(positive_word_list))
    nkeys = nkeys + 1
selected_positive_set =\
    set(unsup_freq.keys()[:nkeys]) & set(positive_word_list)
selected_positive_words = list(selected_positive_set)
selected_positive_words.sort()
print('\nSelected Positive Words:', selected_positive_words)

# identify the most frequent unsup words from the negative word list
# here we use set intersection to find a list of the top 25 negative words
length_test = 0  # initialize test length
nkeys = 0  # slicing index for frequency table extent
while (length_test < 25):
    length_test =\
        len(set(unsup_freq.keys()[:nkeys]) & set(negative_word_list))
    nkeys = nkeys + 1
selected_negative_set =\
    set(unsup_freq.keys()[:nkeys]) & set(negative_word_list)
# list is actually 26 items and contains both 'problem' and 'problems'
# so we will eliminate 'problems' from the selected negative words
selected_negative_set.remove('problems')
selected_negative_words = list(selected_negative_set)
selected_negative_words.sort()
print('\nSelected Negative Words:', selected_negative_words)

# use the complete word lists for POSITIVE and NEGATIVE measures/scores
positive, negative = compute_scores(unsup_corpus)
```

138

```
# create data frame to explore POSITIVE and NEGATIVE measures
unsup_data = {'file': unsup_file_names,\
    'POSITIVE': positive, 'NEGATIVE': negative}
unsup_data_frame = pd.DataFrame(unsup_data)

# summary of distributions of POSITIVE and NEGATIVE scores for unsup corpus
print(unsup_data_frame.describe())

print('\nCorrelation between POSITIVE and NEGATIVE',\
    round(unsup_data_frame['POSITIVE'].corr(unsup_data_frame['NEGATIVE']),3))

# scatter plot of POSITIVE and NEGATIVE scores for unsup corpus
ax = plt.axes()
ax.scatter(unsup_data_frame['NEGATIVE'], unsup_data_frame['POSITIVE'],\
    facecolors = 'none', edgecolors = 'blue')
ax.set_xlabel('NEGATIVE')
ax.set_ylabel('POSITIVE')
plt.savefig('fig_sentiment_text_measures_scatter_plot.pdf',
    bbox_inches = 'tight', dpi=None, facecolor='none', edgecolor='blue',
    orientation='portrait', papertype=None, format=None,
    transparent=True, pad_inches=0.25, frameon=None)

# work on the directory of training files-----------------------------------
# Perhaps POSITIVE and NEGATIVE can be combined in a way to yield effective
# predictions of movie ratings. Let us move to a set of movie reviews for
# supervised learning.  We select the 500 records from a set of positive
# reviews (ratings between 7 and 10) and 500 records from a set of negative
# reviews (ratings between 1 and 4). We begin with the training data.

# /reviews/train/pos/ training set positive reviews
train_pos_file_names, train_pos_corpus, train_pos_aggregate_corpus = \
    corpus_creator(input_directory_path = 'reviews/train/pos/',\
        output_directory_path = 'reviews/train/pos_parsed/')
# use the complete word lists for POSITIVE and NEGATIVE measures/scores
positive, negative = compute_scores(train_pos_corpus)
rating = []
for file_name in train_pos_file_names:
    rating.append(get_rating(str(file_name)))

# create data frame to explore POSITIVE and NEGATIVE measures
train_pos_data = {'train_test':['TRAIN'] * len(train_pos_file_names),\
    'pos_neg': ['POS'] * len(train_pos_file_names),\
    'file_name': train_pos_file_names,\
    'POSITIVE': positive, 'NEGATIVE': negative,\
    'rating': rating}
train_pos_data_frame = pd.DataFrame(train_pos_data)

# /reviews/train/neg/ training set negative reviews
train_neg_file_names, train_neg_corpus, train_neg_aggregate_corpus = \
    corpus_creator(input_directory_path = 'reviews/train/neg/',\
        output_directory_path = 'reviews/train/neg_parsed/')
# use the complete word lists for POSITIVE and NEGATIVE measures/scores
positive, negative = compute_scores(train_neg_corpus)
rating = []
for file_name in train_neg_file_names:
    rating.append(get_rating(str(file_name)))

# create data frame to explore POSITIVE and NEGATIVE measures
train_neg_data = {'train_test':['TRAIN'] * len(train_neg_file_names),\
    'pos_neg': ['NEG'] * len(train_neg_file_names),\
    'file_name': train_neg_file_names,\
    'POSITIVE': positive, 'NEGATIVE': negative,\
    'rating': rating}
train_neg_data_frame = pd.DataFrame(train_neg_data)
```

139

140

```
# merge the positive and negative training data frames
train_data_frame = pd.concat([train_pos_data_frame, train_neg_data_frame],\
    axis = 0, ignore_index = True)
# determining thumbs up or down based on rating
train_data_frame['thumbsupdown'] = \
    train_data_frame['rating'].map(map_to_thumbsupdown)
# compute simple measure of sentiment as POSITIVE - NEGATIVE
train_data_frame['simple'] = \
    train_data_frame['POSITIVE'] - train_data_frame['NEGATIVE']
# examine the data frame
print(pd.crosstab(train_data_frame['pos_neg'],\
    train_data_frame['thumbsupdown']))
print(train_data_frame.head())
print(train_data_frame.tail())
print(train_data_frame.describe())
ratings_grouped = train_data_frame['simple'].\
    groupby(train_data_frame['rating'])
print('\nTraining Data Simple Difference Means by Ratings:',\
    ratings_grouped.mean())
thumbs_grouped = \
    train_data_frame['simple'].groupby(train_data_frame['thumbsupdown'])
print('\nTraining Data Simple Difference Means by Thumbs UP/DOWN:',\
    thumbs_grouped.mean())
# repeat methods for the test data ------------------------------
# /reviews/test/pos/ testing set positive reviews
test_pos_file_names, test_pos_corpus, test_pos_aggregate_corpus = \
    corpus_creator(input_directory_path = 'reviews/test/pos/',\
        output_directory_path = 'reviews/test/pos_parsed/')
# use the complete word lists for POSITIVE and NEGATIVE measures/scores
positive, negative = compute_scores(test_pos_corpus)
rating = []
for file_name in test_pos_file_names:
    rating.append(get_rating(str(file_name)))

# create data frame to explore POSITIVE and NEGATIVE measures
test_pos_data = {'train_test':['TEST'] * len(test_pos_file_names),\
    'pos_neg': ['POS'] * len(test_pos_file_names),\
    'file_name': test_pos_file_names,\
    'POSITIVE': positive, 'NEGATIVE': negative,\
    'rating': rating}
test_pos_data_frame = pd.DataFrame(test_pos_data)

# /reviews/test/neg/ testing set negative reviews
test_neg_file_names, test_neg_corpus, test_neg_aggregate_corpus = \
    corpus_creator(input_directory_path = 'reviews/test/neg/',\
        output_directory_path = 'reviews/test/neg_parsed/')
# use the complete word lists for POSITIVE and NEGATIVE measures/scores
positive, negative = compute_scores(test_neg_corpus)
rating = []
for file_name in test_neg_file_names:
    rating.append(get_rating(str(file_name)))

# create data frame to explore POSITIVE and NEGATIVE measures
test_neg_data = {'train_test':['TEST'] * len(test_neg_file_names),\
    'pos_neg': ['NEG'] * len(test_neg_file_names),\
    'file_name': test_neg_file_names,\
    'POSITIVE': positive, 'NEGATIVE': negative,\
    'rating': rating}
test_neg_data_frame = pd.DataFrame(test_neg_data)

# merge the positive and negative testing data frames
test_data_frame = pd.concat([test_pos_data_frame, test_neg_data_frame],\
    axis = 0, ignore_index = True)

# determining thumbs up or down based on rating
```

```
test_data_frame['thumbsupdown'] = \
    test_data_frame['rating'].map(map_to_thumbsupdown)
# compute simple measure of sentiment as POSITIVE - NEGATIVE
test_data_frame['simple'] = \
    test_data_frame['POSITIVE'] - test_data_frame['NEGATIVE']
# examine the data frame
print(pd.crosstab(test_data_frame['pos_neg'],\
    test_data_frame['thumbsupdown']))
print(test_data_frame.head())
print(test_data_frame.tail())
print(test_data_frame.describe())
ratings_grouped = test_data_frame['simple'].\
    groupby(test_data_frame['rating'])
print('\nTest Data Simple Difference Means by Ratings:',\
    ratings_grouped.mean())
thumbs_grouped = \
    test_data_frame['simple'].groupby(test_data_frame['thumbsupdown'])
print('\nTest Data Simple Difference Means by Thumbs UP/DOWN:',\
    thumbs_grouped.mean())

# repeat methods for the Tom's movie reviews -----------------------------
# /reviews/test/tom/ testing set directory path
test_tom_file_names, test_tom_corpus, test_tom_aggregate_corpus = \
    corpus_creator(input_directory_path = 'reviews/test/tom/',\
        output_directory_path = 'reviews/test/tom_parsed/')

# word counts for Tom's reviews
positive_words = []
negative_words = []
total_words = []
for file in test_tom_corpus:
    positive_words.append(count_positive(file))
    negative_words.append(count_negative(file))
    total_words.append(count_total(file))
# POSITIVE and NEGATIVE measures/scores for Tom's reviews
positive, negative = compute_scores(test_tom_corpus)
rating = []
for file_name in test_tom_file_names:
    rating.append(get_rating(str(file_name)))

# create data frame to check calculations of counts and scores
test_tom_data = {'train_test':['TOM'] * len(test_tom_file_names),\
    'pos_neg': ['POS', 'POS', 'NEG', 'POS', 'NEG', 'NEG', 'POS', 'NEG'],\
    'file_name': test_tom_file_names,\
    'movie': ['Marigolds',\
    'Blade Runner',\
    'Vinny',\
    'Mars Attacks',
    'Fight Club',\
    'Congeniality',\
    'Find Me Guilty',\
    'Moneyball'],\
    'positive_words' : positive_words,\
    'negative_words' : negative_words,\
    'total_words' : total_words,\
    'POSITIVE': positive, 'NEGATIVE': negative,\
    'rating': rating}
test_tom_data_frame = pd.DataFrame(test_tom_data)
# determing thumbs up or down based upon rating
test_tom_data_frame['thumbsupdown'] = \
    test_tom_data_frame['rating'].map(map_to_thumbsupdown)
# compute simple measure of sentiment as POSITIVE - NEGATIVE
test_tom_data_frame['simple'] = \
    test_tom_data_frame['POSITIVE'] - test_tom_data_frame['NEGATIVE']
```

141

142

```
# examine the data frame
print(test_tom_data_frame)
print(test_tom_data_frame.describe())
ratings_grouped = test_tom_data_frame['simple'].\
    groupby(test_tom_data_frame['rating'])
print('\nTom Simple Difference Means by Ratings:',ratings_grouped.mean())
thumbs_grouped = \
    test_tom_data_frame['simple'].groupby(test_tom_data_frame['thumbsupdown'])
print('\nTom Simple Difference Means by Thumbs UP/DOWN:',\
    thumbs_grouped.mean())

# develop predictive models using the training data
# --------------------------------------
# Simple difference method
# --------------------------------------
# use the median of the simple difference between POSITIVE and NEGATIVE
simple_cut_point = train_data_frame['simple'].median()

# algorithm for simple difference method based on training set median
def predict_simple(value):
    if (value > simple_cut_point):
        return('UP')
    else:
        return('DOWN')

train_data_frame['pred_simple'] = \
    train_data_frame['simple'].apply(lambda d: predict_simple(d))
print(train_data_frame.head())

print('\n Simple Difference Training Set Performance\n',\
    'Percentage of Reviews Correctly Classified:',\
    100 * round(evaluate_classifier(train_data_frame['pred_simple'],\
    train_data_frame['thumbsupdown'])[4], 3),'\n')

# evaluate simple difference method in the test set
# using algorithm developed with the training set
test_data_frame['pred_simple'] = \
    test_data_frame['simple'].apply(lambda d: predict_simple(d))
print(test_data_frame.head())
print('\n Simple Difference Test Set Performance\n',\
    'Percentage of Reviews Correctly Classified:',\
    100 * round(evaluate_classifier(test_data_frame['pred_simple'],\
    test_data_frame['thumbsupdown'])[4], 3), '\n')

# --------------------------------------
# Regression difference method
# --------------------------------------
# regression method for determining weights on POSITIVE AND NEGATIVE
# fit a regression model to the training data
regression_model = str('rating ~ POSITIVE + NEGATIVE')
# fit the model to the training set
train_regression_model_fit = smf.ols(regression_model,\
    data = train_data_frame).fit()
# summary of model fit to the training set
print(train_regression_model_fit.summary())

# because we are using predicted rating we use the midpoint
# rating of 5 as the cut-point for making thumbs up or down predictions
regression_cut_point = 5

# algorithm for simple difference method based on training set median
def predict_regression(value):
    if (value > regression_cut_point):
        return('UP')
    else:
```

143

```
        return('DOWN')

# training set predictions from the model fit to the training set
train_data_frame['pred_regression_rating'] =\
    train_regression_model_fit.fittedvalues

# predict thumbs up or down based upon the predicted rating
train_data_frame['pred_regression'] = \
    train_data_frame['pred_regression_rating'].\
        apply(lambda d: predict_regression(d))
print(train_data_frame.head())

print('\n Regression Difference Training Set Performance\n',\
    'Percentage of Reviews Correctly Classified:',\
    100 * round(evaluate_classifier(train_data_frame['pred_regression'],\
    train_data_frame['thumbsupdown'])[4], 3),'\n')

# evaluate regression difference method in the test set
# using algorithm developed with the training set
# predict thumbs up or down based upon the predicted rating

# test set predictions from the model fit to the training set
test_data_frame['pred_regression_rating'] =\
    train_regression_model_fit.predict(test_data_frame)

test_data_frame['pred_regression'] = \
    test_data_frame['pred_regression_rating'].\
        apply(lambda d: predict_regression(d))
print(test_data_frame.head())

print('\n Regression Difference Test Set Performance\n',\
    'Percentage of Reviews Correctly Classified:',\
    100 * round(evaluate_classifier(test_data_frame['pred_regression'],\
    test_data_frame['thumbsupdown'])[4], 3), '\n')

# --------------------------------------------
# Compute text measures for each corpus
# --------------------------------------------
# return to score the document collections with get_text_measures
# for each of the selected words from the sentiment lists
# these new variables will be given the names of the words
# to keep things simple.... there are 50 such variables/words
# identified from our analysis of the unsup corpus above
# start with the training document collection
working_corpus = train_pos_corpus + train_neg_corpus
add_corpus_data = get_text_measures(working_corpus)
add_corpus_data_frame = pd.DataFrame(add_corpus_data)
# merge the new text measures with the existing data frame
train_data_frame =\
    pd.concat([train_data_frame,add_corpus_data_frame],axis=1)
# examine the expanded training data frame
print('\n xtrain_data_frame (rows, cols):',train_data_frame.shape,'\n')
print(train_data_frame.describe())
print(train_data_frame.head())

# start with the test document collection
working_corpus = test_pos_corpus + test_neg_corpus
add_corpus_data = get_text_measures(working_corpus)
add_corpus_data_frame = pd.DataFrame(add_corpus_data)
# merge the new text measures with the existing data frame
test_data_frame = pd.concat([test_data_frame,add_corpus_data_frame],axis=1)
# examine the expanded testing data frame
print('\n xtest_data_frame (rows, cols):',test_data_frame.shape,'\n')
print(test_data_frame.describe())
print(test_data_frame.head())
```

144

```
# end with Tom's reviews as a document collection
working_corpus = test_tom_corpus
add_corpus_data = get_text_measures(working_corpus)
add_corpus_data_frame = pd.DataFrame(add_corpus_data)
# merge the new text measures with the existing data frame
tom_data_frame =\
    pd.concat([test_tom_data_frame,add_corpus_data_frame],axis=1)
# examine the expanded testing data frame
print('\n xtom_data_frame (rows, cols):',tom_data_frame.shape,'\n')
print(tom_data_frame.describe())
print(tom_data_frame.head())

# ---------------------------------------------
# Word/item analysis method for training set
# ---------------------------------------------
# item-rating correlations for all 50 words

item_list = selected_positive_words + selected_negative_words
item_rating_corr = []
for item in item_list:
    item_rating_corr.\
        append(train_data_frame['rating'].corr(train_data_frame[item]))
item_analysis_data_frame =\
    pd.DataFrame({'item': item_list, 'item_rating_corr': item_rating_corr})

# absolute value of item correlation with rating
item_analysis_data_frame['abs_item_rating_corr'] =\
    item_analysis_data_frame['item_rating_corr'].apply(lambda d: abs(d))
# look at sort by absolute value
print(item_analysis_data_frame.sort_index(by = ['abs_item_rating_corr'],\
    ascending = False))

# select subset of items with absolute correlations > 0.05
selected_item_analysis_data_frame =\
    item_analysis_data_frame\
        [item_analysis_data_frame['abs_item_rating_corr'] > 0.05]

# identify the positive items for word/item analysis measure
selected_positive_item_df =\
    selected_item_analysis_data_frame\
        [selected_item_analysis_data_frame['item_rating_corr'] > 0]
possible_positive_items = selected_positive_item_df['item']
print('Possible positive items:',possible_positive_items,'\n')
# note some surprises in the list of positive items
# select list consitent with initial list of positive words
selected_positive_items =\
    list(set(possible_positive_items) & set(positive_word_list))
print('Selected positive items:',selected_positive_items,'\n')

# identify the negative items for word/item analysis measure
selected_negative_item_df =\
    selected_item_analysis_data_frame\
        [selected_item_analysis_data_frame['item_rating_corr'] < 0]
possible_negative_items = selected_negative_item_df['item']
print('Possible negative items:',possible_negative_items,'\n')
# select list consitent with initial list of negative words
selected_negative_items =\
    list(set(possible_negative_items) & set(negative_word_list))
print('Selected negative items:',selected_negative_items,'\n')
# the word "funny" remains a mystery... kept in negative list for now

# selected positive and negative items entered into function
# for obtaining word/item analysis summative score in which
# postive items get +1 point and negative items get -1 point
```

145

```
# ... implemented in imported Python utility get_summative_scores
# start with the training set... identify a cut-off
working_corpus = train_pos_corpus + train_neg_corpus
add_corpus_data = get_summative_scores(working_corpus)
add_corpus_data_frame = pd.DataFrame(add_corpus_data)
# merge the new text measures with the existing data frame
train_data_frame = pd.concat([train_data_frame,add_corpus_data_frame],axis=1)
# examine the expanded training data frame and summative_scores
print('\n train_data_frame (rows, cols):',train_data_frame.shape,'\n')
print(train_data_frame['summative_score'].describe())
print('\nCorrelation of ratings and summative scores:'\
    ,round(train_data_frame['rating'].\
        corr(train_data_frame['summative_score']),3))
ratings_grouped = train_data_frame['summative_score'].\
    groupby(train_data_frame['rating'])
print('\nTraining Data Summative Score Means by Ratings:',\
    ratings_grouped.mean())
thumbs_grouped = \
    train_data_frame['summative_score'].\
        groupby(train_data_frame['thumbsupdown'])
print('\nTraining Data Summative Score Means by Thumbs UP/DOWN:',\
    thumbs_grouped.mean())

# analyses suggest a simple positive/negative cut on summative scores
# algorithm for word/item method based on training set summative_scores
def predict_by_summative_score(value):
    if (value > 0):
        return('UP')
    else:
        return('DOWN')

# evaluate word/item analysis method on training set
train_data_frame['pred_summative_score'] = \
    train_data_frame['summative_score'].\
    apply(lambda d: predict_by_summative_score(d))

print('\n Word/item Analysis Training Set Performance\n',\
    'Percentage of Reviews Correctly Classified by Summative Scores:',\
    100 * round(evaluate_classifier(train_data_frame['pred_summative_score'],\
    train_data_frame['thumbsupdown'])[4], 3),'\n')

# compute summative scores on test data frame
working_corpus = test_pos_corpus + test_neg_corpus
add_corpus_data = get_summative_scores(working_corpus)
add_corpus_data_frame = pd.DataFrame(add_corpus_data)

# merge the new text measures with the existing data frame
test_data_frame = pd.concat([test_data_frame,add_corpus_data_frame],axis=1)

# evaluate word/item analysis method (summative score method) on test set
# using algorithm developed with the training set
test_data_frame['pred_summative_score'] = \
    test_data_frame['summative_score'].\
        apply(lambda d: predict_by_summative_score(d))

print('\n Word/item Analysis Test Set Performance\n',\
    'Percentage of Reviews Correctly Classified by Summative Scores:',\
    100 * round(evaluate_classifier(test_data_frame['pred_summative_score'],\
    test_data_frame['thumbsupdown'])[4], 3), '\n')

# ---------------------------------------
# Logistic regression method
# ---------------------------------------
# translate thumbsupdown into a binary indicator variable y
# here we let thumbs up have the higher value of 1
```

146

[147]

```
thumbsupdown_to_binary = {'UP':1,'DOWN':0}

train_data_frame['y'] =\
    train_data_frame['thumbsupdown'].map(thumbsupdown_to_binary)
# model specification in R-like formula syntax
text_classification_model = 'y ~  beautiful +\
    best + better + classic + enjoy + enough +\
    entertaining + excellent +\
    fans +  fun + good + great + interesting + like +\
    love +  nice + perfect + pretty + right +\
    top + well + won + wonderful + work + worth +\
    bad + boring + creepy + dark + dead+\
    death + evil + fear + funny + hard + kill +\
    killed + lack + lost + mystery +\
    plot + poor + problem + sad + scary +\
    slow + terrible + waste + worst + wrong'

# convert R-like formula into design matrix needed for statsmodels
y,x = patsy.dmatrices(text_classification_model,\
    train_data_frame, return_type = 'dataframe')

# define the logistic regression algorithm
my_logit_model = sm.Logit(y,x)
# fit the model to training set
my_logit_model_fit = my_logit_model.fit()
print(my_logit_model_fit.summary())

# predicted probability of thumbs up for training set
train_data_frame['pred_logit_prob'] =\
    my_logit_model_fit.predict(linear = False)

# map from probability to thumbsupdown with simple 0.5 cut-off
def prob_to_updown(x):
    if(x > 0.5):
        return('UP')
    else:
        return('DOWN')

train_data_frame['pred_logit'] =\
    train_data_frame['pred_logit_prob'].apply(lambda d: prob_to_updown(d))
print('\n Logistic Regression Training Set Performance\n',\
    'Percentage of Reviews Correctly Classified:',\
    100 * round(evaluate_classifier(train_data_frame['pred_logit'],\
    train_data_frame['thumbsupdown'])[4], 3),'\n')

# use the model developed on the training set to predict
# thumbs up or down reviews in the test set
# assume that y is not known... only x used from patsy
y,x = patsy.dmatrices(text_classification_model,\
        test_data_frame, return_type = 'dataframe')
y = []  # ignore known thumbs up/down from test set...
# we want to predict thumbs up/down from the model fit to
# the training set... my_logit_model_fit
test_data_frame['pred_logit_prob'] =\
    my_logit_model_fit.predict(exog = x, linear = False)
test_data_frame['pred_logit'] =\
    test_data_frame['pred_logit_prob'].apply(lambda d: prob_to_updown(d))
print('\n Logistic Regression Test Set Performance\n',\
    'Percentage of Reviews Correctly Classified:',\
    100 * round(evaluate_classifier(test_data_frame['pred_logit'],\
    test_data_frame['thumbsupdown'])[4], 3),'\n')

# ---------------------------------------
# Support vector machines
```

[148]

```
# --------------------------------------
# fit the model to the training set
y,x = patsy.dmatrices(text_classification_model,\
    train_data_frame, return_type = 'dataframe')
my_svm = svm.SVC()
my_svm_fit = my_svm.fit(x, np.ravel(y))
train_data_frame['pred_svm_binary'] = my_svm_fit.predict(x)
binary_to_thumbsupdown = {0: 'DOWN', 1: 'UP'}
train_data_frame['pred_svm'] =\
    train_data_frame['pred_svm_binary'].map(binary_to_thumbsupdown)
print('\n Support Vector Machine Training Set Performance\n',\
    'Percentage of Reviews Correctly Classified:',\
    100 * round(evaluate_classifier(train_data_frame['pred_svm'],\
    train_data_frame['thumbsupdown'])[4], 3),'\n')
# use the model developed on the training set to predict
# thumbs up or down reviews in the test set
# assume that y is not known... only x used from patsy
y,x = patsy.dmatrices(text_classification_model,\
        test_data_frame, return_type = 'dataframe')
y = []  # ignore known thumbs up/down from test set...
test_data_frame['pred_svm_binary'] = my_svm_fit.predict(x)
test_data_frame['pred_svm'] =\
    test_data_frame['pred_svm_binary'].map(binary_to_thumbsupdown)
print('\n Support Vector Machine Test Set Performance\n',\
    'Percentage of Reviews Correctly Classified:',\
    100 * round(evaluate_classifier(test_data_frame['pred_svm'],\
    test_data_frame['thumbsupdown'])[4], 3),'\n')

# --------------------------------------
# Random forests
# --------------------------------------
# fit random forest model to the training data
y,x = patsy.dmatrices(text_classification_model,\
    train_data_frame, return_type = 'dataframe')

# for reproducibility set random number seed with random_state
my_rf_model = RandomForestClassifier(n_estimators = 10, random_state = 9999)
my_rf_model_fit = my_rf_model.fit(x, np.ravel(y))
train_data_frame['pred_rf_binary'] = my_rf_model_fit.predict(x)
train_data_frame['pred_rf'] =\
    train_data_frame['pred_rf_binary'].map(binary_to_thumbsupdown)

print('\n Random Forest Training Set Performance\n',\
    'Percentage of Reviews Correctly Classified:',\
    100 * round(evaluate_classifier(train_data_frame['pred_rf'],\
    train_data_frame['thumbsupdown'])[4], 3),'\n')
# use the model developed on the training set to predict
# thumbs up or down reviews in the test set
# assume that y is not known... only x used from patsy
y,x = patsy.dmatrices(text_classification_model,\
        test_data_frame, return_type = 'dataframe')
y = []  # ignore known thumbs up/down from test set...
test_data_frame['pred_rf_binary'] = my_rf_model_fit.predict(x)
test_data_frame['pred_rf'] =\
    test_data_frame['pred_rf_binary'].map(binary_to_thumbsupdown)

print('\n Random Forest Test Set Performance\n',\
    'Percentage of Reviews Correctly Classified:',\
    100 * round(evaluate_classifier(test_data_frame['pred_rf'],\
    test_data_frame['thumbsupdown'])[4], 3),'\n')

# Suggestions for the student:
# Employ stemming prior to the creation of terms-by-document matrices.
# Try alternative positive and negative word sets for sentiment scoring.
```

149

```
# Try word sets that relate to a wider variety of emotional or opinion states.
# Better still, move beyond a bag-of-words approach to sentiment. Use
# the tools of natural language processing and define text features
# based upon combinations of words such as bigrams (pairs of words)
# and taking note of parts of speech.  Yet another approach would be
# to define ignore negative and positive word lists and work directly
# with identified text features that correlate with movie review ratings or
# do a good job of classifying reviews into positive and negative groups.
# Text features within text classification problems may be defined
# on term document frequency alone or on measures of term document
# frequency adjusted by term corpus frequency. Using alternative
# features and text measures as well as alternative classification methods,
# run a true benchmark within a loop, using hundreds or thousands of iterations.
# See if you can improve upon the performance of modeling methods by
# modifying the values of arguments to algorithms used here.
# Use various methods of classifier performance to evaluate classifiers.
# Try text classification for the movie reviews without using initial
# lists of positive an negative words. That is, identify text features
# for thumbs up/down text classification directly from the training set.
```

150

代码 8-2　情感分析与影视排名分类（R 语言）

```
# Sentiment Analysis Using the Movie Ratings Data (R)

# Note. Results from this program may differ from those published
#       in the book due to changes in the tm package.
#       The original analysis used the tm Dictionary() function,
#       which is no longer available in tm. This function has
#       been replaced by c(as.character()) to set the dictionary
#       as a character vector. Another necessary change concerns
#       the tolower() function, which must now be embedded within
#       the tm content_transformer() function.
#
# Despite changes in the tm functions, we have retained the
# earlier positive and negative word lists for scoring, as
# implemented in the code and utilities appendix under the file
# name <R_utility_program_5.R>, which is brought in by source().

# install these packages before bringing them in by library()
library(tm)  # text mining and document management
library(stringr)  # character manipulation with regular expressions
library(grid)  # grid graphics utilities
library(ggplot2)  # graphics
library(latticeExtra) # package used for text horizon plot
library(caret)  # for confusion matrix function
library(rpart)  # tree-structured modeling
library(e1071)  # support vector machines
library(randomForest)  # random forests
library(rpart.plot)  # plot tree-structured model information

# R preliminaries to get the user-defined utilities for plotting
# place the plotting code file <R_utility_program_3.R>
# in your working directory and execute it by
#     source("R_utility_program_3.R")
# Or if you have the R binary file in your working directory, use
#     load("mtpa_split_plotting_utilities.Rdata")
load("mtpa_split_plotting_utilities.Rdata")
# standardization needed for text measures
standardize <- function(x) {(x - mean(x)) / sd(x)}
# convert to bytecodes to avoid "invalid multibyte string" messages
bytecode.convert <- function(x) {iconv(enc2utf8(x), sub = "byte")}

# read in positive and negative word lists from Hu and Liu (2004)
positive.data.frame <- read.table(file = "Hu_Liu_positive_word_list.txt",
```

```
  header = FALSE, colClasses = c("character"), row.names = NULL,
  col.names = "positive.words")
positive.data.frame$positive.words <-
  bytecode.convert(positive.data.frame$positive.words)
negative.data.frame <- read.table(file = "Hu_Liu_negative_word_list.txt",
  header = FALSE, colClasses = c("character"), row.names = NULL,
  col.names = "negative.words")
negative.data.frame$negative.words <-
  bytecode.convert(negative.data.frame$negative.words)
```

151

```
# we use movie ratings data from Mass et al. (2011)
# available at http://ai.stanford.edu/~amaas/data/sentiment/
# we set up a directory under our working directory structure
# /reviews/train/unsup/ for the unsupervised reviews

directory.location <-
  paste(getwd(),"/reviews/train/unsup/",sep = "")
unsup.corpus <- Corpus(DirSource(directory.location, encoding = "UTF-8"),
  readerControl = list(language = "en_US"))
print(summary(unsup.corpus))

document.collection <- unsup.corpus

# strip white space from the documents in the collection
document.collection <- tm_map(document.collection, stripWhitespace)

# convert uppercase to lowercase in the document collection
document.collection <- tm_map(document.collection, content_transformer(tolower))

# remove numbers from the document collection
document.collection <- tm_map(document.collection, removeNumbers)

# remove punctuation from the document collection
document.collection <- tm_map(document.collection, removePunctuation)

# using a standard list, remove English stopwords from the document collection
document.collection <- tm_map(document.collection,
  removeWords, stopwords("english"))

# there is more we could do in terms of data preparation
# stemming... looking for contractions... possessives...
# previous analysis of a list of top terms showed a number of word
# contractions which we might like to drop from further analysis,
# recognizing them as stop words to be dropped from the document collection
initial.tdm <- TermDocumentMatrix(document.collection)
examine.tdm <- removeSparseTerms(initial.tdm, sparse = 0.96)
top.words <- Terms(examine.tdm)
print(top.words)

more.stop.words <- c("cant","didnt","doesnt","dont","goes","isnt","hes",
  "shes","thats","theres","theyre","wont","youll","youre","youve")
document.collection <- tm_map(document.collection,
  removeWords, more.stop.words)

some.proper.nouns.to.remove <-
  c("dick","ginger","hollywood","jack","jill","john","karloff",
    "kudrow","orson","peter","tcm","tom","toni","welles","william","wolheim")
document.collection <- tm_map(document.collection,
  removeWords, some.proper.nouns.to.remove)

# there is still more we could do in terms of data preparation
# but we will work with the bag of words we have for now
```

152

```
# the workhorse technique will be TermDocumentMatrix()
# for creating a terms-by-documents matrix across the document collection
```

```
# in previous text analytics with the taglines data we let the data
# guide us to the text measures... with sentiment analysis we have
# positive and negative dictionaries (to a large extent) defined in
# advance of looking at the data...
# positive.words and negative.words lists were read in earlier
# these come from the work of Hu and Liu (2004)
# positive.words = list of  positive words
# negative.words = list of  negative words
# we will start with these lists to build dictionaries
# that seem to make sense for movie reviews analysis
# Hu.Liu.positive.dictionary <- Dictionary(positive.data.frame$positive.words)
Hu.Liu.positive.dictionary <-
    c(as.character(positive.data.frame$positive.words))
reviews.tdm.Hu.Liu.positive <- TermDocumentMatrix(document.collection,
  list(dictionary = Hu.Liu.positive.dictionary))
examine.tdm <- removeSparseTerms(reviews.tdm.Hu.Liu.positive, 0.95)
top.words <- Terms(examine.tdm)
print(top.words)
Hu.Liu.frequent.positive <- findFreqTerms(reviews.tdm.Hu.Liu.positive, 25)
# this provides a list positive words occurring at least 25 times
# a review of this list suggests that all make sense (have content validity)
# test.positive.dictionary <- Dictionary(Hu.Liu.frequent.positive)
test.positive.dictionary <- c(as.character(Hu.Liu.frequent.positive))

# .... now for the negative words
# Hu.Liu.negative.dictionary <- Dictionary(negative.data.frame$negative.words)
Hu.Liu.negative.dictionary <-
    c(as.character(negative.data.frame$negative.words))
reviews.tdm.Hu.Liu.negative <- TermDocumentMatrix(document.collection,
  list(dictionary = Hu.Liu.negative.dictionary))
examine.tdm <- removeSparseTerms(reviews.tdm.Hu.Liu.negative, 0.97)
top.words <- Terms(examine.tdm)
print(top.words)
Hu.Liu.frequent.negative <- findFreqTerms(reviews.tdm.Hu.Liu.negative, 15)
# this provides a short list negative words occurring at least 15 times
# across the document collection... one of these words seems out of place
# as they could be thought of as positive: "funny"
test.negative <- setdiff(Hu.Liu.frequent.negative,c("funny"))
# test.negative.dictionary <- Dictionary(test.negative)
test.negative.dictionary <- c(as.character(test.negative))

# we need to evaluate the text measures we have defined
# for each of the documents count the total words
# and the number of words that match the positive and negative dictionaries
total.words <- integer(length(names(document.collection)))
positive.words <- integer(length(names(document.collection)))
negative.words <- integer(length(names(document.collection)))
other.words <- integer(length(names(document.collection)))
```

153

```
reviews.tdm <- TermDocumentMatrix(document.collection)
for(index.for.document in seq(along=names(document.collection))) {
  positive.words[index.for.document] <-
    sum(termFreq(document.collection[[index.for.document]],
    control = list(dictionary = test.positive.dictionary)))
  negative.words[index.for.document] <-
    sum(termFreq(document.collection[[index.for.document]],
    control = list(dictionary = test.negative.dictionary)))
  total.words[index.for.document] <-
    length(reviews.tdm[,index.for.document][["i"]])
  other.words[index.for.document] <- total.words[index.for.document] -
    positive.words[index.for.document] - negative.words[index.for.document]
  }
document <- names(document.collection)
text.measures.data.frame <- data.frame(document,total.words,
  positive.words, negative.words, other.words, stringsAsFactors = FALSE)
```

```
rownames(text.measures.data.frame) <- paste("D",as.character(0:499),sep="")
# compute text measures as percentages of words in each set
text.measures.data.frame$POSITIVE <-
  100 * text.measures.data.frame$positive.words /
  text.measures.data.frame$total.words
text.measures.data.frame$NEGATIVE <-
  100 * text.measures.data.frame$negative.words /
    text.measures.data.frame$total.words
# let us look at the resulting text measures we call POSITIVE and NEGATIVE
# to see if negative and positive dimensions appear to be on a common scale
# that is... is this a single dimension in the document space
# we use principal component biplots to explore text measures
# here we can use the technique to check on POSITIVE and NEGATIVE
principal.components.solution <-
  princomp(text.measures.data.frame[,c("POSITIVE","NEGATIVE")], cor = TRUE)
print(summary(principal.components.solution))
# biplot rendering of text measures and documents by year
pdf(file = "fig_sentiment_text_measures_biplot.pdf", width = 8.5, height = 11)
biplot(principal.components.solution, xlab = "First Pricipal Component",
  xlabs = rep("o", times = length(names(document.collection))),
  ylab = "Second Principal Component", expand = 0.7)
dev.off()
# results... the eigenvalues suggest that there are two underlying dimensions
# POSITIVE and NEGATIVE vectors rather than pointing in opposite directions
# they appear to be othogonal to one another... separate dimensions
# here we see the scatter plot for the two measures...
# if they were on the same dimension, they would be negatively correlated
# in fact they are correlated negatively but the correlation is very small
with(text.measures.data.frame, print(cor(POSITIVE, NEGATIVE)))
pdf(file = "fig_sentiment_text_measures_scatter_plot.pdf",
  width = 8.5, height = 8.5)
ggplot.object <- ggplot(data = text.measures.data.frame,
  aes(x = NEGATIVE, y = POSITIVE)) +
    geom_point(colour = "darkblue", shape = 1)
ggplot.print.with.margins(ggplot.object.name = ggplot.object,
  left.margin.pct=10, right.margin.pct=10,
  top.margin.pct=10,bottom.margin.pct=10)
dev.off()
# Perhaps POSITIVE and NEGATIVE can be combined in a way to yield effective
# predictions of movie ratings. Let us move to a set of movie reviews for
# supervised learning.  We select the 500 records from a set of positive
# reviews (ratings between 7 and 10) and 500 records from a set of negative
# reviews (ratings between 1 and 4).

# a set of 500 positive reviews... part of the training set
directory.location <-
  paste(getwd(),"/reviews/train/pos/",sep = "")
pos.train.corpus <- Corpus(DirSource(directory.location, encoding = "UTF-8"),
  readerControl = list(language = "en_US"))
print(summary(pos.train.corpus))
# a set of 500 negative reviews... part of the training set
directory.location <-
  paste(getwd(),"/reviews/train/neg/",sep = "")
neg.train.corpus <- Corpus(DirSource(directory.location, encoding = "UTF-8"),
  readerControl = list(language = "en_US"))
print(summary(neg.train.corpus))

# combine the positive and negative training sets
train.corpus <- c(pos.train.corpus, neg.train.corpus)

# strip white space from the documents in the collection
train.corpus <- tm_map(train.corpus, stripWhitespace)

# convert uppercase to lowercase in the document collection
train.corpus <- tm_map(train.corpus, content_transformer(tolower))
```

154

```
# remove numbers from the document collection
train.corpus <- tm_map(train.corpus, removeNumbers)

# remove punctuation from the document collection
train.corpus <- tm_map(train.corpus, removePunctuation)

# using a standard list, remove English stopwords from the document collection
train.corpus <- tm_map(train.corpus,
  removeWords, stopwords("english"))

# there is more we could do in terms of data preparation
# stemming... looking for contractions... possessives...
# previous analysis of a list of top terms showed a number of word
# contractions which we might like to drop from further analysis,
# recognizing them as stop words to be dropped from the document collection
initial.tdm <- TermDocumentMatrix(train.corpus)
examine.tdm <- removeSparseTerms(initial.tdm, sparse = 0.96)
top.words <- Terms(examine.tdm)
print(top.words)

more.stop.words <- c("cant","didnt","doesnt","dont","goes","isnt","hes",
  "shes","thats","theres","theyre","wont","youll","youre","youve")
train.corpus <- tm_map(train.corpus,
  removeWords, more.stop.words)

some.proper.nouns.to.remove <-
  c("dick","ginger","hollywood","jack","jill","john","karloff",
    "kudrow","orson","peter","tcm","tom","toni","welles","william","wolheim")
train.corpus <- tm_map(train.corpus,
  removeWords, some.proper.nouns.to.remove)

# compute list-based text measures for the training corpus
# for each of the documents count the total words
# and the number of words that match the positive and negative dictionaries
total.words <- integer(length(names(train.corpus)))
positive.words <- integer(length(names(train.corpus)))
negative.words <- integer(length(names(train.corpus)))
other.words <- integer(length(names(train.corpus)))

reviews.tdm <- TermDocumentMatrix(train.corpus)

for(index.for.document in seq(along=names(train.corpus))) {
  positive.words[index.for.document] <-
    sum(termFreq(train.corpus[[index.for.document]],
    control = list(dictionary = test.positive.dictionary)))
  negative.words[index.for.document] <-
    sum(termFreq(train.corpus[[index.for.document]],
    control = list(dictionary = test.negative.dictionary)))
  total.words[index.for.document] <-
    length(reviews.tdm[,index.for.document][["i"]])
  other.words[index.for.document] <- total.words[index.for.document] -
    positive.words[index.for.document] - negative.words[index.for.document]
  }

document <- names(train.corpus)
train.data.frame <- data.frame(document,total.words,
  positive.words, negative.words, other.words, stringsAsFactors = FALSE)
rownames(train.data.frame) <- paste("D",as.character(0:999),sep="")
# compute text measures as percentages of words in each set
train.data.frame$POSITIVE <-
  100 * train.data.frame$positive.words /
  train.data.frame$total.words
train.data.frame$NEGATIVE <-
  100 * train.data.frame$negative.words /
```

155

```
    train.data.frame$total.words

# rating is embedded in the document name... extract with regular expressions
for(index.for.document in seq(along = train.data.frame$document)) {
  first_split <- strsplit(train.data.frame$document[index.for.document],
    split = "[_]")
  second_split <- strsplit(first_split[[1]][2], split = "[.]")
  train.data.frame$rating[index.for.document] <- as.numeric(second_split[[1]][1])
  } # end of for-loop for defining ratings and thumbsupdown
train.data.frame$thumbsupdown <- ifelse((train.data.frame$rating > 5), 2, 1)
train.data.frame$thumbsupdown <-
  factor(train.data.frame$thumbsupdown, levels = c(1,2),
    labels = c("DOWN","UP"))
```

156

```
# a set of 500 positive reviews... part of the test set
directory.location <-
  paste(getwd(),"/reviews/test/pos/",sep = "")

pos.test.corpus <- Corpus(DirSource(directory.location, encoding = "UTF-8"),
  readerControl = list(language = "en_US"))
print(summary(pos.test.corpus))

# a set of 500 negative reviews... part of the test set
directory.location <-
  paste(getwd(),"/reviews/test/neg/",sep = "")

neg.test.corpus <- Corpus(DirSource(directory.location, encoding = "UTF-8"),
  readerControl = list(language = "en_US"))
print(summary(neg.test.corpus))

# combine the positive and negative testing sets
test.corpus <- c(pos.test.corpus, neg.test.corpus)

# strip white space from the documents in the collection
test.corpus <- tm_map(test.corpus, stripWhitespace)

# convert uppercase to lowercase in the document collection
test.corpus <- tm_map(test.corpus, content_transformer(tolower))

# remove numbers from the document collection
test.corpus <- tm_map(test.corpus, removeNumbers)

# remove punctuation from the document collection
test.corpus <- tm_map(test.corpus, removePunctuation)

# using a standard list, remove English stopwords from the document collection
test.corpus <- tm_map(test.corpus,
  removeWords, stopwords("english"))

# there is more we could do in terms of data preparation
# stemming... looking for contractions... possessives...
# previous analysis of a list of top terms showed a number of word
# contractions which we might like to drop from further analysis,
# recognizing them as stop words to be dropped from the document collection
initial.tdm <- TermDocumentMatrix(test.corpus)
examine.tdm <- removeSparseTerms(initial.tdm, sparse = 0.96)
top.words <- Terms(examine.tdm)
print(top.words)
more.stop.words <- c("cant","didnt","doesnt","dont","goes","isnt","hes",
  "shes","thats","theres","theyre","wont","youll","youre","youve")
test.corpus <- tm_map(test.corpus,
  removeWords, more.stop.words)
some.proper.nouns.to.remove <-
  c("dick","ginger","hollywood","jack","jill","john","karloff",
    "kudrow","orson","peter","tcm","tom","toni","welles","william","wolheim")
test.corpus <- tm_map(test.corpus,
```

157

```
    removeWords, some.proper.nouns.to.remove)

# compute list-based text measures for the testing corpus
# for each of the documents count the total words
# and the number of words that match the positive and negative dictionaries
total.words <- integer(length(names(test.corpus)))
positive.words <- integer(length(names(test.corpus)))
negative.words <- integer(length(names(test.corpus)))
other.words <- integer(length(names(test.corpus)))

reviews.tdm <- TermDocumentMatrix(test.corpus)

for(index.for.document in seq(along=names(test.corpus))) {
  positive.words[index.for.document] <-
    sum(termFreq(test.corpus[[index.for.document]],
    control = list(dictionary = test.positive.dictionary)))
  negative.words[index.for.document] <-
    sum(termFreq(test.corpus[[index.for.document]],
    control = list(dictionary = test.negative.dictionary)))
  total.words[index.for.document] <-
    length(reviews.tdm[,index.for.document][["i"]])
  other.words[index.for.document] <- total.words[index.for.document] -
    positive.words[index.for.document] - negative.words[index.for.document]
  }
document <- names(test.corpus)
test.data.frame <- data.frame(document,total.words,
  positive.words, negative.words, other.words, stringsAsFactors = FALSE)
rownames(test.data.frame) <- paste("D",as.character(0:999),sep="")

# compute text measures as percentages of words in each set
test.data.frame$POSITIVE <-
  100 * test.data.frame$positive.words /
  test.data.frame$total.words
 test.data.frame$NEGATIVE <-
  100 * test.data.frame$negative.words /
    test.data.frame$total.words

# rating is embedded in the document name... extract with regular expressions
for(index.for.document in seq(along = test.data.frame$document)) {
  first_split <- strsplit(test.data.frame$document[index.for.document],
    split = "[_]")
  second_split <- strsplit(first_split[[1]][2], split = "[.]")
  test.data.frame$rating[index.for.document] <- as.numeric(second_split[[1]][1])
  } # end of for-loop for defining

test.data.frame$thumbsupdown <- ifelse((test.data.frame$rating > 5), 2, 1)
test.data.frame$thumbsupdown <-
  factor(test.data.frame$thumbsupdown, levels = c(1,2),
    labels = c("DOWN","UP"))

# a set of 4 positive and 4 negative reviews... testing set of Tom's reviews
directory.location <-
  paste(getwd(),"/reviews/test/tom/",sep = "")

tom.corpus <- Corpus(DirSource(directory.location, encoding = "UTF-8"),
  readerControl = list(language = "en_US"))
print(summary(tom.corpus))

# strip white space from the documents in the collection
tom.corpus <- tm_map(tom.corpus, stripWhitespace)

# convert uppercase to lowercase in the document collection
tom.corpus <- tm_map(tom.corpus, content_transformer(tolower))

# remove numbers from the document collection
tom.corpus <- tm_map(tom.corpus, removeNumbers)
```

158

```
# remove punctuation from the document collection
tom.corpus <- tm_map(tom.corpus, removePunctuation)

# using a standard list, remove English stopwords from the document collection
tom.corpus <- tm_map(tom.corpus,
  removeWords, stopwords("english"))

# there is more we could do in terms of data preparation
# stemming... looking for contractions... possessives...
# previous analysis of a list of top terms showed a number of word
# contractions which we might like to drop from further analysis,
# recognizing them as stop words to be dropped from the document collection

initial.tdm <- TermDocumentMatrix(tom.corpus)
examine.tdm <- removeSparseTerms(initial.tdm, sparse = 0.96)
top.words <- Terms(examine.tdm)
print(top.words)

more.stop.words <- c("cant","didnt","doesnt","dont","goes","isnt","hes",
  "shes","thats","theres","theyre","wont","youll","youre","youve")
tom.corpus <- tm_map(tom.corpus,
  removeWords, more.stop.words)

some.proper.nouns.to.remove <-
  c("dick","ginger","hollywood","jack","jill","john","karloff",
    "kudrow","orson","peter","tcm","tom","toni","welles","william","wolheim")
tom.corpus <- tm_map(tom.corpus,
  removeWords, some.proper.nouns.to.remove)

# compute list-based text measures for tom corpus
# for each of the documents count the total words
# and the number of words that match the positive and negative dictionaries

total.words <- integer(length(names(tom.corpus)))
positive.words <- integer(length(names(tom.corpus)))
negative.words <- integer(length(names(tom.corpus)))
other.words <- integer(length(names(tom.corpus)))

reviews.tdm <- TermDocumentMatrix(tom.corpus)
for(index.for.document in seq(along=names(tom.corpus))) {
  positive.words[index.for.document] <-
    sum(termFreq(tom.corpus[[index.for.document]],
    control = list(dictionary = test.positive.dictionary)))
  negative.words[index.for.document] <-
    sum(termFreq(tom.corpus[[index.for.document]],
    control = list(dictionary = test.negative.dictionary)))
  total.words[index.for.document] <-
    length(reviews.tdm[,index.for.document][["i"]])
  other.words[index.for.document] <- total.words[index.for.document] -
    positive.words[index.for.document] - negative.words[index.for.document]
  }

document <- names(tom.corpus)
tom.data.frame <- data.frame(document,total.words,
  positive.words, negative.words, other.words, stringsAsFactors = FALSE)
rownames(tom.data.frame) <- paste("D",as.character(0:7),sep="")

# compute text measures as percentages of words in each set
tom.data.frame$POSITIVE <-
  100 * tom.data.frame$positive.words /
  tom.data.frame$total.words

tom.data.frame$NEGATIVE <-
  100 * tom.data.frame$negative.words /
```

159

```
    tom.data.frame$total.words

# rating is embedded in the document name... extract with regular expressions

for(index.for.document in seq(along = tom.data.frame$document)) {
  first_split <- strsplit(tom.data.frame$document[index.for.document],
    split = "[_]")
  second_split <- strsplit(first_split[[1]][2], split = "[.]")
  tom.data.frame$rating[index.for.document] <- as.numeric(second_split[[1]][1])
  } # end of for-loop for defining

tom.data.frame$thumbsupdown <- ifelse((tom.data.frame$rating > 5), 2, 1)
tom.data.frame$thumbsupdown <-
  factor(tom.data.frame$thumbsupdown, levels = c(1,2),
    labels = c("DOWN","UP"))

tom.movies <- data.frame(movies =
  c("The Effect of Gamma Rays on Man-in-the-Moon Marigolds",
    "Blade Runner","My Cousin Vinny","Mars Attacks",
    "Fight Club","Miss Congeniality 2","Find Me Guilty","Moneyball"))

# check out the measures on Tom's ratings
tom.data.frame.review <-
  cbind(tom.movies,tom.data.frame[,names(tom.data.frame)[2:9]])
print(tom.data.frame.review)
```

160

```
# develop predictive models using the training data
# --------------------------------------
# Simple difference method
# --------------------------------------
train.data.frame$simple <-
     train.data.frame$POSITIVE - train.data.frame$NEGATIVE
# check out simple difference method... is there a correlation with ratings?
with(train.data.frame, print(cor(simple, rating)))
# we use the training data to define an optimal cutoff...
# trees can help with finding the optimal split point for simple.difference
try.tree <- rpart(thumbsupdown ~ simple, data = train.data.frame)
print(try.tree)  # note that the first split value
# an earlier analysis had this value as -0.7969266
# create a user-defined function for the simple difference method
predict.simple <- function(x) {
  if (x >= -0.7969266) return("UP")
  if (x < -0.7969266) return("DOWN")
  }
# evaluate predictive accuracy in the training data
train.data.frame$pred.simple <- character(nrow(train.data.frame))
for (index.for.review in seq(along = train.data.frame$pred.simple)) {
  train.data.frame$pred.simple[index.for.review] <-
    predict.simple(train.data.frame$simple[index.for.review])
  }
train.data.frame$pred.simple <-
  factor(train.data.frame$pred.simple)
train.pred.simple.performance <-
  confusionMatrix(data = train.data.frame$pred.simple,
  reference = train.data.frame$thumbsupdown, positive = "UP")
# report full set of statistics relating to predictive accuracy
print(train.pred.simple.performance)
cat("\n\nTraining set percentage correctly predicted by",
  " simple difference method = ",
  sprintf("%1.1f",train.pred.simple.performance$overall[1]*100)," Percent",sep="")
# evaluate predictive accuracy in the test data
# SIMPLE DIFFERENCE METHOD
test.data.frame$simple <-
     test.data.frame$POSITIVE - train.data.frame$NEGATIVE
```

```
test.data.frame$pred.simple <- character(nrow(test.data.frame))
for (index.for.review in seq(along = test.data.frame$pred.simple)) {
  test.data.frame$pred.simple[index.for.review] <-
    predict.simple(test.data.frame$simple[index.for.review])
  }

test.data.frame$pred.simple <-
  factor(test.data.frame$pred.simple)
test.pred.simple.performance <-
  confusionMatrix(data = test.data.frame$pred.simple,
  reference = test.data.frame$thumbsupdown, positive = "UP")
# report full set of statistics relating to predictive accuracy
print(test.pred.simple.performance)
cat("\n\nTest set percentage correctly predicted = ",
  sprintf("%1.1f",test.pred.simple.performance$overall[1]*100),"
    Percent",sep="")

# ---------------------------------------
# Regression difference method
# ---------------------------------------
# regression method for determining weights on POSITIVE AND NEGATIVE
# fit a regression model to the training data
regression.model <- lm(rating ~ POSITIVE + NEGATIVE, data = train.data.frame)
print(regression.model)  # provides 5.5386 + 0.2962(POSITIVE) -0.3089(NEGATIVE)
train.data.frame$regression <-
  predict(regression.model, newdata = train.data.frame)
# determine the cutoff for regression.difference
  try.tree <- rpart(thumbsupdown ~ regression, data = train.data.frame)
print(try.tree)  # note that the first split is at 5.264625
# create a user-defined function for the simple difference method
predict.regression <- function(x) {
  if (x >= 5.264625) return("UP")
  if (x < 5.264625) return("DOWN")
  }
train.data.frame$pred.regression <-  character(nrow(train.data.frame))
for (index.for.review in seq(along = train.data.frame$pred.simple)) {
  train.data.frame$pred.regression[index.for.review] <-
    predict.regression(train.data.frame$regression[index.for.review])
  }
train.data.frame$pred.regression <-
  factor(train.data.frame$pred.regression)
train.pred.regression.performance <-
  confusionMatrix(data = train.data.frame$pred.regression,
  reference = train.data.frame$thumbsupdown, positive = "UP")
# report full set of statistics relating to predictive accuracy
print(train.pred.regression.performance)  # result 67.3 percent
cat("\n\nTraining set percentage correctly predicted by regression = ",
  sprintf("%1.1f",train.pred.regression.performance$overall[1]*100),
    " Percent",sep="")
# regression method for determining weights on POSITIVE AND NEGATIVE
# for the test set we use the model developed on the training set
test.data.frame$regression <-
  predict(regression.model, newdata = test.data.frame)

test.data.frame$pred.regression <-  character(nrow(test.data.frame))
for (index.for.review in seq(along = test.data.frame$pred.simple)) {
  test.data.frame$pred.regression[index.for.review] <-
    predict.regression(test.data.frame$regression[index.for.review])
  }

test.data.frame$pred.regression <-
  factor(test.data.frame$pred.regression)
test.pred.regression.performance <-
  confusionMatrix(data = test.data.frame$pred.regression,
  reference = test.data.frame$thumbsupdown, positive = "UP")
```

161

162

```
# report full set of statistics relating to predictive accuracy
print(test.pred.regression.performance)  # result 67.3 percent
cat("\n\nTest set percentage correctly predicted = ",
  sprintf("%1.1f",test.pred.regression.performance$overall[1]*100),
    " Percent",sep="")

# ---------------------------------------------
# Word/item analysis method for train.corpus
# ---------------------------------------------
# return to the training corpus to develop simple counts
# for each of the words in the sentiment list
# these new variables will be given the names of the words
# to keep things simple.... there are 50 such variables/words
# identified from an earlier analysis, as published in the book
working.corpus <- train.corpus
# run common code from utilities for scoring the working corpus
# this common code uses 25 positive and 25 negative words
# identified in an earlier analysis of these data
source("R_utility_program_5.R")
add.data.frame <- data.frame(amazing,beautiful,classic,enjoy,
  enjoyed,entertaining,excellent,fans,favorite,fine,fun,humor,
  lead,liked,love,loved,modern,nice,perfect,pretty,
  recommend,strong,top,wonderful,worth,bad,boring,cheap,creepy,dark,dead,
  death,evil,hard,kill,killed,lack,lost,miss,murder,mystery,plot,poor,
  sad,scary,slow,terrible,waste,worst,wrong)
train.data.frame <- cbind(train.data.frame,add.data.frame)
# ---------------------------------------------
# Word/item analysis method for test.corpus
# ---------------------------------------------
# return to the testing corpus to develop simple counts
# for each of the words in the sentiment list
# these new variables will be given the names of the words
# to keep things simple.... there are 50 such variables/words
working.corpus <- test.corpus
# run common code from utilities for scoring the working corpus
source("R_utility_program_5.R")
add.data.frame <- data.frame(amazing,beautiful,classic,enjoy,
  enjoyed,entertaining,excellent,fans,favorite,fine,fun,humor,
  lead,liked,love,loved,modern,nice,perfect,pretty,
  recommend,strong,top,wonderful,worth,bad,boring,cheap,creepy,dark,dead,
  death,evil,hard,kill,killed,lack,lost,miss,murder,mystery,plot,poor,
  sad,scary,slow,terrible,waste,worst,wrong)
 test.data.frame <- cbind(test.data.frame,add.data.frame)
# ---------------------------------------------
# Word/item analysis method for tom.corpus
# ---------------------------------------------
# return to the toming corpus to develop simple counts
# for each of the words in the sentiment list
# these new variables will be given the names of the words
# to keep things simple.... there are 50 such variables/words
working.corpus <- tom.corpus
# run common code from utilities for scoring the working corpus
source("R_utility_program_5.R")
add.data.frame <- data.frame(amazing,beautiful,classic,enjoy,
  enjoyed,entertaining,excellent,fans,favorite,fine,fun,humor,
  lead,liked,love,loved,modern,nice,perfect,pretty,
  recommend,strong,top,wonderful,worth,bad,boring,cheap,creepy,dark,dead,
  death,evil,hard,kill,killed,lack,lost,miss,murder,mystery,plot,poor,
  sad,scary,slow,terrible,waste,worst,wrong)
tom.data.frame <- cbind(tom.data.frame,add.data.frame)

# use phi coefficient... correlation with rating as index of item value
# again we draw upon the earlier positive and negative lists
phi <- numeric(50)
item <- c("amazing","beautiful","classic","enjoy",
```

163

```
  "enjoyed","entertaining","excellent","fans","favorite","fine","fun","humor",
  "lead","liked","love","loved","modern","nice","perfect","pretty",
  "recommend","strong","top","wonderful","worth",
  "bad","boring","cheap","creepy","dark","dead",
  "death","evil","hard","kill","killed","lack",
  "lost","miss","murder","mystery","plot","poor",
  "sad","scary","slow","terrible","waste","worst","wrong")
item.analysis.data.frame <- data.frame(item,phi)
item.place <- 14:63
for (index.for.column in 1:50) {
  item.analysis.data.frame$phi[index.for.column] <-
    cor(train.data.frame[, item.place[index.for.column]],train.data.frame[,8])
  }

# sort by absolute value of the phi coefficient with the rating
item.analysis.data.frame$absphi <- abs(item.analysis.data.frame$phi)
item.analysis.data.frame <-
  item.analysis.data.frame[sort.list(item.analysis.data.frame$absphi,
    decreasing = TRUE),]

# subset of words with phi coefficients greater than 0.05 in absolute value
selected.items.data.frame <-
  subset(item.analysis.data.frame, subset = (absphi > 0.05))

# use the sign of the phi coefficient as the item weight
selected.positive.data.frame <-
  subset(selected.items.data.frame, subset = (phi > 0.0))
selected.positive.words <- as.character(selected.positive.data.frame$item)

selected.negative.data.frame <-
  subset(selected.items.data.frame, subset = (phi < 0.0))
selected.negative.words <- as.character(selected.negative.data.frame$item)

# these lists define new dictionaries for scoring

reviews.tdm <- TermDocumentMatrix(train.corpus)

temp.positive.score <- integer(length(names(train.corpus)))
temp.negative.score <- integer(length(names(train.corpus)))
for(index.for.document in seq(along=names(train.corpus))) {
  temp.positive.score[index.for.document] <-
    sum(termFreq(train.corpus[[index.for.document]],
    control = list(dictionary = selected.positive.words)))
  temp.negative.score[index.for.document] <-
    sum(termFreq(train.corpus[[index.for.document]],
    control = list(dictionary = selected.negative.words)))
  }
train.data.frame$item.analysis.score <-
  temp.positive.score - temp.negative.score

# use the training set and tree-structured modeling to determine the cutoff
  try.tree<-rpart(thumbsupdown ~ item.analysis.score, data = train.data.frame)
print(try.tree)  # note that the first split is at -0.5
# create a user-defined function for the simple difference method
predict.item.analysis <- function(x) {
  if (x >= -0.5) return("UP")
  if (x < -0.5) return("DOWN")
  }
train.data.frame$pred.item.analysis <-  character(nrow(train.data.frame))
for (index.for.review in seq(along = train.data.frame$pred.simple)) {
  train.data.frame$pred.item.analysis[index.for.review] <-
  predict.item.analysis(train.data.frame$item.analysis.score[index.for.review])
  }
train.data.frame$pred.item.analysis <-
  factor(train.data.frame$pred.item.analysis)
```

164

```
train.pred.item.analysis.performance <-
  confusionMatrix(data = train.data.frame$pred.item.analysis,
  reference = train.data.frame$thumbsupdown, positive = "UP")
# report full set of statistics relating to predictive accuracy
print(train.pred.item.analysis.performance)  # result 73.9 percent
cat("\n\nTraining set percentage correctly predicted by item analysis = ",
  sprintf("%1.1f",train.pred.item.analysis.performance$overall[1]*100),
    " Percent",sep="")

# use item analysis method of scoring with the test set

reviews.tdm <- TermDocumentMatrix(test.corpus)

temp.positive.score <- integer(length(names(test.corpus)))
temp.negative.score <- integer(length(names(test.corpus)))
for(index.for.document in seq(along=names(test.corpus))) {
  temp.positive.score[index.for.document] <-
    sum(termFreq(test.corpus[[index.for.document]],
    control = list(dictionary = selected.positive.words)))
  temp.negative.score[index.for.document] <-
    sum(termFreq(test.corpus[[index.for.document]],
    control = list(dictionary = selected.negative.words)))
  }

test.data.frame$item.analysis.score <-
  temp.positive.score - temp.negative.score

test.data.frame$pred.item.analysis <-  character(nrow(test.data.frame))
for (index.for.review in seq(along = test.data.frame$pred.simple)) {
  test.data.frame$pred.item.analysis[index.for.review] <-
  predict.item.analysis(test.data.frame$item.analysis.score[index.for.review])
  }
test.data.frame$pred.item.analysis <-
  factor(test.data.frame$pred.item.analysis)
test.pred.item.analysis.performance <-
  confusionMatrix(data = test.data.frame$pred.item.analysis,
  reference = test.data.frame$thumbsupdown, positive = "UP")

# report full set of statistics relating to predictive accuracy
print(test.pred.item.analysis.performance)  # result 74 percent

cat("\n\nTest set percentage correctly predicted by item analysis = ",
  sprintf("%1.1f",test.pred.item.analysis.performance$overall[1]*100),
    " Percent",sep="")

# --------------------------------------
# Logistic regression method
# --------------------------------------
text.classification.model <- {thumbsupdown ~ amazing + beautiful +
  classic + enjoy + enjoyed +
  entertaining + excellent +
  fans + favorite + fine + fun + humor + lead + liked +
  love + loved + modern + nice + perfect + pretty +
  recommend + strong + top + wonderful + worth +
  bad + boring + cheap + creepy + dark + dead +
  death + evil + hard + kill +
  killed + lack + lost + miss + murder + mystery +
  plot + poor + sad + scary +
  slow + terrible + waste + worst + wrong}

# full logistic regression model
logistic.regression.fit <- glm(text.classification.model,
  family=binomial(link=logit), data = train.data.frame)
print(summary(logistic.regression.fit))
```

165

```
# obtain predicted probability values for training set
logistic.regression.pred.prob <-
  as.numeric(predict(logistic.regression.fit, newdata = train.data.frame,
  type="response"))

train.data.frame$pred.logistic.regression <-
  ifelse((logistic.regression.pred.prob > 0.5),2,1)

train.data.frame$pred.logistic.regression <-
  factor(train.data.frame$pred.logistic.regression, levels = c(1,2),
    labels = c("DOWN","UP"))

train.pred.logistic.regression.performance <-
  confusionMatrix(data = train.data.frame$pred.logistic.regression,
  reference = train.data.frame$thumbsupdown, positive = "UP")

# report full set of statistics relating to predictive accuracy
print(train.pred.logistic.regression.performance)  # result 75.2 percent
cat("\n\nTraining set percentage correct by logistic regression = ",
  sprintf("%1.1f",train.pred.logistic.regression.performance$overall[1]*100),
    " Percent",sep="")
```

166

```
# now we use the model developed on the training set with the test set
# obtain predicted probability values for test set
logistic.regression.pred.prob <-
  as.numeric(predict(logistic.regression.fit, newdata = test.data.frame,
  type="response"))

test.data.frame$pred.logistic.regression <-
  ifelse((logistic.regression.pred.prob > 0.5),2,1)

test.data.frame$pred.logistic.regression <-
  factor(test.data.frame$pred.logistic.regression, levels = c(1,2),
    labels = c("DOWN","UP"))

test.pred.logistic.regression.performance <-
  confusionMatrix(data = test.data.frame$pred.logistic.regression,
  reference = test.data.frame$thumbsupdown, positive = "UP")

# report full set of statistics relating to predictive accuracy
print(test.pred.logistic.regression.performance)  # result 72.6 percent

cat("\n\nTest set percentage correctly predicted by logistic regression = ",
  sprintf("%1.1f",test.pred.logistic.regression.performance$overall[1]*100),
    " Percent",sep="")
# ----------------------------------------
# Support vector machines
# ----------------------------------------
# determine tuning parameters prior to fitting model
train.tune <- tune(svm, text.classification.model, data = train.data.frame,
                   ranges = list(gamma = 2^(-8:1), cost = 2^(0:4)),
                   tunecontrol = tune.control(sampling = "fix"))
# display the tuning results (in text format)
print(train.tune)
# fit the support vector machine to the training data using tuning parameters
train.data.frame.svm <- svm(text.classification.model, data = train.data.frame,
  cost=4, gamma=0.00390625, probability = TRUE)
train.data.frame$pred.svm <- predict(train.data.frame.svm, type="class",
newdata=train.data.frame)
train.pred.svm.performance <-
  confusionMatrix(data = train.data.frame$pred.svm,
  reference = train.data.frame$thumbsupdown, positive = "UP")

# report full set of statistics relating to predictive accuracy
print(train.pred.svm.performance)  # result 79.0 percent
```

```
cat("\n\nTraining set percentage correctly predicted by SVM = ",
  sprintf("%1.1f",train.pred.svm.performance$overall[1]*100),
    " Percent",sep="")

# use the support vector machine model identified in the training set
# to do text classification on the test set
test.data.frame$pred.svm <- predict(train.data.frame.svm, type="class",
newdata=test.data.frame)
test.pred.svm.performance <-
  confusionMatrix(data = test.data.frame$pred.svm,
  reference = test.data.frame$thumbsupdown, positive = "UP")

# report full set of statistics relating to predictive accuracy
print(test.pred.svm.performance)  # result 71.6 percent

cat("\n\nTest set percentage correctly predicted by SVM = ",
  sprintf("%1.1f",test.pred.svm.performance$overall[1]*100),
    " Percent",sep="")

# ----------------------------------------
# Random forests
# ----------------------------------------
# fit random forest model to the training data
set.seed (9999)  # for reproducibility
train.data.frame.rf <- randomForest(text.classification.model,
  data=train.data.frame, mtry=3, importance=TRUE, na.action=na.omit)

# review the random forest solution
print(train.data.frame.rf)

# check importance of the individual explanatory variables
pdf(file = "fig_sentiment_random_forest_importance.pdf",
width = 11, height = 8.5)
varImpPlot(train.data.frame.rf, main = "")
dev.off()

train.data.frame$pred.rf <- predict(train.data.frame.rf, type="class",
  newdata = train.data.frame)

train.pred.rf.performance <-
  confusionMatrix(data = train.data.frame$pred.rf,
  reference = train.data.frame$thumbsupdown, positive = "UP")

# report full set of statistics relating to predictive accuracy
print(train.pred.rf.performance)  # result 82.2 percent

cat("\n\nTraining set percentage correctly predicted by random forests = ",
  sprintf("%1.1f",train.pred.rf.performance$overall[1]*100),
    " Percent",sep="")

# use the model fit to the training data to predict the the test data
test.data.frame$pred.rf <- predict(train.data.frame.rf, type="class",
  newdata = test.data.frame)

test.pred.rf.performance <-
  confusionMatrix(data = test.data.frame$pred.rf,
  reference = test.data.frame$thumbsupdown, positive = "UP")

# report full set of statistics relating to predictive accuracy
print(test.pred.rf.performance)  # result 74.0 percent

cat("\n\nTest set percentage correctly predicted by random forests = ",
  sprintf("%1.1f",test.pred.rf.performance$overall[1]*100),
    " Percent",sep="")
```

167

168

```
# measurement model performance summary

methods <- c("Simple difference","Regression difference",
  "Word/item analysis","Logistic regression",
  "Support vector machines","Random forests")

methods.performance.data.frame <- data.frame(methods)

methods.performance.data.frame$training <-
  c(train.pred.simple.performance$overall[1]*100,
    train.pred.regression.performance$overall[1]*100,
    train.pred.item.analysis.performance$overall[1]*100,
    train.pred.logistic.regression.performance$overall[1]*100,
    train.pred.svm.performance$overall[1]*100,
    train.pred.rf.performance$overall[1]*100)

methods.performance.data.frame$test <-
  c(test.pred.simple.performance$overall[1]*100,
    test.pred.regression.performance$overall[1]*100,
    test.pred.item.analysis.performance$overall[1]*100,
    test.pred.logistic.regression.performance$overall[1]*100,
    test.pred.svm.performance$overall[1]*100,
    test.pred.rf.performance$overall[1]*100)

# random forest predictions for Tom's movie reviews

tom.data.frame$pred.rf <- predict(train.data.frame.rf, type="class",
  newdata = tom.data.frame)

print(tom.data.frame[,c("thumbsupdown","pred.rf")])

tom.pred.rf.performance <-
  confusionMatrix(data = tom.data.frame$pred.rf,
  reference = tom.data.frame$thumbsupdown, positive = "UP")

# report full set of statistics relating to predictive accuracy

print(tom.pred.rf.performance)  # result 74.0 percent

cat("\n\nTraining set percentage correctly predicted by random forests = ",
  sprintf("%1.1f",tom.pred.rf.performance$overall[1]*100),
    "Percent",sep="")

# building a simple tree to classify reviews

simple.tree <- rpart(text.classification.model,
  data=train.data.frame,)

# plot the regression tree result from rpart

pdf(file = "fig_sentiment_simple_tree_classifier.pdf", width = 8.5, height = 8.5)
prp(simple.tree, main="",
  digits = 3,  # digits to display in terminal nodes
  nn = TRUE,  # display the node numbers
  fallen.leaves = TRUE,  # put the leaves on the bottom of the page
  branch = 0.5,  # change angle of branch lines
  branch.lwd = 2,  # width of branch lines
  faclen = 0,  # do not abbreviate factor levels
  trace = 1,  # print the automatically calculated cex
  shadow.col = 0,  # no shadows under the leaves
  branch.lty = 1,  # draw branches using dotted lines
  split.cex = 1.2,  # make the split text larger than the node text
  split.prefix = "is ",  # put "is" before split text
  split.suffix = "?",  # put "?" after split text
  split.box.col = "blue",  # lightgray split boxes (default is white)
```

169

```
  split.col = "white",  # color of text in split box
  split.border.col = "blue",  # darkgray border on split boxes
  split.round = .25)  # round the split box corners a tad
dev.off()

# simple tree predictions for Tom's movie reviews
tom.data.frame$pred.simple.tree <- predict(simple.tree, type="class",
  newdata = tom.data.frame)

print(tom.data.frame[,c("thumbsupdown","pred.rf","pred.simple.tree")])
```

170

第9章
发现共同主题

Curly：“你知道生活的秘诀是什么吗？”（Curly 举起一根手指）。

Curly：“这就是。”

Mitch：“你的手指？”

Curly：“一件事，只有一件事。你只要坚持那件事，其余的都不重要。”

Mitch：“可是，这到底是一件什么事呢？”

Curly：“这就是你必须要去寻找的答案。”

Curly（Jack Palance 饰）、Mitch Robbins（Billy Crystal 饰）

电影《城市乡巴佬》（1991 年）

我选择安静的生活。我没有什么政治野心或对于出名的欲望。我在政治上的参与程度仅限于在全国的选举中去投票而已。然而，即使是这样一个安静的人也不得不关心世界上发生的事情以及看上去似乎是永无止境的国与国之间的冲突。

也许我们可以通过历史去学习如何更好地理解外交事务的本质。也许我们可以找到共同的主题，也就是如何定义一个国家以及我们为什么会像我们已经做的那样去做事情的思路。

共同的历史主题可以从我们选择的一系列与我们所关心的问题相关的文档中去寻找。使用任何文本分析，我们可以客观地去开展研究。我们让数据引导我们去寻找主题。在本章里，我们介绍将多维标度法和聚类分析应用于从网络中获取的文本数据的整个过程。

为了找到共同的政治主题，我们选择了美国总统所做的国情咨文（POTUS），也就是在C.7 中介绍的那些演讲稿。我们关注的是近代政治史，因此就选择了自艾森豪威尔担任总统后的相关文件。我们所关注的时间段是 1960～2014 年，在此期间共有 52 篇美国总统所做的国情咨文。这些国情咨文构成了一组文档或者是用于分析的文本语料库。

对于文本分析来说，分析成功往往是通过判断而不是使用算法。在整个分析过程中的每一个阶段，我们要决定保留什么，舍弃什么。

英文中有标准的停止词。这些往往是代词和介词，它们对于识别出共同主题或者区分文档能够起的作用微乎其微。我们可以基于 Natural Language Took Kit 提供的数据来确定这些停止词（Bird, Klein, and Loper 2009）。

通过对 POTUS 数据进行分析，我们会注意到在国情咨文中还有一些其他的共同词也对识别共同主题或者区分不同演讲起不了任何作用。这些演讲向美国国会发表，因此演讲中国会以及国会议员的词没有什么作用，同样没有作用的还有美国和美洲这些词。我们把这些词加入到停止词列表中。

对原始的文本文档进行解析涉及删除标点符号、将大写字母转换成小写字母，以及舍弃停止词。得到的结果就是一个由 52 篇演讲稿构成的文档集或语料库，其中 25 篇出自于民主党总统、27 篇出自于共和党总统。

我们为语料库构建一个由术语和文档构成的矩阵。同样，我们通过判断来决定保留多少词，即该矩阵的行数。通过对这 52 个文档进行解析，在除去停止词后，依然还有 10 938 个172不同的词。我们需要决定将多少数量的词包含在我们的研究中。是使用矩阵中出现频率最高的前 50、100、150，还是 200 个词呢？这个问题的答案将会影响所有需要将这个矩阵作为输入进行的分析。

分析的目标是通过使用对区分分档十分有用的特征或单词将 POTUS 文本转化成为一个简洁的表述。一种方式是通过找到在一个文档中出现频率很高但是在其他文档中出现频繁都不高的那些词。也就是说，我们可以使用一个词频矩阵，即矩阵单元的值为相应的词在相对应的文档中出现的频率。常用的选择是词出现的频率乘以文档在整个语料库中出现频率的倒数。这种技术使用缩写 TF-IDF 来表达。我们随后都将采用这种方法。

我们将 TF-IDF 矩阵中排名前 200 个的词赋予距离度量值，找到所有演讲之间的距离。然后将这个距离矩阵作为多维标度算法的一个输入。我们针对这个问题找到一个二维解决方案，得到一个如图 9-1 所示的总统演讲图。我们可以看到，总统演讲可以很自然地在空间上进行组合。肯尼迪总统和约翰逊总统的很多演讲彼此间都很接近。奥巴马总统的演讲彼此间也非常接近，等等。

这个多维标度解决方案使我们对 TF-IDF 矩阵能够为识别总统的演讲提供有用的特征建立了信心。但是我们还不能就此结束。回顾一下，我们最初的目标是找到共同主题。为此，我们将重心转移到聚类分析。特别需要指出的是，我们寻求的是对词进行聚类，而不是对演讲进行聚类。

我们使用统计的方式来确定所聚类的个数，由此而获得基于词的两个聚类。使用与聚类中心的距离作为条件，我们找到 8 个与第一个聚类关联性很强的词，另外 8 个与第二个聚类关联性很强的词。

对于第一个我们称之为“保持强大（Stay Strong）”的聚类，以下词最接近它的中心：“自从（since）”“保持（keep）”“远（far）”“天（day）”“强大（strong）”“保护（protect）”“高（high）”和“重要（important）”。对于第二个我们称之为“帮助国家（Help Nation）”的聚类，以下的词最接近它的中心：“使得（make）”“时间（time）”“帮助（help）”“国家（nation）”“年份（years）”“工作（work）”“每一个（every）”和“今晚（tonight）”。此外，我173们还可以考虑外部防御与保护或重点在帮助他人及提供就业机会的国内政策。我们可以使用

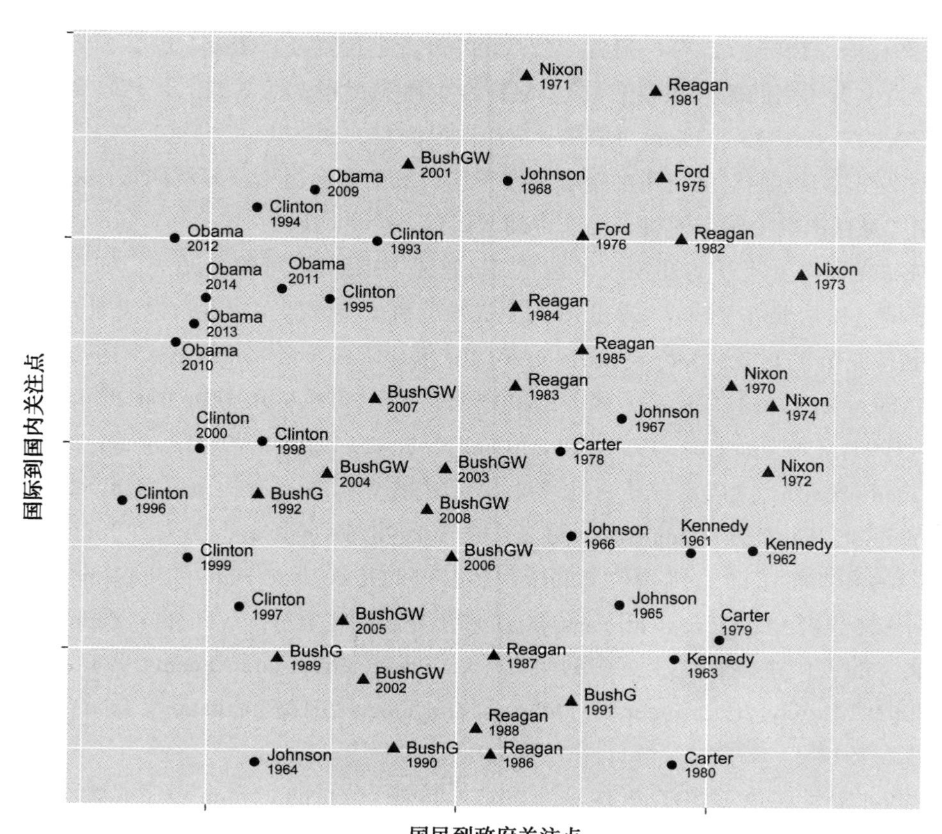

图 9-1　根据语言对总统进行映射

174

这些单词列表为个人演讲或者为总统建立文本度量。我们同样也可以观察从肯尼迪到奥巴马 52 年间主题的走向。

以下这个实例用来说明处理大量文本并找到共同主题的方法。这些方法适用于所有类型的文本和内容。

认为只需要通过算法就能够得知文本所包含的故事是不正确的。在顺利找到共同主题中，分析判断与算法同样重要。我们是应用自然语言进行演讲的专家，不应该期待计算机在不久的将来就能够取代我们。

对政治性文本进行分析是一个热门的研究领域，具有代表性的工作是 Roderick P. Hart (2000a, 2000b, 2001) 对 1948 年～1996 年期间的 13 届美国总统参与竞选活动的政治演讲进行的研究。Hart 提出了一个通用的文本度量，以确定性、乐观主义、活跃性、现实性和共同性（共同的价值观）作为维度对涉及文档的基调进行了归纳。每项度量都包括基于选定的单词列表对词进行统计。某些词获得正加权，另外一些词则获得负加权。对于乐观主义度量，与赞美、满足和鼓励相关的词获得正加权，而与指责、困难和否定相关的词获得负加权。现实性度量则以熟知度这个概念作为基础。

作为一个独立的文本度量，熟知度的计算以英语中被公认为最常见的 44 个单词作为依据。对三个主要的群体（政客、记者和公众）的声音进行复查，Hart观察到，随着时间的推移，复杂度上升（更低的熟知度分数）。横穿整个研究过程，在基调上，与政客以及公众相比较，记者的声音毫无疑问是最负面的（乐观性很低）。

只要不在政策方面，演讲中的常态化在美国政坛上都会得到很好的回报。对于政治候选人，使用常见用语的中间派演讲人比非中间派候选人会更加成功。

面向随意文本分析学的建模技术包括基于差异或距离度量的多元法，如多维标度分析和聚类分析，和基于协方差或关联矩阵的主成份分析。差异与距离度量以及聚类分析法在以下文献中有充分的讨论：Meyer (2014a, 2014b), Kaufman and Rousseeuw (1990) 和 Izenman (2008)，Maechler (2014a) 提供了相关的R语言代码。多维标度分析方面的工作可以参考 Davison (1992), Cox and Cox (1994), Everitt and Rabe-Hesketh (1997), Izenman (2008) 和 Borg and Groenen (2010)。主成份分析在很多对多元法的介绍中都有提及（Manly 1994; Sharma 1996; Gnanadesikan 1997; Johnson and Wichern 1998; Everitt and Dunn 2001; Seber 2000; Izenman 2008），相应的二维图在 Gabriel (1971) 和 Gower and Hand (1996) 中都有相关讨论。潜在语义分析（latent semantic analysis）和潜在狄利克雷分配（latent Dirichlet allocation）为识别文本的共同主题提供了更多的工具（Landauer, Foltz, and Laham 1998; Blei, Ng, and Jordan 2003; Murphy 2012; Ingersoll, Morton, and Farris 2013; Landauer, McNamara, Dennis, and Kintsch 2014)）。

代码9-1展示的是对POTUS演讲的数据文本进行分析的Python程序。该程序来源于很多用于文本解析、自然语言处理和多元法的软件包。

代码 9-1 发现共同主题：POTUS 演讲（Python）

```
# Discovering Common Themes: POTUS Speeches (Python)

# prepare for Python version 3x features and functions
from __future__ import division, print_function

# import packages for text processing and multivariate analysis
import os  # operating system functions
from fnmatch import fnmatch   # character string matching
import re  # regular expressions
import nltk  # draw on the Python natural language toolkit
import pandas as pd  # DataFrame structure and operations
import numpy as np  # arrays and numerical processing
import scipy
import matplotlib.pyplot as plt  # 2D plotting

# terms-by-documents matrix
from sklearn.feature_extraction.text import CountVectorizer
from sklearn.feature_extraction.text import TfidfTransformer

# alternative distance metrics for multidimensional scaling
from sklearn.metrics import euclidean_distances
from sklearn.metrics.pairwise import linear_kernel as cosine_distances
from sklearn.metrics.pairwise import manhattan_distances as manhattan_distances
from sklearn.metrics import silhouette_score as silhouette_score

from sklearn import manifold  # multidimensional scaling
from sklearn.cluster import KMeans  # cluster analysis by partitioning
```

```
# function for walking and printing directory structure
def list_all(current_directory):
    for root, dirs, files in os.walk(current_directory):
        level = root.replace(current_directory, '').count(os.sep)
        indent = ' ' * 4 * (level)
        print('{}{}/'.format(indent, os.path.basename(root)))
        subindent = ' ' * 4 * (level + 1)
        for f in files:
            print('{}{}'.format(subindent, f))

# define list of codes to be dropped from documents
# carriage-returns, line-feeds, tabs
codelist = ['\r', '\n', '\t']

# we will drop the standard stop words
standard_stopwords = nltk.corpus.stopwords.words('english')
print(map(lambda t: t.encode('ascii'), standard_stopwords))
# special words associated with the occasion to be dropped from analysis
# along with the usual English stopwords
more_stopwords = ['speaker','president', 'mr', 'ms', 'mrs', 'th', 'house',\
    'representative', 'representatives', 'senate', 'senator',\
    'americans', 'american', 'america', \
     'one', 'two', 'three', 'four', 'five', 'six', 'seven', 'eight', 'nine',\
    'united', 'states', 'us', 'we', 'applause', 'ladies', 'gentlemen',\
    'congress', 'country' 'weve', 'youve', 'member', 'members',\
    'with', 'without', 'also', 'yet', 'half', 'also', 'many', 'see', 'said'\
    'years', 'year', 'even', 'ever', 'use', 'well', 'much', 'know', 'let', 'less']
print(map(lambda t: t.encode('ascii'), more_stopwords))

# start with the initial list and add to it for POTUS text work
stoplist = standard_stopwords + more_stopwords
print(map(lambda t: t.encode('ascii'), stoplist))

# employ regular expressions to parse documents
# here we are working with characters and character sequences
def text_parse(string):
    # replace non-alphanumeric with space
    temp_string = re.sub('[^a-zA-Z]', '  ', string)
    # replace codelist codes with space
    for i in range(len(codelist)):
        stopstring = ' ' + codelist[i] + '  '
        temp_string = re.sub(stopstring, '  ', temp_string)
    # replace single-character words with space
    temp_string = re.sub('\s.\s', ' ', temp_string)
    # convert uppercase to lowercase
    temp_string = temp_string.lower()
    # replace selected character strings/stop-words with space
    for i in range(len(stoplist)):
        stopstring = ' ' + str(stoplist[i]) + ' '
        temp_string = re.sub(stopstring, ' ', temp_string)
    # replace multiple blank characters with one blank character
    temp_string = re.sub('\s+', ' ', temp_string)
    return(temp_string)

# word stemming... looking for contractions... possessives...
# if we want to do stemming at a later time, we could use
#     porter = nltk.PorterStemmer()
# in a construction like this
#     words_stemmed =  [porter.stem(word) for word in initial_words]
# we examined stemming but found the results to be undesirable
```

183[⊖]

⊖ 为适应中文版阅读习惯，这里原书中 P176～P177 的文字分别排在代码 9-1、代码 9-2 和代码 9-3 前面，P178～P182 的图 9-3～图 9-11 排在代码 9-1 后面。——编辑注

184

```
# data for POTUS example are in one directory called POTUS
# all oral State of the Union addresses after Dwight D. Eisenhower
# 52 files (25 Democratic speeches, 27 Republican speeches)
# there is no scraping requirement for working with these text data
# but there are some comments with the text files, such as [Applause]
# because the speeches were delivered orally. Such comments need to
# be deleted from the files before additional text processing is done

# examine the directory structure to ensure POTUS is present
current_directory = os.getcwd()
list_all(current_directory)
# identify text file names of files/documents... should be 52 files
POTUS_file_names =\
    [name for name in os.listdir('POTUS') if fnmatch(name, '*.txt')]
print('\n\nNumber of files/documents:',len(POTUS_file_names))
# extract metadata from file names for labels on plots/reports
party_label = []  # initialize list
pres_label = []  # initialize list
year_label = []  # initialize list
for file_name in POTUS_file_names:
    file_name_less_extension = file_name.split('.')[0]
    party_label.append(file_name_less_extension.split('_')[0])
    pres_label.append(file_name_less_extension.split('_')[2])
    year_label.append(file_name_less_extension.split('_')[3])

# for numeric year in plots
year = []
for y in year_label: year.append(int(y))

# make working directory for a text corpus of parsed documents
# used for review of files and subsequent processing from files
os.mkdir('WORK_POTUS')

noutfiles = 0  # intialize count of working files for corpus
working_corpus = []  # initialize corpus with simple list structure
# work on files one at a time parsing and saving to new directory
for input_file_name in POTUS_file_names:
    # read in file
    this_file_name = input_file_name
    with open('POTUS/' + this_file_name, 'rt') as finput:
        text = finput.read()  # text string
        clean_text = text_parse(text)
        # output file name will be the same as input file name
        # but we store the file in the new directory WORK_POTUS
        output_file_name = "WORK_POTUS/" + input_file_name
        with open(output_file_name, 'wt') as foutput:
            foutput.write(str(clean_text))
            noutfiles = noutfiles + 1
            working_corpus.append(clean_text)  # build list-structured corpus
print('\nParsing complete: ', str(noutfiles) + ' files written to WORK_POTUS')
print('\nworking_corpus list: ', str(len(working_corpus)) + ' items')

# terms-by-documents matrix method to be employed
# with various numbers of top words
# check on top 200 words
tdm_method = CountVectorizer(max_features = 200, binary = True)
# employ simple term frequency
tdm_method = CountVectorizer(max_features = 200, binary = True)
examine_POTUS_tdm = tdm_method.fit(working_corpus)
top_200_words = examine_POTUS_tdm.get_feature_names()
# get clean printing of the top words
print('\nTop 200 words in POTUS corpus\n')
print(map(lambda t: t.encode('ascii'), top_200_words))  # print sans unicode

# check on top 100 words
tdm_method = CountVectorizer(max_features = 100, binary = True)
```

185

```
examine_POTUS_tdm = tdm_method.fit(working_corpus)
top_100_words = examine_POTUS_tdm.get_feature_names()
# get clean printing of the top words
print('\nTop 100 words in POTUS corpus\n')
print(map(lambda t: t.encode('ascii'), top_100_words))  # print sans unicode

# check on top 50 words
tdm_method = CountVectorizer(max_features = 50, binary = True)
examine_POTUS_tdm = tdm_method.fit(working_corpus)
top_100_words = examine_POTUS_tdm.get_feature_names()
# get clean printing of the top words
print('\nTop 100 words in POTUS corpus\n')
print(map(lambda t: t.encode('ascii'), top_100_words))  # print sans unicode

# ------------------------------------
# Full solution with all words TF-IDF
# ------------------------------------
# use the term frequency inverse document frequency matrix
# . . . begin by computing simple term frequency
count_vectorizer = CountVectorizer(min_df = 1)
term_freq_matrix = count_vectorizer.fit_transform(working_corpus)
# print vocabulary without unicode symbols... big vocabulary commented out
# print(map(lambda t: t.encode('ascii'), count_vectorizer.vocabulary_))
# . . . then transform to term frequency times inverse document frequency
# To get TF-IDF weighted word vectors tf times idf
tfidf = TfidfTransformer()  # accept default settings
tfidf.fit(term_freq_matrix)
tfidf_matrix = tfidf.transform(term_freq_matrix)
print('Shape of term frequency/inverse document frequency matrix',\
    str(tfidf_matrix.shape))

# -----------------------------
# Try top 200 words TF-IDF
# -----------------------------
# use the term frequency inverse document frequency matrix
# . . . begin by computing simple term frequency
count_vectorizer = CountVectorizer(min_df = 1, max_features = 200)
term_freq_matrix = count_vectorizer.fit_transform(working_corpus)
# print the feature names without unicode symbols...
top_200_words = count_vectorizer.get_feature_names()
print(map(lambda t: t.encode('ascii'), top_200_words))

# . . . then transform to term frequency times inverse document frequency
# To get TF-IDF weighted word vectors tf times idf
tfidf = TfidfTransformer()  # accept default settings
tfidf.fit(term_freq_matrix)
tfidf_matrix = tfidf.transform(term_freq_matrix)
print('Shape of term frequency/inverse document frequency matrix',\
    str(tfidf_matrix.shape))
# for input to subsequent analyses we need to choose
# either the standard terms x documents frequency matrix
# or the term frequency/inverse document frequency matrix
# we choose the latter with 200 words
```

186

```
# ------------------------------
# Multidimensional Scaling
# ------------------------------
# dissimilarity measures and multidimensional scaling
# consider alternative pairwise distance metrics from sklearn modules
# euclidean_distances, cosine_distances, manhattan_distances (city-block)
# note that different metrics provide different solutions
POTUS_distance_matrix = euclidean_distances(tfidf_matrix)
# POTUS_distance_matrix = manhattan_distances(tfidf_matrix)
# POTUS_distance_matrix = cosine_distances(tfidf_matrix)
```

```
mds_method = manifold.MDS(n_components = 2, random_state = 9999,\
    dissimilarity = 'precomputed')
mds_fit = mds_method.fit(POTUS_distance_matrix)
mds_coordinates = mds_method.fit_transform(POTUS_distance_matrix)

# plot mds solution in two dimensions using party labels
# defined by multidimensional scaling
plt.figure()
plt.scatter(mds_coordinates[:,0],mds_coordinates[:,1],\
    facecolors = 'none', edgecolors = 'none')  # points in white (invisible)
labels = party_label
for label, x, y in zip(labels, mds_coordinates[:,0], mds_coordinates[:,1]):
    plt.annotate(label, (x,y), xycoords = 'data')
plt.xlabel('First Dimension')
plt.ylabel('Second Dimension')
plt.show()
plt.savefig('fig_text_mds_POTUS_party.pdf',
    bbox_inches = 'tight', dpi=None, facecolor='w', edgecolor='b',
    orientation='landscape', papertype=None, format=None,
    transparent=True, pad_inches=0.25, frameon=None)

# plot mds solution in two dimensions using President names as labels
# defined by multidimensional scaling
plt.figure()
plt.scatter(mds_coordinates[:,0],mds_coordinates[:,1],\
    facecolors = 'none', edgecolors = 'none')  # points in white (invisible)
labels = pres_label
for label, x, y in zip(labels, mds_coordinates[:,0], mds_coordinates[:,1]):
    plt.annotate(label, (x,y), xycoords = 'data')
plt.xlabel('First Dimension')
plt.ylabel('Second Dimension')
plt.show()
plt.savefig('fig_text_mds_POTUS_pres.pdf',
    bbox_inches = 'tight', dpi=None, facecolor='w', edgecolor='b',
    orientation='landscape', papertype=None, format=None,
    transparent=True, pad_inches=0.25, frameon=None)

# plot mds solution in two dimensions using years as labels
# defined by multidimensional scaling
plt.figure()
plt.scatter(mds_coordinates[:,0],mds_coordinates[:,1],\
    facecolors = 'none', edgecolors = 'none')  # points in white (invisible)
labels = year_label
for label, x, y in zip(labels, mds_coordinates[:,0], mds_coordinates[:,1]):
    plt.annotate(label, (x,y), xycoords = 'data')
plt.xlabel('First Dimension')
plt.ylabel('Second Dimension')
plt.show()
plt.savefig('fig_text_mds_POTUS_year.pdf',
    bbox_inches = 'tight', dpi=None, facecolor='w', edgecolor='b',
    orientation='landscape', papertype=None, format=None,
    transparent=True, pad_inches=0.25, frameon=None)

# ------------------------------
# Cluster Analysis
# ------------------------------
# investigate alternative numbers of clusters using silhouette score
silhouette_value = []
k = range(2,10)
for i in k:
    kmeans_model = KMeans(n_clusters = i, random_state = 9999).\
        fit(np.transpose(tfidf_matrix))
    labels = kmeans_model.labels_
    silhouette_value.append(silhouette_score(np.transpose(tfidf_matrix),\
        labels, metric = 'euclidean'))
```

187

```
# highest silhouette score is for two clusters

# classification of words into groups for further analysis
# use transpose of the terms-by-document matrix and cluster analysis
# try two clusters/groups of words

clustering_method = KMeans(n_clusters = 2, random_state = 9999)
clustering_solution = clustering_method.fit(np.transpose(tfidf_matrix))
cluster_membership = clustering_method.predict(np.transpose(tfidf_matrix))
word_distance_to_center = clustering_method.transform(np.transpose(tfidf_matrix))

# top words data frame for reporting k-means clustering results
top_words_data = {'word': top_200_words, 'cluster': cluster_membership,\
    'dist_to_0': word_distance_to_center[0:,0],\
    'dist_to_1': word_distance_to_center[0:,1]}
distance_name_list = ['dist_to_0','dist_to_1']
top_words_data_frame = pd.DataFrame(top_words_data)
for cluster in range(2):
    words_in_cluster =\
        top_words_data_frame[top_words_data_frame['cluster'] == cluster]
    sorted_data_frame =\
        top_words_data_frame.sort_index(by = distance_name_list[cluster],\
        ascending = True)
    print('\n Top Words in Cluster :',cluster,'------------------------------')
    print(sorted_data_frame[:8])  # top 8 words in each cluster

# ----------------------------------------------
# Cluster Analysis Results Suggest Common Themes
# ----------------------------------------------

# Top Words in Cluster : 0 ------------------------------
#     cluster  dist_to_0  dist_to_1       word
#           0   0.150253   0.862167       since
#           0   0.154762   0.789895       keep
#           0   0.160067   0.862156       far
#           0   0.160653   0.801106       day
#           0   0.168222   0.808046       strong
#           0   0.169240   0.843582       protect
#           0   0.171769   0.777582       high
#           0   0.173527   0.848002       important

# Top Words in Cluster : 1 ------------------------------
#     cluster  dist_to_0  dist_to_1     word
#           1   0.602360   0.369731     make
#           1   0.575850   0.425409     time
#           1   0.623849   0.446006     help
#           1   0.727506   0.446941     nation
#           1   0.875372   0.454196     years
#           1   0.645027   0.508289     work
#           1   0.679918   0.508682     every
#           1   0.650285   0.532537     tonight

# a two-cluster solution seems to make sense with words
# toward the center of each cluster fitting together
# let's use pairs of top words from each cluster to name the clusters
# cluster index 0: "Stay Strong"
# cluster index 1: "Help Nation"

# name the clusters in the top words data frame
cluster_to_name = {0:'Stay Strong',1:'Help Nation'}
top_words_data_frame['cluster_name'] =\
    top_words_data_frame['cluster'].map(cluster_to_name)

# ----------------------------------------------
# Output Results of MDS and Cluster Analysis
```

188

```
# ----------------------------------------------

# write multidimensional scaling solution to comma-delimited text file
mds_data = {'party': party_label, 'pres': pres_label, 'year': year_label,\
    'first_dimension': list(mds_coordinates[:,0]),\
    'second_dimension': list(mds_coordinates[:,1])}
mds_data_frame = pd.DataFrame(mds_data)
mds_data_frame.to_csv('POTUS_mds.csv')

# write cluster analysis solution to comma-delimited text file
top_words_data_frame.to_csv('POTUS_top_words_clustering.csv')

# ----------------------------------------------
# Create Aggregate Text Files for Presidents
# ----------------------------------------------
# aggregate State of the Union addresses for each of
# ten presidents formed by simple concatenation
# working from the WORK_POTUS directory of parsed text
Kennedy = ''  # initialize aggregate text string
Johnson = ''
Nixon = ''
Ford = ''
Carter = ''
Reagan = ''
BushG = ''
Clinton = ''
BushGW = ''
Obama = ''

for input_file_name in POTUS_file_names:
    # read in file
    this_file_name = input_file_name
    with open('WORK_POTUS/' + this_file_name, 'rt') as finput:
        text = finput.read()  # text string
        file_name_less_extension = input_file_name.split('.')[0]
        pres = file_name_less_extension.split('_')[2]  # President
        # append to the appropriate President string with ' ' separator
        if pres == 'Kennedy':
            Kennedy = Kennedy + ' ' + text
        elif pres == 'Johnson':
            Johnson = Johnson + ' ' + text
        elif pres == 'Nixon':
            Nixon = Nixon + ' ' + text
        elif pres == 'Ford':
            Ford = Ford + ' ' + text
        elif pres == 'Carter':
            Carter = Carter + ' ' + text
        elif pres == 'Reagan':
            Reagan = Reagan + ' ' + text
        elif pres == 'BushG':
            BushG = BushG + ' ' + text
        elif pres == 'Clinton':
            Clinton = Clinton + ' ' + text
        elif pres == 'BushGW':
            BushGW = BushGW + ' ' + text
        elif pres == 'Obama':
            Obama = Obama + ' ' + text
        else:
            print('\n\nError in processing file:',this_file_name,'\n\n')

# store the strings as text files in a new directory ALL_POTUS
os.mkdir('ALL_POTUS')
with open('ALL_POTUS/Kennedy.txt', 'wt')  as foutput:
    foutput.write(str(Kennedy))
```

189

```
with open('ALL_POTUS/Johnson.txt', 'wt') as foutput:
    foutput.write(str(Johnson))
with open('ALL_POTUS/Nixon.txt', 'wt') as foutput:
    foutput.write(str(Nixon))
with open('ALL_POTUS/Ford.txt', 'wt') as foutput:
    foutput.write(str(Ford))
with open('ALL_POTUS/Carter.txt', 'wt') as foutput:
    foutput.write(str(Carter))
with open('ALL_POTUS/Reagan.txt', 'wt') as foutput:
    foutput.write(str(Reagan))
with open('ALL_POTUS/BushG.txt', 'wt') as foutput:
    foutput.write(str(BushG))
with open('ALL_POTUS/Clinton.txt', 'wt') as foutput:
    foutput.write(str(Clinton))
with open('ALL_POTUS/BushGW.txt', 'wt') as foutput:
    foutput.write(str(BushGW))
with open('ALL_POTUS/Obama.txt', 'wt') as foutput:
    foutput.write(str(Obama))

# Suggestions for the student: Use word clusters to define text
# measures that vary across addresses, Presidents, and years.
# Try word stemming prior to the definition of a
# terms-by-documents matrix. Try longer lists of words
# for the identified clusters. Try alternative numbers
# of top words or alternative numbers of clusters.
# Repeat the multidimensional scaling and cluster analysis
# using the aggregate text files for the ten Presidents.
# Try other methods for identifying common themes, such as
# latent semantic analysis or latent Dirichlet allocation.
```

190

对于绘制词云图和使用散点图展示非重叠文本，目前已经有现成的 R 语言程序可以使用（Fellows 2014a; Fellows 2014b）。我们使用这些工具程序来给国情咨文的文本绘制图形，见图 9-2～图 9-11。代码 9-2 提供了生成这些词云的代码。

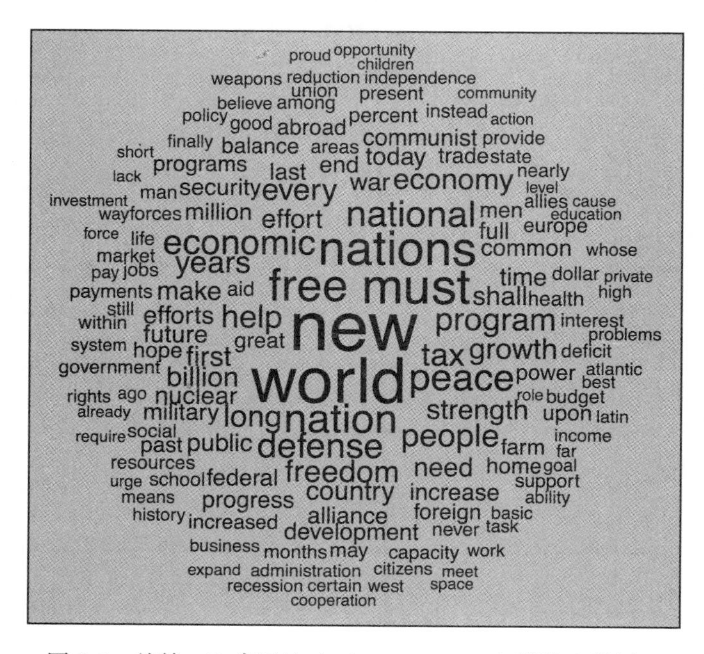

图 9-2　约翰　F. 肯尼迪（John F. Kennedy）讲演中的词云

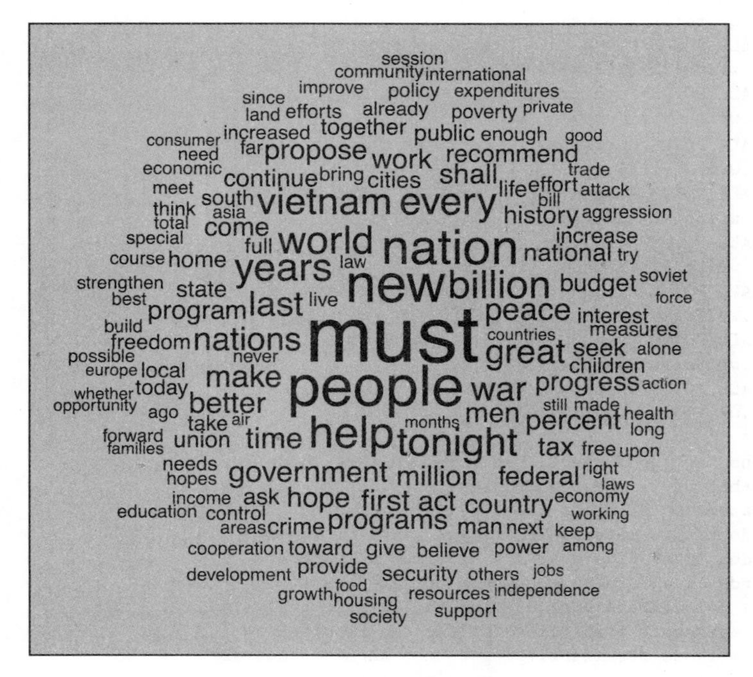

图 9-3 林登 B. 约翰逊（Lyndon B. Johnson）讲演中的词云

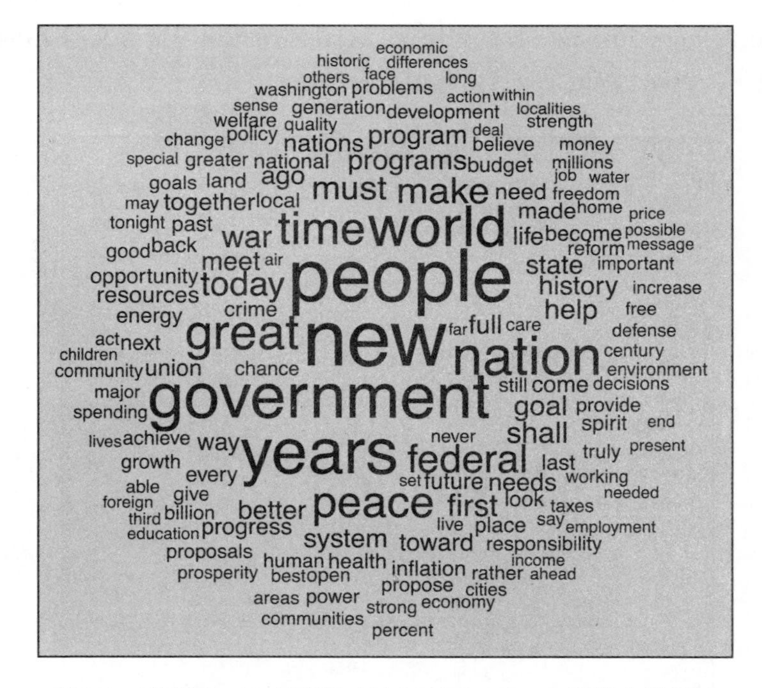

图 9-4 理查德 M. 尼克松（Richard M. Nixon）演讲中的词云

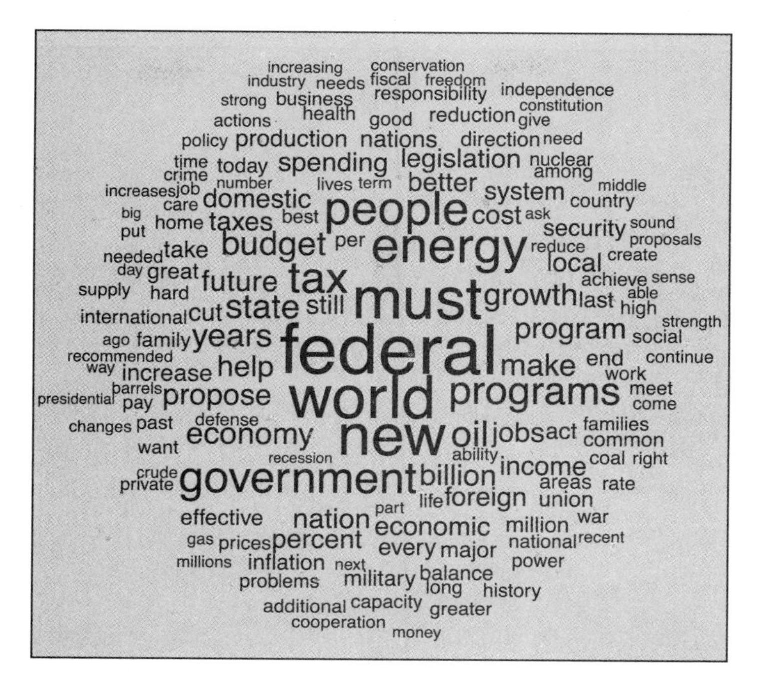

图 9-5　杰拉尔德　R. 福特（Gerald R. Ford）演讲中的词云

图 9-6　吉米 · 卡特（Jimmy Carter）演讲中的词云

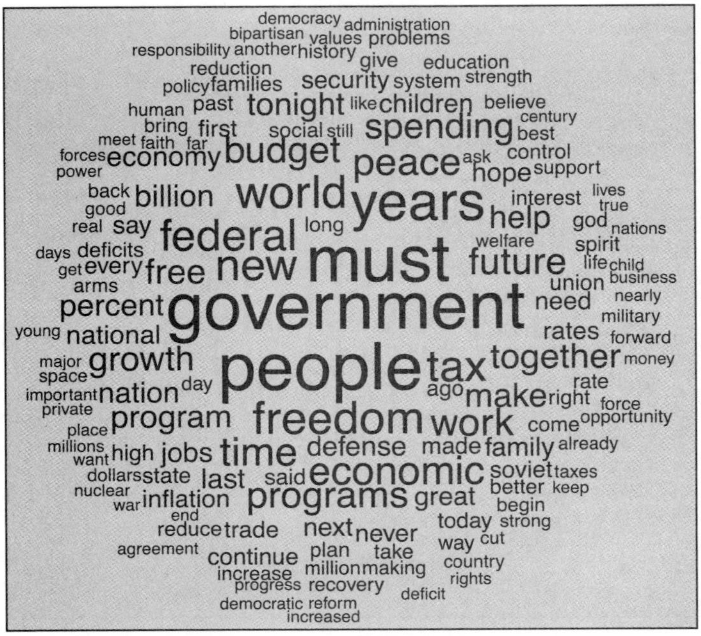

democracy administration
bipartisan values problems
responsibility another history
reduction give education
policy families security system strength
human past tonight like children believe
bring first social still spending century
meet faith far best
forces economy budget peace ask control
power hope support
back billion world years interest lives
good true
real say federal long help god nations
days deficits welfare spirit
get every free new must future life child
arms union business
percent government need nearly
young national military
rates forward
major growth people tax together money
space day rate
important nation ago make right force
private
place program freedom work come opportunity
millions want high jobs time defense made family already
dollars state last said economic soviet taxes
nuclear better keep
war inflation programs great begin
end
reduce trade next never today strong
agreement plan take way cut
continue country
increase million making
progress recovery rights
democratic reform deficit
increased

图 9-7 罗纳德·里根（Ronald Reagan）演讲中的词云

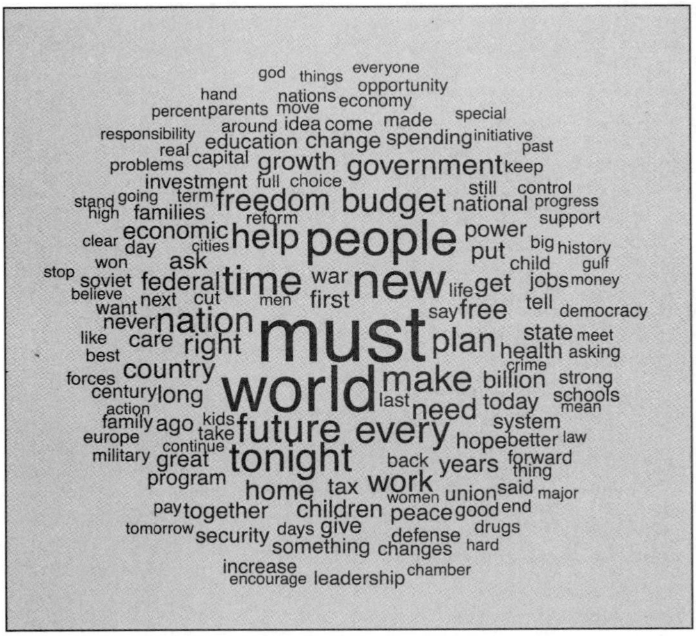

god things everyone
opportunity
hand nations economy
percent parents move
around idea come made special
responsibility education change spending initiative
real past
problems capital growth government keep
investment full choice still control
stand going term national progress
high families freedom budget support
reform
clear economic help people power
day cities big history
won ask put child gulf
stop soviet federal time war new life get jobs money
believe next cut men first tell
want say free democracy
never nation
like care right must plan state meet
best health asking
crime
forces country
century long world make billion strong
action last schools
family ago kids need today mean
europe take future every system
continue hope better law
military great tonight back years forward
thing
program home tax work union said major
women
pay together children peace good end
tomorrow security days give defense drugs
something changes hard
increase
encourage leadership chamber

图 9-8 乔治·布什（George Bush）演讲中的词云

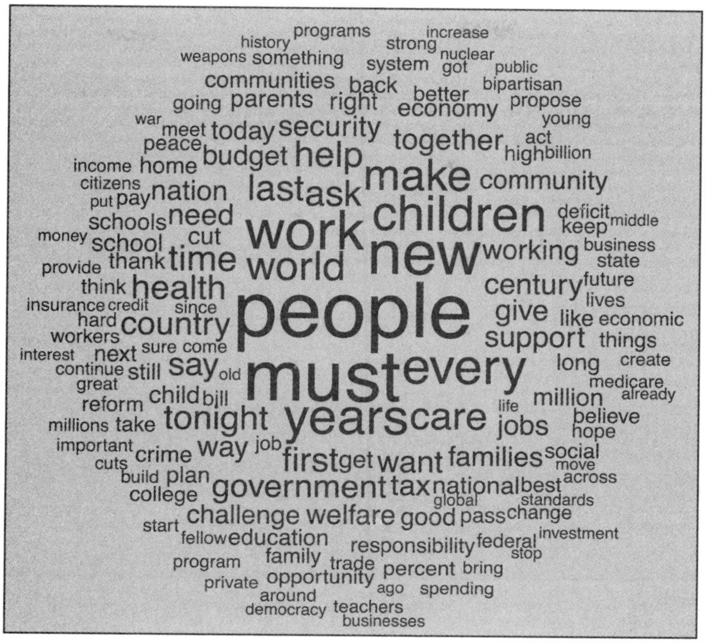

图 9-9　威廉　J. 克林顿（William J. Clinton）演讲中的词云

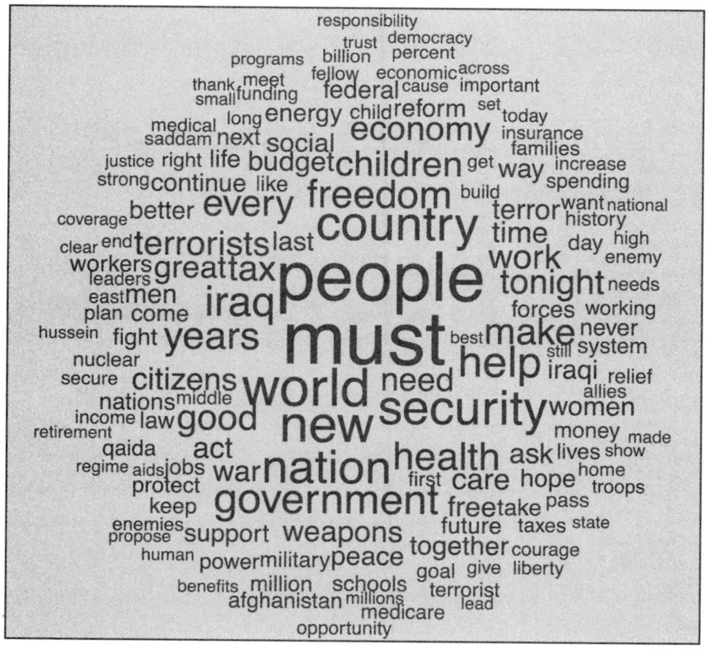

图 9-10　乔治　W. 布什（George W. Bush）演讲中的词云

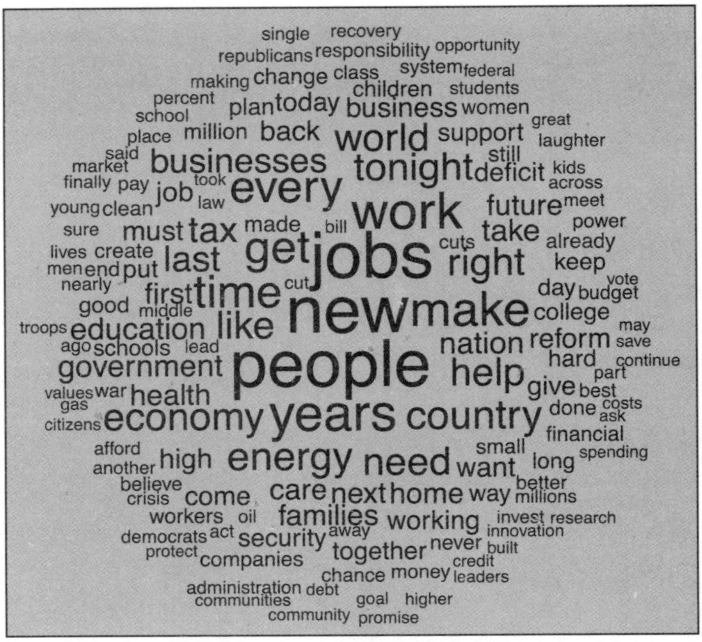

191（其中图为178～182）

图 9-11 贝拉克·奥巴马（Barack Obama）演讲中的词云

代码 9-2 生成词云：POTUS 演讲（R 语言）

```
# Making Word Clouds: POTUS Speeches (R)
# install R wordcloud package
library(wordcloud)

# ----------------------------------
# word cloud for John F. Kennedy
# ----------------------------------
Kennedy.text <- scan("ALL_POTUS/Kennedy.txt", what = "char", sep = "\n")
# replace uppercase with lowercase letters
Kennedy.text <- tolower(Kennedy.text)
# strip out all non-letters and return vector
Kennedy.text.preword.vector <- unlist(strsplit(Kennedy.text, "\\W"))
# drop all empty words
Kennedy.text.vector <-
 Kennedy.text.preword.vector[which(nchar(Kennedy.text.preword.vector) > 0)]
pdf(file = "fig_text_wordcloud_of_Kennedy_speeches.pdf", width = 8.5, height = 8.5)
set.seed(1234)
wordcloud(Kennedy.text.vector, min.freq = 5,
  max.words = 150,
  random.order=FALSE,
  random.color=FALSE,
  rot.per=0.0,
  colors="black",
  ordered.colors=FALSE,
  use.r.layout=FALSE,
  fixed.asp=TRUE)
dev.off()

# ----------------------------------
# word cloud for Lyndon B. Johnson
# ----------------------------------
Johnson.text <- scan("ALL_POTUS/Johnson.txt", what = "char", sep = "\n")
# replace uppercase with lowercase letters
Johnson.text <- tolower(Johnson.text)
```

```
# strip out all non-letters and return vector
Johnson.text.preword.vector <- unlist(strsplit(Johnson.text, "\\W"))
# drop all empty words
Johnson.text.vector <-
 Johnson.text.preword.vector[which(nchar(Johnson.text.preword.vector) > 0)]
pdf(file = "fig_text_wordcloud_of_Johnson_speeches.pdf", width = 8.5, height = 8.5)
set.seed(1234)
wordcloud(Johnson.text.vector, min.freq = 5,
  max.words = 150,
  random.order=FALSE,
  random.color=FALSE,
  rot.per=0.0,
  colors="black",
  ordered.colors=FALSE,
  use.r.layout=FALSE,
  fixed.asp=TRUE)
dev.off()
# -----------------------------------
# word cloud for Richard M Nixon
# -----------------------------------
Nixon.text <- scan("ALL_POTUS/Nixon.txt", what = "char", sep = "\n")
# replace uppercase with lowercase letters
Nixon.text <- tolower(Nixon.text)
# strip out all non-letters and return vector
Nixon.text.preword.vector <- unlist(strsplit(Nixon.text, "\\W"))
# drop all empty words
Nixon.text.vector <-
 Nixon.text.preword.vector[which(nchar(Nixon.text.preword.vector) > 0)]
pdf(file = "fig_text_wordcloud_of_Nixon_speeches.pdf", width = 8.5, height = 8.5)
set.seed(1234)
wordcloud(Nixon.text.vector, min.freq = 5,
  max.words = 150,
  random.order=FALSE,
  random.color=FALSE,
  rot.per=0.0,
  colors="black",
  ordered.colors=FALSE,
  use.r.layout=FALSE,
  fixed.asp=TRUE)
dev.off()

# -----------------------------------
# word cloud for Gerald R. Ford
# -----------------------------------
Ford.text <- scan("ALL_POTUS/Ford.txt", what = "char", sep = "\n")
# replace uppercase with lowercase letters
Ford.text <- tolower(Ford.text)
# strip out all non-letters and return vector
Ford.text.preword.vector <- unlist(strsplit(Ford.text, "\\W"))
# drop all empty words
Ford.text.vector <-
 Ford.text.preword.vector[which(nchar(Ford.text.preword.vector) > 0)]
pdf(file = "fig_text_wordcloud_of_Ford_speeches.pdf", width = 8.5, height = 8.5)
set.seed(1234)
wordcloud(Ford.text.vector, min.freq = 5,
  max.words = 150,
  random.order=FALSE,
  random.color=FALSE,
  rot.per=0.0,
  colors="black",
  ordered.colors=FALSE,
  use.r.layout=FALSE,
  fixed.asp=TRUE)
dev.off()
```

192

```
# ------------------------------------
# word cloud for Jimmy Carter
# ------------------------------------
Carter.text <- scan("ALL_POTUS/Carter.txt", what = "char", sep = "\n")
# replace uppercase with lowercase letters
Carter.text <- tolower(Carter.text)
# strip out all non-letters and return vector
Carter.text.preword.vector <- unlist(strsplit(Carter.text, "\\W"))
# drop all empty words
Carter.text.vector <-
 Carter.text.preword.vector[which(nchar(Carter.text.preword.vector) > 0)]
pdf(file = "fig_text_wordcloud_of_Carter_speeches.pdf", width = 8.5, height = 8.5)
set.seed(1234)
wordcloud(Carter.text.vector, min.freq = 5,
  max.words = 150,
  random.order=FALSE,
  random.color=FALSE,
  rot.per=0.0,
  colors="black",
  ordered.colors=FALSE,
  use.r.layout=FALSE,
  fixed.asp=TRUE)
dev.off()

# ------------------------------------
# word cloud for Ronald Reagan
# ------------------------------------
Reagan.text <- scan("ALL_POTUS/Reagan.txt", what = "char", sep = "\n")
# replace uppercase with lowercase letters
Reagan.text <- tolower(Reagan.text)
# strip out all non-letters and return vector
Reagan.text.preword.vector <- unlist(strsplit(Reagan.text, "\\W"))
# drop all empty words
Reagan.text.vector <-
 Reagan.text.preword.vector[which(nchar(Reagan.text.preword.vector) > 0)]
pdf(file = "fig_text_wordcloud_of_Reagan_speeches.pdf", width = 8.5, height = 8.5)
set.seed(1234)
wordcloud(Reagan.text.vector, min.freq = 5,
  max.words = 150,
  random.order=FALSE,
  random.color=FALSE,
  rot.per=0.0,
  colors="black",
  ordered.colors=FALSE,
  use.r.layout=FALSE,
  fixed.asp=TRUE)
dev.off()

# ------------------------------------
# word cloud for George Bush
# ------------------------------------
BushG.text <- scan("ALL_POTUS/BushG.txt", what = "char", sep = "\n")
# replace uppercase with lowercase letters
BushG.text <- tolower(BushG.text)
# strip out all non-letters and return vector
BushG.text.preword.vector <- unlist(strsplit(BushG.text, "\\W"))
# drop all empty words
BushG.text.vector <-
 BushG.text.preword.vector[which(nchar(BushG.text.preword.vector) > 0)]
pdf(file = "fig_text_wordcloud_of_BushG_speeches.pdf", width = 8.5, height = 8.5)
set.seed(1234)
wordcloud(BushG.text.vector, min.freq = 5,
  max.words = 150,
  random.order=FALSE,
```

193

194

```
   random.color=FALSE,
   rot.per=0.0,
   colors="black",
   ordered.colors=FALSE,
   use.r.layout=FALSE,
   fixed.asp=TRUE)
dev.off()

# -----------------------------------
# word cloud for William J. Clinton
# -----------------------------------
Clinton.text <- scan("ALL_POTUS/Clinton.txt", what = "char", sep = "\n")
# replace uppercase with lowercase letters
Clinton.text <- tolower(Clinton.text)
# strip out all non-letters and return vector
Clinton.text.preword.vector <- unlist(strsplit(Clinton.text, "\\W"))
# drop all empty words
Clinton.text.vector <-
 Clinton.text.preword.vector[which(nchar(Clinton.text.preword.vector) > 0)]
pdf(file = "fig_text_wordcloud_of_Clinton_speeches.pdf", width = 8.5, height = 8.5)
set.seed(1234)
wordcloud(Clinton.text.vector, min.freq = 5,
  max.words = 150,
  random.order=FALSE,
  random.color=FALSE,
  rot.per=0.0,
  colors="black",
  ordered.colors=FALSE,
  use.r.layout=FALSE,
  fixed.asp=TRUE)
dev.off()

# -----------------------------------
# word cloud for George W. Bush
# -----------------------------------
BushGW.text <- scan("ALL_POTUS/BushGW.txt", what = "char", sep = "\n")
# replace uppercase with lowercase letters
BushGW.text <- tolower(BushGW.text)
# strip out all non-letters and return vector
BushGW.text.preword.vector <- unlist(strsplit(BushGW.text, "\\W"))
# drop all empty words
BushGW.text.vector <-
 BushGW.text.preword.vector[which(nchar(BushGW.text.preword.vector) > 0)]
pdf(file = "fig_text_wordcloud_of_BushGW_speeches.pdf", width = 8.5, height = 8.5)
set.seed(1234)

wordcloud(BushGW.text.vector, min.freq = 5,
  max.words = 150,
  random.order=FALSE,
  random.color=FALSE,
  rot.per=0.0,
  colors="black",
  ordered.colors=FALSE,
  use.r.layout=FALSE,
  fixed.asp=TRUE)
dev.off()

# -----------------------------------
# word cloud for Barack Obama
# -----------------------------------
Obama.text <- scan("ALL_POTUS/Obama.txt", what = "char", sep = "\n")
# replace uppercase with lowercase letters
Obama.text <- tolower(Obama.text)
# strip out all non-letters and return vector
```

195

```
Obama.text.preword.vector <- unlist(strsplit(Obama.text, "\\W"))
# drop all empty words
Obama.text.vector <-
 Obama.text.preword.vector[which(nchar(Obama.text.preword.vector) > 0)]
pdf(file = "fig_text_wordcloud_of_Obama_speeches.pdf", width = 8.5, height = 8.5)
set.seed(1234)
wordcloud(Obama.text.vector, min.freq = 5,
  max.words = 150,
  random.order=FALSE,
  random.color=FALSE,
  rot.per=0.0,
  colors="black",
  ordered.colors=FALSE,
  use.r.layout=FALSE,
  fixed.asp=TRUE)
dev.off()
```

196

在许多场景下，我们只不过是将词云看作是一种分散、量化的艺术。但是对于 *POTUS* 演讲来说，词云有助于我们识别词在多维标度分析中维度的含义。换言之，我们可以利用那些关于总统的词云来指导我们对图的坐标轴进行标记。

在给横坐标轴进行标记时，我们将克林顿作为左端、将尼克松作为右端来关注词云之间的差异。对纵坐标轴进行辨别会更加困难些，这是因为基于每一位总统进行标识会难以识别。有些总统在从上到下的全部区域内都分布着。需要注意的是约翰逊总统和乔治 W·布什总统在图中的任何区域都没有标示出来。对于纵坐标轴，我们可以考虑将尼克松作为最顶端、肯尼迪作为最底端，或者将奥巴马作为最顶端、乔治·布什作为最底端的差异。虽然与通用文本分析学一样，对维度进行标记也依赖于研究者的主观判断，我们的主观直觉会更加全面地受到词云的影响。

对 *POTUS* 演讲的研究结果让我们把横坐标轴标记为"国民到政府关注点"、纵坐标轴标记为"国际到国内关注点"。这些坐标轴与我们传统上所认为的自由与保守政治哲学并不一致。通常被认为是自由派的肯尼迪和卡特在图的右侧出现，与尼克松、福特和里根出现在同一侧。

与标注着红色州和蓝色州的选举夜地区分布图更有说服力，用总统自己的语言绘制出的图勾画出美国 52 年来的共同主题。那就是展示在图 9-1 中的多维标度分析结果。代码 9-2 是生成这个多维标度分析结果以及文本图的代码。

我们从文本，也就是一个文档集开始。然后，我们确定文本中重要的词或特征，给这些词赋予值，即被某些人称为是给文本注释的过程。我们运用预测分析学对这些数字进行分析，其中包括多元法。最后，我们用文字和图的方式来报告我们的发现。文本分析学将我们从词带向数字（文本度量），然后又回到词本身。有时，我们也把文本度量转化成为图，这是一个我们称之为文本映射的过程。

好奇的读者也许会问："这些政治方面的研究与 Web 和网络数据科学有什么关系？"我们的回答是：这些都与 Web 和网络数据科学有很大的关系。POTUS 演讲是一个非常容易得到的公开文本来源，用以展示我们广泛应用于研究网络数据的过程。

我们可以用本章中介绍的方法对任何文档集进行分析，包括网页、博客、微博、维基、搜索查询与列表、研究报告以及顾客反馈。文本分析学的方法也可以应用于来源于焦点小组和在线评论的定性的研究数据进行分析，附录 B 将进一步进行介绍。网络数据在很大程度上就是文本，因此，学习如何进行文本分析以及如何确定共同主题非常重要。

代码 9-3 From Text Measures to Text Maps: POTUS Speeches (R)

```
# From Text Measures to Text Maps: POTUS Speeches (R)

# install R wordcloud package

library(ggplot2)  # grammar of graphics plotting system

# --------------------
#  Presidents Studied
# --------------------
#  John F. Kennedy
#  Lyndon B. Johnson
#  Richard M Nixon
#  Gerald R. Ford
#  Jimmy Carter
#  Ronald Reagan
#  George Bush
#  William J. Clinton
#  George W. Bush
#  Barack Obama
# --------------------

# read in results from multidimensional scaling analysis
mds_data_frame <- read.csv("POTUS_mds.csv", stringsAsFactors = FALSE)
print(str(mds_data_frame))

mds_data_frame$year_label <- as.character(mds_data_frame$year)
mds_data_frame$party_label <- rep("", length = nrow(mds_data_frame))
for (i in seq(along = mds_data_frame$party)) {
    if(mds_data_frame$party[i] == "D")
        mds_data_frame$party_label[i] <- "Democrat"
    if(mds_data_frame$party[i] == "R")
        mds_data_frame$party_label[i] <- "Republican"
    }

# direct manipulation of text position to avoid overlapping text
mds_data_frame$pres_x <- mds_data_frame$first_dimension + 0.025
mds_data_frame$pres_y <- mds_data_frame$second_dimension + 0.015
mds_data_frame$year_x <- mds_data_frame$first_dimension + 0.025
mds_data_frame$year_y <- mds_data_frame$second_dimension - 0.015

# Carter 1979 up and to the left
mds_data_frame$pres_x[2] <- mds_data_frame$first_dimension[2] + 0.025 -0.0225
mds_data_frame$pres_y[2] <- mds_data_frame$second_dimension[2] + 0.015 + 0.035
mds_data_frame$year_x[2] <- mds_data_frame$first_dimension[2] + 0.025 -0.0225
mds_data_frame$year_y[2] <- mds_data_frame$second_dimension[2] - 0.015 + 0.035

# Clinton 1994 move down
mds_data_frame$pres_x[5] <- mds_data_frame$first_dimension[5] + 0.025
mds_data_frame$pres_y[5] <- mds_data_frame$second_dimension[5] + 0.015 - 0.0125
mds_data_frame$year_x[5] <- mds_data_frame$first_dimension[5] + 0.025
mds_data_frame$year_y[5] <- mds_data_frame$second_dimension[5] - 0.015 - 0.0125

# Clinton 2000 up and to the left
mds_data_frame$pres_x[11] <- mds_data_frame$first_dimension[11] + 0.025 -0.0325
mds_data_frame$pres_y[11] <- mds_data_frame$second_dimension[11] + 0.015 + 0.045
mds_data_frame$year_x[11] <- mds_data_frame$first_dimension[11] + 0.025 -0.0325
mds_data_frame$year_y[11] <- mds_data_frame$second_dimension[11] - 0.015 + 0.045
```

197

```
# Kennedy 1961 move up and to the left
mds_data_frame$pres_x[17] <- mds_data_frame$first_dimension[17] + 0.025 -0.0455
mds_data_frame$pres_y[17] <- mds_data_frame$second_dimension[17] + 0.015 + 0.04
mds_data_frame$year_x[17] <- mds_data_frame$first_dimension[17] + 0.025 - 0.0455
mds_data_frame$year_y[17] <- mds_data_frame$second_dimension[17] - 0.015 + 0.04

# Kennedy 1963 move down
mds_data_frame$pres_x[19] <- mds_data_frame$first_dimension[19] + 0.025
mds_data_frame$pres_y[19] <- mds_data_frame$second_dimension[19] + 0.015 - 0.0125
mds_data_frame$year_x[19] <- mds_data_frame$first_dimension[19] + 0.025
mds_data_frame$year_y[19] <- mds_data_frame$second_dimension[19] - 0.015 - 0.0125

# Obama 2009 move up
mds_data_frame$pres_x[20] <- mds_data_frame$first_dimension[20] + 0.025
mds_data_frame$pres_y[20] <- mds_data_frame$second_dimension[20] + 0.015 + 0.0125
mds_data_frame$year_x[20] <- mds_data_frame$first_dimension[20] + 0.025
mds_data_frame$year_y[20] <- mds_data_frame$second_dimension[20] - 0.015 + 0.0125

# Obama 2010 move down and to the left
mds_data_frame$pres_x[21] <- mds_data_frame$first_dimension[21] + 0.025 -0.0125
mds_data_frame$pres_y[21] <- mds_data_frame$second_dimension[21] + 0.015 - 0.03
mds_data_frame$year_x[21] <- mds_data_frame$first_dimension[21] + 0.025 -0.0125
mds_data_frame$year_y[21] <- mds_data_frame$second_dimension[21] - 0.015 - 0.03

# Obama 2011 move up and to the left
mds_data_frame$pres_x[22] <- mds_data_frame$first_dimension[22] + 0.025 -0.0125
mds_data_frame$pres_y[22] <- mds_data_frame$second_dimension[22] + 0.015 + 0.0355
mds_data_frame$year_x[22] <- mds_data_frame$first_dimension[22] + 0.025 -0.0125
mds_data_frame$year_y[22] <- mds_data_frame$second_dimension[22] - 0.015 + 0.0355

# Obama 2014 move up and to the left
mds_data_frame$pres_x[25] <- mds_data_frame$first_dimension[25] + 0.025 -0.0255
mds_data_frame$pres_y[25] <- mds_data_frame$second_dimension[25] + 0.015 + 0.0415
mds_data_frame$year_x[25] <- mds_data_frame$first_dimension[25] + 0.025 - 0.0255
mds_data_frame$year_y[25] <- mds_data_frame$second_dimension[25] - 0.015 + 0.0415

# BushG 1990 move down
mds_data_frame$pres_x[27] <- mds_data_frame$first_dimension[27] + 0.025
mds_data_frame$pres_y[27] <- mds_data_frame$second_dimension[27] + 0.015 - 0.0125
mds_data_frame$year_x[27] <- mds_data_frame$first_dimension[27] + 0.025
mds_data_frame$year_y[27] <- mds_data_frame$second_dimension[27] - 0.015 - 0.0125

# BushG 1992 move down
mds_data_frame$pres_x[29] <- mds_data_frame$first_dimension[29] + 0.025
mds_data_frame$pres_y[29] <- mds_data_frame$second_dimension[29] + 0.015 - 0.0125
mds_data_frame$year_x[29] <- mds_data_frame$first_dimension[29] + 0.025
mds_data_frame$year_y[29] <- mds_data_frame$second_dimension[29] - 0.015 - 0.0125

# Nixon 1970 move up
mds_data_frame$pres_x[40] <- mds_data_frame$first_dimension[40] + 0.025
mds_data_frame$pres_y[40] <- mds_data_frame$second_dimension[40] + 0.015 + 0.0125
mds_data_frame$year_x[40] <- mds_data_frame$first_dimension[40] + 0.025
mds_data_frame$year_y[40] <- mds_data_frame$second_dimension[40] - 0.015 + 0.0125

# Reagan 1988 move up
mds_data_frame$pres_x[52] <- mds_data_frame$first_dimension[52] + 0.025
mds_data_frame$pres_y[52] <- mds_data_frame$second_dimension[52] + 0.015 + 0.0125
mds_data_frame$year_x[52] <- mds_data_frame$first_dimension[52] + 0.025
mds_data_frame$year_y[52] <- mds_data_frame$second_dimension[52] - 0.015 + 0.0125

pdf(file = "fig_mds_solution_POTUS_Speeches.pdf", width = 8.5, height = 8.5)
ggplot_object <- ggplot(data = mds_data_frame,
    aes(x = first_dimension, y = second_dimension,
        shape = party_label, colour = party_label)) +
```

198

```
    geom_point(size = 3) +
    scale_colour_manual(values =
        c(Democrat = "darkblue", Republican = "darkred")) +
    geom_text(aes(x = pres_x , y = pres_y, label = pres),
        size = 4, hjust = 0, colour = "black") +
    geom_text(aes(x = year_x , y = year_y, label = year_label),
        size = 3, hjust = 0, colour = "black") +
    xlim(-0.70, 0.85) +
    xlab("People-to-Government Focus") +
    ylab("International-to-Domestic Focus") +
    theme(axis.title.x = element_text(size = 15, colour = "black")) +
    theme(axis.title.y = element_text(size = 15, colour = "black")) +
    theme(axis.text.x = element_text(colour = "white")) +
    theme(axis.text.y = element_text(colour = "white")) +
    theme(legend.position = "bottom") +
    theme(legend.title = element_text(size = 0.000001)) +
    theme(legend.text = element_text(size = 15))
print(ggplot_object)
dev.off()
```

199

第 10 章

推　　荐

"这些人都是谁呀?"

Butch Cassidy（Paul Newman 饰）
电影《虎豹小霸王》(1969 年)

作为 Amazon.com 网站的常客，我常常收到来自该网站的购书推荐。2013 年 5 月，也就是我所撰写的《Modeling Techniques》(建模技术）第 1 版发行之前，亚马逊（Amazon）给我发来了一封邮件，其中有推荐给我的十本书，排名第一的正是我当时正在写作的那一本。像其他好的出版商一样，我的出版商往往会接受新书预订，因此亚马逊就询问了我是否需要预订那本书。

那些对我如此了解的人都是谁? 在图书方面是亚马逊，在影视方面是 Netflix，在音乐方面是 Pandora，等等，他们似乎已经知道了关于我的很多偏好，然而我并没有告诉过他们很多信息。

推荐系统建立在*稀疏矩阵*基础之上。在 Netflix 曾经组织过的一个奖金为一百万美元的竞赛中（2006.5.29～2009.9.21），Netflix 为一个知名的推荐系统问题提供了包含 17 770 部电影和 480 189 位客户的数据。每一位客户最多只能租看一至两百部电影，因此如果我们构造一个用于表达客户 - 电影租赁之间关系的矩阵，当一位客户与一部电影之间存在租赁关系时，则相应的矩阵单元中的值设为 1，不存在此关系则设为 0，那么我们由此而得到的矩阵将有 99% 的单元中的值会是 0，也就是说这个矩阵 99% 稀疏。

推荐系统所使用的算法，有时又称作推荐引擎，必须能够有效处理稀疏矩阵。目前已经提出了很多可行的推荐算法。其中一类推荐系统被称为*基于内容的推荐系统*，这些系统会利用客户自身的特征、过去的订单以及表露出的偏好。基于内容的推荐系统可能也会依赖于产品本身的特性。

另外一类推荐系统是*协同过滤*（也被称为社交过滤或分组过滤），这一类系统建立在以下前提上：对某一类产品的评级相似的客户，也会对其他类的产品做相似的评级。那么，要想预测 Janiya 喜欢哪方面的音乐，就去寻找与 Janiya 最相近的邻居，看他们喜欢哪方面的音乐。

假设 Brit 已经观看过 10 部电影，现在一个影片服务供应商想要给 Brit 推荐一部新电影。

找到这样一部新电影的一个方法就是先找到那些观看过很多Brit已经看过的影片的客户。然后，服务供应商遍历这些近邻客户，找出他们看过的相同的电影，其中共享率最高的那些电影就成为向Brit进行推荐的基础。

有趣的是，Kaggle公司举办的一场比赛要求数据分析师为R语言软件包的用户构建一个推荐引擎。Conway and White（2012）使用比赛本身的数据展示了如何构建一个近邻推荐系统。我们发现，通过“依赖”和“反向依赖”链接，R语言环境本身的信息可以对Kaggle公司的这次R语言竞赛本身的数据提供补充。我们可以将任意的程序设计环境表示成一个由软件包、函数或模块相互间进行调用而组成的网络。通过网络拓扑，我们就有很大的把握辨识出一个给定节点的最近邻居节点。

关联规则建模（也称为亲和性分析或购物篮分析）是另外一种构建推荐系统的方法。关联规则建模关心的问题是：*哪些事物通常会一起出现？哪些产品通常会被一起预订或购买？哪些活动通常会同时发生？哪些网站通常会一起被访问？*

关联规则模型涉及大规模的列联表统计分析，而我们的工作就是要决定需要查找哪些列联表。关联规则模型面临的一个关键挑战就是所产生的数量十分庞大的规则。

202

*条目集合*是指从仓库的所有条目中选出一些条目构成的一个集合。条目集合的大小是指这个集合中所包含的条目数。条目集合可以由两个、三个或者更多个条目构成。即使对于微软公司来说，不同的条目集合的个数也非常庞大。

*关联规则*会将一个条目集合分解为两个子集合，其中，*前驱*子集合排在*后继*子集合之前。关联规则的个数多于条目集合的个数⊖。Apriori算法Agraval et al. (1996)通过使用能够反映关联规则潜在实用性的选择准则来应对关联规则数量庞大的问题。

第一个准则是关于条目集合的*支持度*或普及率。每一个条目集合都会被评估，以确定这个条目集合在仓库数据集合中所出现次数的比例。如果这个比例超过了一个最小支持度阈值或准则，那么这个条目集合就将进入下一阶段的分析。支持度标准为0.01表示每100个市场购物篮中的一个必须包含这个条目集合，支持度标准为0.001表示每1000个市场购物篮中的一个必须包含这个条目集合。

第二个准则是关于关联规则的*可信度*或可预测性。可信度通过将一个条目集合的支持度除以前驱中条目子集的支持度而得到。这是一个在给定前驱的前提下，对后继的条件概率进行的估算。在选择关联规则时，我们设定的可信度准则要远远大于支持度准则⊜。支持度和

⊖ 对微软这个数据集来说，如果它有 $K=294$ 个网站区域以及相对应的二进制数据矩阵，可以计算出不同数据集合的数量为

$$2K=2^{294}\approx 3.182\,868\times 10^{88}$$

⊜ 对于两个条目子集 A 和 B，存在两个可能的关联规则：（$A\Rightarrow B$）和（$B\Rightarrow A$）。我们只需要考虑这两个规则中的其中一个而已，因为我们更倾向于选择可信度高的规则。就拿规则（$A\Rightarrow B$）来说，它是在已知 A 的前提下，出现 B 的条件概率。同样，规则（$B\Rightarrow A$）是在已知 B 的前提下，出现 A 的条件概率：

$$P(B|A)=\frac{P(AB)}{P(A)} \qquad P(A|B)=\frac{P(AB)}{P(B)}$$

因此，如果子集 A 的支持度高于子集 B 的支持度，那么，$P(A)>P(B)$，$P(A|B)>P(B|A)$。由于我们倾向于选择可信度高的规则，那么很显然，具有高支持度的条目子集将会承担作为后继的角色。

203 可信度准则由研究人员任意设定。

为了演示一个基于关联规则的系统如何进行推荐，我们借鉴微软匿名网络数据（Anonymous Microsoft Web Data）案例。此案例的数据包含了用户标识符和网络区域标识符。访问者的选择揭示了对微软网站进行访问的用户的偏好。

我们为这些数据构造一个访客 - 区域矩阵。在这个矩阵中，我们将矩阵单元的值设为 1 代表访问，设为 0 代表未访问，因此，得到的矩阵中大多数单元的值都将是 0，也就是一个稀疏矩阵。设计的关联规则算法就是应用于这样的矩阵。

支持度和可信度的设置因具体问题而异。对于微软的这个数据集合，我们将初始的支持度准则设为 0.025，然后绘制一个反映对网站各个区域访问频率的图，以显示对每个区域的访问是否满足这个准则。参看图 10-1 以及表 10-1 中的目录名与说明。

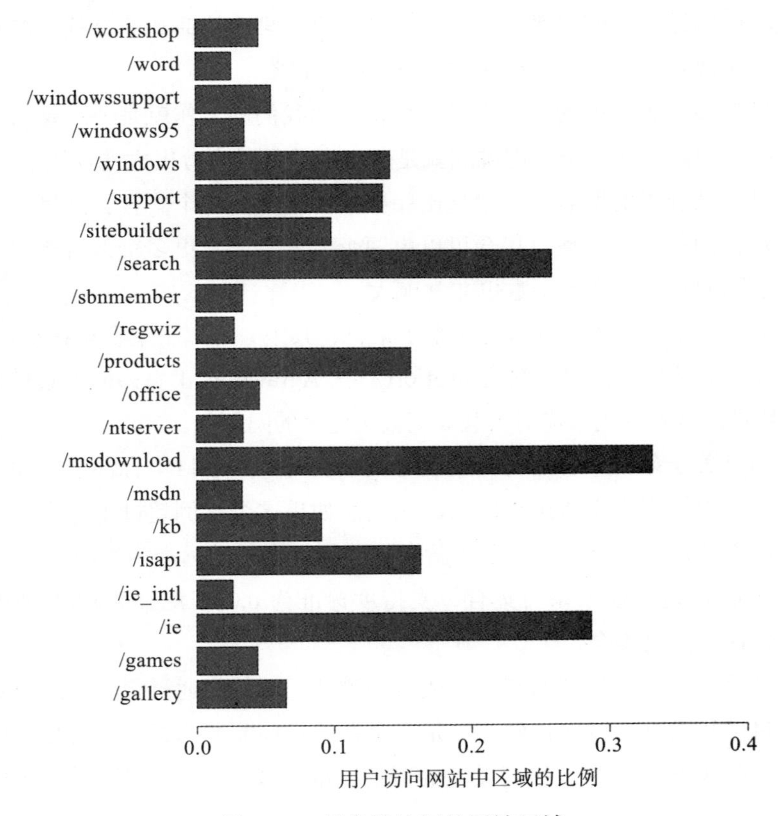

图 10-1 最常被访问的网站区域

表 10-1 最常被访问的网站区域与说明

目录名	说明	目录名	说明
/gallery	网站构建者介绍	/kb	知识库
/games	游戏	/msdn	开发者网络
/ie	IE 浏览器	/msdownload	免费下载
/ie_intl	国际版 IE 浏览器内容	/ntserver	Windows NT 服务器
/isapi	网络服务器应用编程接口	/office	微软 Office 信息

（续）

目 录 名	说 明	目 录 名	说 明
/products	产品（一个通用类别）	/windows	Windows 操作系统大全
/regwiz	注册向导	/windows95	Windows 95
/sbnmember	SiteBuilder 网络会员	/windowssupport	Windows 95 支持
/search	搜索	/word	微软 Word 新闻
/sitebuilder	面向开发者的互联网站构建	/workshop	开发者研讨会
/support	支持桌面		

将支持度和可信度的阈值分别设置为 0.01 和 0.025 将会产生一个包含 268 条关联规则的集合。图 10-2 展示的是面向这 268 条关联规则，以支持度作为横坐标轴、可信度作为纵坐标轴的散点图。每个点的颜色编码代表相对应规则的提升度，这是一种对相对预测可信度的度量。[⊖]

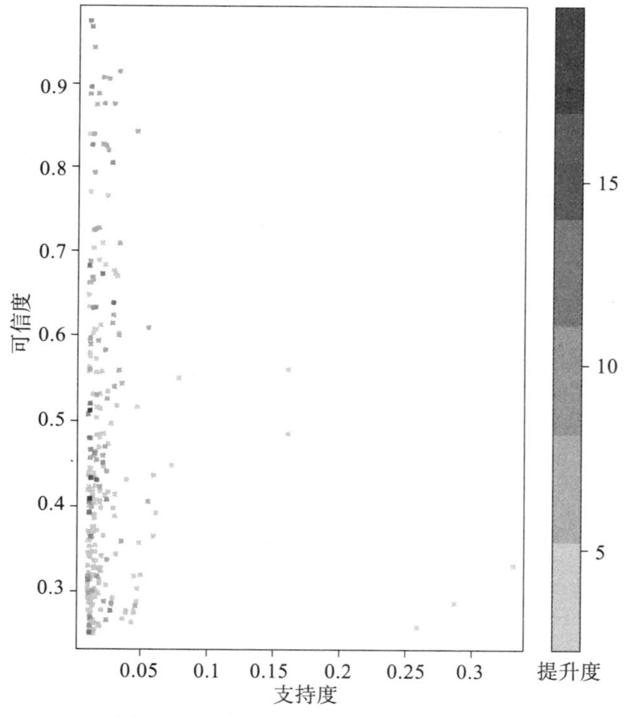

图 10-2 关联规则的支持度与可信度

⊖ 就像支持度和可信度一样，提升度也是通过概率公式进行表达的。提升度可以理解为是在已知关联规则 $(A\Rightarrow B)$ 的前提下，对后继 B 进行的可信度除以在没有这个关联规则的前提下对 B 进行预测的可信度。在不知道 A 和关联规则 $(A\Rightarrow B)$ 的前提下，我们观测到条目子集 B 的可信度是 $P(B)$。在已知 A 和关联规则 $(A\Rightarrow B)$ 的前提下，根据之前的定义，我们观测到条目子集 B 的可信度是 $P(B|A)$。以上这两个值的比值就是提升度：

$$\frac{P(B|A)}{P(B)}=\left(\frac{P(AB)}{P(A)}\right)\left(\frac{1}{P(B)}\right)=\frac{P(AB)}{P(A)P(B)}$$

通过查看以上这个等式最右侧比值的分子和分母，我们可以发现，在独立性假设的前提下，它们是相等的。也就是说，如果子集 A 和子集 B 之间不存在任何关系，那么 A 和 B 的联合概率就等于它们各自的概率相乘：$P(AB)=P(A)P(B)$。因此，提升度就是用于度量关联规则中子集之间偏离独立的程度。

图 10-3 是对同样一组关联规则的又一种展示方法：气泡矩阵图。处于前驱子集（也就是关联规则左侧 LHS）中的一个条目用于标识这个矩阵的顶部，而处于后继子集（也就是关联规则右侧 RHS）中的一个条目则用于标识这个矩阵的右侧。支持度通过气泡的大小来体现，而提升度则通过气泡的颜色深度来体现，如图 10-3 所示。

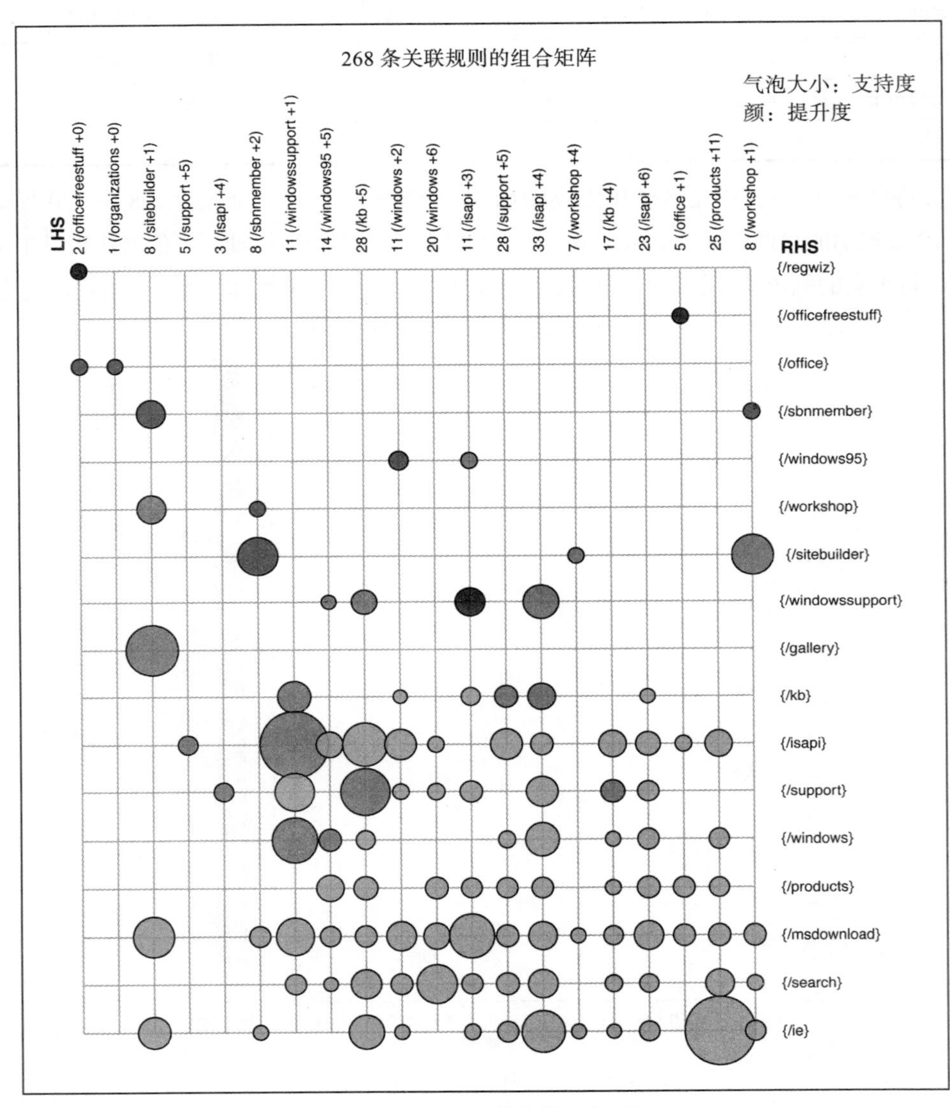

204
~
208

图 10-3 关联规则的前驱与后继

让我们回顾一下前面讨论过的那几个对关联规则的度量。支持度代表着普及率。作为一个对相对频率或概率的估计，它的取值应为 0～1。支持度具有较低的值也是可以接受的，但是如果某个区域的值远低于其他区域的值，那么可能表明这个区域对网站浏览者来说不重要。

可信度代表在已知前驱的前提下，预测后继出现的概率。作为一个条件概率，可信度的取值也在 0～1 之间，并且越高越好。作为非零概率值的一个比值，提升度的取值范围为实数轴的正数部分。提升度大于1.00对于管理来说才有意义，并且关联规则的提升度越高越好。

在解决市场或顾客分类问题时，使用事务聚类法可能会有帮助。也就是说，在使用关联规则分析之前，我们可以先使用聚类分析法来识别出网站访客群体中具有相似习惯的访客群体。然后，我们再对这些访客群体分别进行分析。

假如我们可以随意地使用有效的算法，那么解决人人皆知的“干草堆里的细针”问题就成为一项很容易的事情。然而，我们在这个过程中学到了任何有价值的东西吗？我们会知道在下一个干草堆里又如何去寻找吗？我们又该如何将关联规则应用到推荐系统中呢？

到目前为止，我们对这个微软案例所做的均是描述性的而非预测性的。被选出的关联规则只是对网站访客的浏览习惯进行了描述或对网站结构进行了分析。某些区域常被访问，而另外一些则不常被访问。现在我们想看到的是这些关联规则是否提供了价值。我们是否可以利用这些关联规则去预测访客今后将会做什么？

在这个微软案例中，预测访问者的下一步行为就类似于一种推荐。这种推荐的价值在网页场景下很容易理解，这是因为对网页来说，我们可以在访客发出请求前，就提前准备好这些网页的内容。也就是说，一个推荐系统可以利用所期望的访客后续提交的请求事先加载好网页内容，从而改进网站的性能。在服务器端提前加载好的网页随时准备在接收到访客请求后的第一时间就发送出去。这是开展最初研究并最终应用于微软案例的动力之一（Breese, Heckerman, and Kadie 1998）。 209

在这个微软案例中，我们使用了一个持续性的测试来测试模型的预测准确性。关联规则模型在测试中的准确率达到36%。也就是说，假如我们知道访客已经访问过哪个区域，我们就有36%的概率成功地预测访客接下来要访问的区域。

Hastie, Tibshirani, and Friedman (2009) 回顾了关联规则的理论基础，提供了支持度、可信度（可预测性）和提升度（偏离独立性的程度）的正式定义，并对它们的应用以及Agraval et al. (1996) 提出的Apriori算法进行了讨论。机器学习领域文献（Tan, Steinbach, and Kumar 2006; Witten, Frank, and Hall 2011; Harrington 2012; Rajaraman and Ullman 2012）也有一些这方面的讨论。Bruzzese and Davino (2008) 回顾了关联规则的数据可视化。

为了找到一个感兴趣的规则，数据科学家可能需要查阅大量的关联规则。因此，很多研究都关注如何选择关联规则（Hahsler, Buchta, and Hornik 2008）、如何为关联规则模型选择参数（Dippold and Hruschka 2013）以及如何在同一分析中同时应用分割模型和关联规则模型（Boztug and Reutterer 2008），这些研究方向并不会让我们感到意外。

Hahsler, Grün, and Hornik (2005), Hahsler et al. (2011) 和 Hahsler et al. (2014a, 2014b) 讨论了关联规则模型的R语言实现，此外，Hahsler and Chelluboina (2014a, 2014b) 提供了数据可视化方面的支持。更多的关联规则算法可以通过R语言程序提供接口，从Weka获得（Hornik 2014a, 2014b）。对于稀疏矩阵，可以使用一些特殊的算法（Bates and Maechler 2014）。Ricci et al. (2011) 和 Dehuri et al. (2012) 对推荐系统进行了回顾。R语言软件由Hahsler (2014a, 2014b) 提供。

代码10-1是本章的工作实例中使用的程序，来源于Hahsler et al. (2014a) 和 Hahsler and

Chelluboina (2014a)，其中将 Apriori 算法应用于关联规则。这段代码展示了一种面向微软案例来构建网站区域推荐的方法。

210

代码 10-1 从规则到推荐：微软案例（R 语言）

```
# From Rules to Recommendations: The Microsoft Case (R)

# install necessary packages prior to running

library(arules)  # association rules
library(arulesViz)  # data visualization of association rules
library(RColorBrewer)  # color palettes for plots

# set criteria for association rule modeling for this analysis
# set support and confidence settings for input to arules package

low_support_setting <- 0.01
high_support_setting <- 0.025
confidence_setting <- 0.25

# -------------------------------------------------------
# carmap is modified recode function from car package
# to avoid conflict with recode function from arules
# car package from Fox (2014) and Fox and Weisberg (2011)
# -------------------------------------------------------

carmap <- function (var, recodes, as.factor.result = FALSE,
    as.numeric.result = FALSE, levels)
{
    lo <- -Inf
    hi <- Inf
    recodes <- gsub("\n|\t", " ", recodes)
    recode.list <- rev(strsplit(recodes, ";")[[1]])
    is.fac <- is.factor(var)
    if (missing(as.factor.result))
        as.factor.result <- is.fac
    if (is.fac)
        var <- as.character(var)
    result <- var

    for (term in recode.list) {
        if (0 < length(grep(":", term))) {
            range <- strsplit(strsplit(term, "=")[[1]][1], ":")
            low <- eval(parse(text = range[[1]][1]))
            high <- eval(parse(text = range[[1]][2]))
            target <- eval(parse(text = strsplit(term, "=")[[1]][2]))
            result[(var >= low) & (var <= high)] <- target
        }
        else if (0 < length(grep("^else=", term))) {
            target <- eval(parse(text = strsplit(term, "=")[[1]][2]))
            result[1:length(var)] <- target
        }
        else {
            set <- eval(parse(text = strsplit(term, "=")[[1]][1]))
            target <- eval(parse(text = strsplit(term, "=")[[1]][2]))
            for (val in set) {
                if (is.na(val))
                  result[is.na(var)] <- target
                else result[var == val] <- target
            }
        }
    }
    if (as.factor.result) {
```

211

```
        result <- if (!missing(levels))
            factor(result, levels = levels)
        else as.factor(result)
    }
    else if (as.numeric.result && (!is.numeric(result))) {
        result.valid <- na.omit(result)
        opt <- options(warn = -1)
        result.valid <- as.numeric(result.valid)
        options(opt)
        if (!any(is.na(result.valid)))
            result <- as.numeric(result)
    }
    result
}

# ----------------------------------
# Data preparation work
# ----------------------------------
#
# note that the initial data file has seven initial records of data
# that are identifiers only ... these were deleted...
# then the records that define the attributes (A records) are placed
# in a file <000_attribute_data.csv>
# the base training data for the study are the web activity records
# (C and V records) which are in the original data files
#
# from the documentation we have the following information
#
#  Case and Vote Lines:
#  For each user, there is a case line followed by zero or more vote lines.
#
#  For example:
#      C,"10164",10164
#      V,1123,1
#      V,1009,1
#      V,1052,1
#          Where:
#             'C' marks this as a case line,
#             '10164' is the case ID number of a user,
#             'V' marks the vote lines for this case,
#             '1123', 1009', 1052' are the attributes ID's
#                  of Vroots that a user visited.
#                  '1' may be ignored.
#    The datasets record which Vroots [web site areas]
#    each user visited in a one-week timeframe in Feburary 1998.
# note that all C and V data were set as numeric data on input
# for both training and test data files
# this was done in a plain text editor
#
# our editing resulted in files <000_training_data.csv> and
#    <000_test_data.csv>
# the former is used to define association rules
# the latter can be used to test predictions/recommendations
# based upon these association rules
#
# we detected what appear to be two miscodings in the attribute data:
# "/security." was changed to "/security" and "/msdownload." to "/msdownload"

# ----------------------------------
# User-defined functions
# ----------------------------------
# function for replacing web area numeric IDs with meaningful area names
map_microsoft <- function(x) {
carmap(var = x,
recodes = '1287 = "/autoroute";1288 = "/library";1289 = "/masterchef";
```

212

```
1297 = "/centroam";1215 = "/developer";1279 = "/msgolf";1239 = "/msconsult";
1282 = "/home";1251 = "/referencesupport";1121 = "/magazine";
1083 = "/msaccesssupport";1145 = "/vfoxprosupport";1276 = "/vtestsupport";
1200 = "/benelux";1259 = "/controls";1155 = "/sidewalk";1092 = "/vfoxpro";
1004 = "/search";1057 = "/powerpoint";1140 = "/netherlands";
1198 = "/pictureit";1147 = "/msft";1005 = "/norge";1026 = "/sitebuilder";
1119 = "/corpinfo";1216 = "/vrml";1218 = "/publishersupport";
1205 = "/hardwaresupport";1269 = "/business";1031 = "/msoffice";1003 = "/kb";
1238 = "/exceldev";1118 = "/sql";1242 = "/msgarden";1171 = "/merchant";
1175 = "/msprojectsupport";1021 = "/visualc";1222 = "/msofc";
1284 = "/partner";1294 = "/bookshelf";1053 = "/visualj";1293 = "/encarta";
1167 = "/hwtest";1202 = "/advtech";1234 = "/off97cat";1054 = "/exchange";
1262 = "/chile";1074 = "/ntworkstation";1027 = "/intdev";1061 = "/promo";
1236 = "/globaldev";1212 = "/worldwide";1204 = "/msscheduleplus";
1196 = "/ie40";1188 = "/korea";1228 = "/vtest";1078 = "/ntserversupport";
1008 = "/msdownload";1052 = "/word";1091 = "/hwdev";1280 = "/music";
1247 = "/wineguide";1064 = "/activeplatform";1065 = "/java";
1133 = "/frontpagesupport";1102 = "/homeessentials";1132 = "/msmoneysupport";
1240 = "/thailand";1225 = "/piracy";1130 = "/syspro";1157 = "/win32dev";
1058 = "/referral";1076 = "/ntwkssupport";1163 = "/opentype";1187 = "/odbc";
1152 = "/rus";1139 = "/k-12";1223 = "/finland";1001 = "/support";
1043 = "/smallbiz";1165 = "/poland";1194 = "/china";1138 = "/mind";
1158 = "/imedia";1094 = "/mshome";1055 = "/kids";1277 = "/stream";
1143 = "/workshoop";1068 = "/vbscript";1229 = "/uruguay";1177 = "/events";
1014 = "/officefreestuff";1019 = "/mspowerpoint";1122 = "/mindshare";
1041 = "/workshop";1033 = "/logostore";1233 = "/vbscripts";
1211 = "/smsmgmtsupport";1199 = "/feedback";1024 = "/iis";1179 = "/colombia";
1067 = "/frontpage";1181 = "/kidssupport";1174 = "/nz";
1162 = "/infoservsupport";1046 = "/iesupport";1197 = "/sqlsupport";
1231 = "/win32devsupport";1141 = "/europe";1120 = "/switch";1112 = "/canada";
1142 = "/southafrica";1250 = "/middleeast";1214 = "/finserv";
1190 = "/repository";1098 = "/devonly";1263 = "/services";
1049 = "/supportnet";1073 = "/taiwan";1166 = "/mexico";
1226 = "/msschedplussupport";1184 = "/msexcelsupport";1025 = "/gallery";
1160 = "/visualcsupport";1156 = "/powered";1268 = "/javascript";
1220 = "/macofficesupport";1060 = "/msword";1203 = "/danmark";
1176 = "/jscript";1168 = "/salesinfo";1066 = "/musicproducer";1128 = "/msf";
1275 = "/security";1136 = "/usa";1146 = "/msp";1237 = "/devdays";
1081 = "/accessdev";1016 = "/excel";1069 = "/windowsce";
1148 = "/channel_resources";1161 = "/workssupport";1013 = "/vbasicsupport";
1116 = "/switzerland";1093 = "/vba";1249 = "/fortransupport";
1095 = "/catalog";1023 = "/spain";1192 = "/visualjsupport";1080 = "/brasil";
1050 = "/macoffice";1255 = "/msmq";1273 = "/mdn";1206 = "/select";
1230 = "/mailsupport";1172 = "/belgium";1011 = "/officedev";1009 = "/windows";
1096 = "/mspress";1235 = "/onlineeval";1070 = "/activex";1154 = "/project";
1099 = "/cio";1186 = "/college";1291 = "/news";1256 = "/sia";
1270 = "/developr";1232 = "/standards";1159 = "/transaction";
1035 = "/windowssupport";1164 = "/smsmgmt";
1077 = "/msofficesupport";1295 = "/train_cert";1056 = "/sports";
1006 = "/misc";1272 = "/softlib";1123 = "/germany";
1151 = "/mspowerpointsupport";1103 = "/works";1243 = "/usability";
1244 = "/devwire";1260 = "/trial";1258 = "/peru";1208 = "/israel";
1106 = "/cze";1124 = "/industry";1114 = "/servad";1012 = "/outlookdev";
1045 = "/netmeeting";1082 = "/access";1261 = "/diyguide";1137 = "/mscorp";
1059 = "/sverige";1037 = "/windows95";1227 = "/argentina";
1281 = "/intellimouse";1134 = "/backoffice";1044 = "/mediadev";
1028 = "/oledev";1248 = "/softimage";1085 = "/exchangesupport";
1131 = "/moneyzone";1079 = "/australia";1048 = "/publisher";
1042 = "/vstudio";1075 = "/jobs";1201 = "/hardware";1105 = "/france";
1153 = "/venezuela";1292 = "/northafrica";1015 = "/msexcel";
1290 = "/devmovies";1017 = "/products";1010 = "/vbasic";
1126 = "/mediamanager";1144 = "/devnews";1191 = "/management";
1002 = "/athome";1213 = "/corporate_solutions";1084 = "/uk";
1178 = "/msdownload";1036 = "/organizations";1257 = "/devvideos";
1180 = "/slovenija";1246 = "/gamesdev";1088 = "/outlook";
```

213

```
1117 = "/sidewinder";1097 = "/latam";1266 = "/licenses";
1072 = "/vinterdev";1169 = "/msproject";1107 = "/slovakia";
1089 = "/officereference";1038 = "/sbnmember";1224 = "/atec";
1086 = "/oem";1108 = "/teammanager";1007 = "/ie_intl";1252 = "/giving";
1283 = "/cinemania";1127 = "/netshow";1189 = "/internet";1110 = "/mastering";
1090 = "/gamessupport";1109 = "/technet";1040 = "/office";1150 = "/infoserv";
1195 = "/portugal";1111 = "/ssafe";1274 = "/pdc";1267 = "/caribbean";
1113 = "/security";1245 = "/ofc";1253 = "/worddev";1087 = "/proxy";
1185 = "/sna";1209 = "/turkey";1063 = "/intranet";1101 = "/oledb";
1264 = "/se_partners";1032 = "/games";1173 = "/moli";1051 = "/scheduleplus";
1278 = "/hed";1062 = "/msaccess";1020 = "/msdn";1104 = "/hk";
1071 = "/automap";1000 = "/regwiz";1135 = "/mswordsupport";1207 = "/icp";
1217 = "/ireland";1254 = "/ie3";1022 = "/truetype";1183 = "/italy";
1170 = "/mail";1241 = "/india";1149 = "/adc";1029 = "/clipgallerylive";
1221 = "/mstv";1115 = "/hun";1125 = "/imagecomposer";1039 = "/isp";
1034 = "/ie";1265 = "/ssafesupport";1271 = "/mdsn";1129 = "/ado";
1018 = "/isapi";1193 = "/offdevsupport";1219 = "/ads";1030 = "/ntserver";
1182 = "/fortran";1100 = "/education";1210 = "/snasupport"')
}
```

214

```
# ------------------------------------------------------------------
# user-defined function to identify antecedent items and
# consequent items lists... as many items as are given in rules

make_lists_for_all_rules <- function(association_rules) {
    # function returns items lists for antecedents and consequents
    # note. default apriori algorithm finds single-item consequents
    #       but we will return lists for consequents as well
    all_antecedent_matrix <- as(lhs(association_rules), "matrix")
    all_consequent_matrix <- as(rhs(association_rules), "matrix")
    # create list of lists for antecedent items in each association rule
    all_antecedent_items <- NULL
    for (irule in 1:nrow(all_antecedent_matrix)) {
        this_rule_antecedent_items <- NULL
        for (jitem in 1:ncol(all_antecedent_matrix)) {
            if (all_antecedent_matrix[irule, jitem] == 1)
            this_rule_antecedent_items <-
                c(this_rule_antecedent_items,
                    colnames(all_antecedent_matrix)[jitem])
         }  # end inner for-loop for gathering up items in this rule
         all_antecedent_items <-
             rbind(all_antecedent_items,list(this_rule_antecedent_items))
     }  # end outter for-loop for rules

    all_consequent_items <- NULL
    for (irule in 1:nrow(all_consequent_matrix)) {
        this_rule_consequent_items <- NULL
        for (jitem in 1:ncol(all_consequent_matrix)) {
            if (all_consequent_matrix[irule, jitem] == 1)
                this_rule_consequent_items <-
                c(this_rule_consequent_items,
                    colnames(all_consequent_matrix)[jitem])
            }  # end inner for-loop for gathering up items in this rule
            all_consequent_items <-
                rbind(all_consequent_items,list(this_rule_consequent_items))
    }  # end outter for-loop for rules

    list(all_antecedent_items,all_consequent_items)

    }  # end of function make.item.lists.for.all rules

# -----------------------------------------------------------------------

make_lists_for_single_item_antecedent_rules <- function(association_rules) {
    # function returns items lists for antecedents and consequents
```

```
    # note. default apriori algorithm finds single-item consequents
    #       but we will return lists for consequents as well
    all_antecedent_matrix <- as(lhs(association_rules), "matrix")
    all_consequent_matrix <- as(rhs(association_rules), "matrix")

    # create list of lists for antecedent items in each association rule
    indices_for_single_antecedent_rules <- NULL
    all_antecedent_items <- NULL
    for (irule in 1:nrow(all_antecedent_matrix)) {
        this_rule_antecedent_items <- NULL
        for (jitem in 1:ncol(all_antecedent_matrix)) {
            if (all_antecedent_matrix[irule, jitem] == 1)
            this_rule_antecedent_items <-
                c(this_rule_antecedent_items,
                    colnames(all_antecedent_matrix)[jitem])
            }  # end inner for-loop for gathering up items in this rule
        all_antecedent_items <-
            rbind(all_antecedent_items,list(this_rule_antecedent_items))
        if (length(this_rule_antecedent_items) == 1)
            indices_for_single_antecedent_rules <-
                c(indices_for_single_antecedent_rules, irule)
        } # end outter for-loop for rules

    all_consequent_items <- NULL
    for (irule in 1:nrow(all_consequent_matrix)) {
        this_rule_consequent_items <- NULL
        for (jitem in 1:ncol(all_consequent_matrix)) {
            if (all_consequent_matrix[irule, jitem] == 1)
                this_rule_consequent_items <-
            c(this_rule_consequent_items,
                colnames(all_consequent_matrix)[jitem])
            }  # end inner for-loop for gathering up items in this rule
        all_consequent_items <-
            rbind(all_consequent_items,list(this_rule_consequent_items))
        }  # end outter for-loop for rules

    antecedent_items <-
        all_antecedent_items[indices_for_single_antecedent_rules,]
    consequent_items <-
        all_consequent_items[indices_for_single_antecedent_rules,]
    list(antecedent_items,consequent_items)

    }  # end of function make_lists_for_single_item_antecedent_rules

# -------------------------------------------
# user-defined function to track processing
# used for long-running loop processing code
multiple_of_one_thousand <- function(x) {
    # return true if x is a multiple of 1000
    returnvalue <- FALSE
    if(trunc(x/1000)==(x/1000)) returnvalue <- TRUE
    returnvalue
    } # end of function multiple_of_one_thousand

# ---------------------------------------------------
# Read in training data... create transactions data
# ---------------------------------------------------
# read in the input data for attribute names
# we do not use these data in the jump-start program
# it could be used later to provide an interpretation of association rules
# the rules are identified by numbers only in training and test data
# the attribute data show what those names mean in terms of the
# subdirectory name (area_name) and its description
# we will use the area_name as our web_area_id because it is
# short enough for listings and more meaningful than numbers
```

215

216

```
# in order to do this, we will need to define the mapping
# much as we would with dictionaries and the map method in Python
attribute_input_data <-
    read.csv(file = "000_attribute_data.csv", header = FALSE,
        col.names = c("A_type", "numeric_value", "ignore_value",
            "description", "area_name"), stringsAsFactors = FALSE)

# read in the input data for users and web areas visited
training_input_data <- read.csv(file = "000_training_data.csv",
    header = FALSE, col.names = c("CV_type","value","ignore_value"),
    stringsAsFactors = FALSE)

# the following code creates a transaction data frame for the training data
# as needed for association rule analysis with the R package arules
# the transaction data frame has two columns, the first being user_id
# and the second being the web_area_id, a web site area visited by the user

n_records <- nrow(training_input_data )  # stopping rule for while-loop

alphanumeric_transaction_data_frame <- NULL  # initialize object
n_record_count <- 0  # intialize record count

# note that this step can take a while to run...
# we set things up to report on the process, but this reporting
# may not work on all operating systems... be patient with this step
# note also that this step may be skipped on subsequent runs
# because you will have the transactions_file_name in place
cat("\n\nTransaction processing percentage complete: ")
while (n_record_count < n_records) {
  n_record_count <- n_record_count + 1
  # report to screen the proportion of job completed
  if(multiple_of_one_thousand(n_record_count))
      cat(" ",trunc(100 * (n_record_count /  n_records)))
  # read one record from input data
  this_record <- training_input_data[n_record_count,]
  if (this_record$CV_type == "C") user_id <-
      this_record$value  # C for user_id
  if (this_record$CV_type == "V") {  # V record provides web_area_id
    web_area_id <- map_microsoft(this_record$value)  # meaningful area name
    # add one record to the transaction.data.frame
    alphanumeric_transaction_data_frame <-
      rbind(alphanumeric_transaction_data_frame, data.frame(user_id, web_area_id))
    }
  }
# write the transactions to a comma-delimited file
transactions_file_name <- "000_training_transactions.csv"
write.csv(alphanumeric_transaction_data_frame,
    file = "000_training_transactions.csv", row.names = FALSE)
cat("\nTransactions sent to ",transactions_file_name,"\n")
# ------------------------------------------------
# Use transactions data to find association rules
# ------------------------------------------------
transactions_object <-
    read.transactions(file = transactions_file_name, cols = c(1,2),
        format = "single", sep = ",", rm.duplicates = TRUE)

cat("\n",dim(transactions_object)[1],"unique user_id values")
cat("\n",dim(transactions_object)[2],"unique web_area_id values")

# examine frequency for each item with support
# greater than high_support_setting
pdf(file = "fig_wnds_recommend_web_area_support_bar_plot.pdf",
  width = 8.5, height = 11)
itemFrequencyPlot(transactions_object, support = high_support_setting,
  cex.names=0.8, xlim = c(0,0.4),
```

217

```
  type = "relative", horiz = TRUE, col = "darkblue", las = 1,
  xlab = "Proportion of Users Viewing the Web Area")
 title("Association Rules Analysis (Areas with Highest Support)")
dev.off()

# obtain large set of association rules for web areas
# this is done by setting very low criteria for support and confidence
# set support at low_support_setting and confidence at confidence_setting
# using our best judgment
association_rules <- apriori(transactions_object,
  parameter =
      list(support = low_support_setting, confidence = confidence_setting))
print(summary(association_rules))

# data visualization of association rules in scatter plot
pdf(file = "fig_wnds_recommend_web_area_rules_scatter_plot.pdf",
    width = 8.5,height = 11)
plot(association_rules,
  control=list(jitter=2, col = rev(brewer.pal(9, "Blues")[4:9])),
  shading = "lift")
dev.off()

# grouped matrix of rules
pdf(file = "fig_wnds_recommend_web_area_bubble_chart.pdf",
    width = 8.5, height = 11)
plot(association_rules, method="grouped",
  control=list(col = rev(brewer.pal(9, "Blues")[4:9])))
dev.off()

# route association rule results to text file
sink(file = "fig_wnds_recommend_partial_list_of_rules.txt")
cat("Web Usage Analysis by Association Rules\n")
cat("\n",dim(transactions_object)[1],"unique user_id values")
cat("\n",dim(transactions_object)[2],"unique web_area_id values")
cat("\n\nCriteria set for discovering association rules:")
cat("\nMinimum support set to:   ",low_support_setting)
cat("\nMinimum confidence set to:",confidence_setting,"\n\n")
print(summary(association_rules))
cat("\n","Resulting Web Area Association Rules\n\n")
inspect(association_rules)
sink()

# ------------------------------------------------
# Select association rules for recommender model
# ------------------------------------------------
# let's focus upon rules with a single-item/area-antecedent rules

selected_rules <-
    make_lists_for_single_item_antecedent_rules(association_rules)
antecedent_areas <- as.numeric(unlist(selected_rules[[1]]))
consequent_areas <- as.numeric(unlist(selected_rules[[2]]))

# ----------------------------------------------------
# Read in test data and create users-by-areas matrix
# ----------------------------------------------------

test_input_data <- read.csv(file = "000_test_data.csv", header = FALSE,
  col.names = c("CV_type","value","ignore_value"), stringsAsFactors = FALSE)

case_test_data <- subset(test_input_data, subset = (CV_type == "C"))
list_of_case_id_names <- as.character(sort(unique(case_test_data$value)))

web_area_test_data <- subset(test_input_data, subset = (CV_type == "V"))
list_of_web_area_id_names <-
    as.character(sort(unique(web_area_test_data$value)))
```

218

```
# the following code creates a test data matrix for the test data
# rows correspond to the user_ids
# columns correspond to the web_area_ids

# initialize test matrix as matrix of zeroes
test_data_matrix <- matrix(0, nrow = length(list_of_case_id_names),
  ncol = length(list_of_web_area_id_names),
  dimnames = list(as.character(list_of_case_id_names),
                  as.character(list_of_web_area_id_names)))

n_records <- nrow(test_input_data )  # used as stopping rule for while-loop

n_record_count <- 0  # intialize record count

# note that this step should go pretty fast...
cat("\n\nTest data processing percentage complete: ")
while (n_record_count < n_records) {
  n_record_count <- n_record_count + 1
  # report to screen the proportion of job completed
  if(multiple_of_one_thousand(n_record_count))
      cat(" ",trunc(100 * (n_record_count /  n_records)))
  # read one record from input data
  this_record <- test_input_data[n_record_count,]

  if (this_record$CV_type == "C")
      name_user_id <- as.character(this_record$value)  # C record for user_id
  if (this_record$CV_type == "V") {  # V record provides web_area_id
    name_web_area_id <- map_microsoft(this_record$value)
    # enter 1 in appropriate cell of test_data_matrix
    test_data_matrix[name_user_id, name_web_area_id] <- 1
    }
  }

# quick check of the test_data_matrix
# there should be as many 1s in test_data_matrix
#    as there are V records in the test data file
if(nrow(web_area_test_data) != sum(test_data_matrix))
    stop("test_data_matrix problem")

# useful rows for testing are those with at least two web areas visited
# sum of rows is number of 1s
user_areas_visited <- apply(test_data_matrix, 1, FUN = sum)

# keep track of the antecedent that is selected for testing
selected_antecedent <- rep(NA, length = nrow(test_data_matrix)) # initialize

# set up binary indicator for a correct prediction
# from any of the association rules
correct_predictions <- rep(0, length = nrow(test_data_matrix)) # initialize

# set up a counter for the number of rules
# that match the antecedent for this user
antecedent_rule_matches <-
    rep(0, length = nrow(test_data_matrix))  # initialize

# set up character string vector for storing consequents correctly predicted
correctly_predicted_consequents <-
    rep("", length = nrow(test_data_matrix))  # initialize

# for each user visiting more than one area
# select an area from his/her visited list at random
# predict from this area other areas likely to have been visited
# using the selected association rules from previous work
# determine if, in fact, these predictions are correct
```

219

```
set.seed(9999)  # for reproducible results
for(iuser in 1:nrow(test_data_matrix)) {
  if(user_areas_visited[iuser] > 1) {
    this_user_row <- test_data_matrix[iuser,]
    this_user_web_areas <-
        as.numeric(names(this_user_row[(this_user_row == 1)]))
    # pick one of the areas at random
    randomized_web_areas <- sample(this_user_web_areas)
    selected_antecedent[iuser] <-
        randomly_selected_antecedent <- randomized_web_areas[1]
    randomly_selected_consequents <-
        randomized_web_areas[2:length(randomized_web_areas)]
    # apply association rules to obtain predicted consequents
    for (iarea in seq(along = antecedent_areas)) {
      if (randomly_selected_antecedent == antecedent_areas[iarea]) {
        antecedent_rule_matches[iuser] <- antecedent_rule_matches[iuser] + 1
        predicted_consequent <- consequent_areas[iarea]
        if (predicted_consequent %in% randomly_selected_consequents) {
          correct_predictions[iuser] <- correct_predictions[iuser] + 1
          correctly_predicted_consequents[iuser] <-
            paste(correctly_predicted_consequents[iuser],
                predicted_consequent, " ")
          }
        }
      } # end for-loop for checking association rules

    } # end of if-block for users with more than one site visited
  } # end of testing for-loop

prediction_data_frame <- na.omit(data.frame(user.id = list_of_case_id_names,
                                    user_areas_visited,
                                    selected_antecedent,
                                    antecedent_rule_matches,
                                    correct_predictions,
                                    correctly_predicted_consequents))

cat("\n\nPercentage predicted accurately by recommendation model: ",
  round((100 * sum(correct_predictions) /
      sum(antecedent_rule_matches)),digits = 3),"\n")

# Suggestions for the student: Try alternative values for support
# and confidence and note changes in the set off association rules.
# Develop a formal assessment strategy for evaluating the accuracy
# of recommendations. How much better are we doing with the model
# than we would expect from the levels of support across website areas?
# We see Percentage predicted accurately by recommendation model: 36.461
# How good is that value compared to what we could get without a model?
# What about recommendation system models other than association rules?
# Could any of those models be used given the data we have here?
```

220

221

第 11 章
网络博弈

“我也比较担心 Walt Waldowski，然而不觉得痛苦。他的牌友们发生了一些争执并让他来评理，但是他说，谁对谁错又怎么样，只不过是一个纸牌游戏而已。”

Father John Patrick Mulcahy（Rene Auberjonois 饰）

电影《MASH》（1970 年）

几乎每天下午，我都要到所居住的公寓的游泳池去游泳。除了几只不会游泳的小飞虫之外，整个泳池通常只有我一个人。如果任何这些小生物侵犯了我的空间，那么它们的生死就掌握在我的手中。然而，往往出于仁慈，我会尽量将它们拨到游泳池外。这就是我的生活。我会花费大量的时间为不同的事情去纠错。

基于代理的模型是另一个可以让我扮演上帝角色的地方。如果我能够定义出一个网络的初始结构、连接的性质以及节点的特征，那么我就可以运用这些信息为这个网络建立起一个行为模型。我可以模拟节点或代理的行为，还有它们之间的相互交互。

使用基于代理的模型，我们构建一个微型世界来模拟现实世界中的现象。这为模拟复杂性提供了一个良好的平台，对于网络数据科学领域尤为重要。首先研究和应用离散事件模拟理论以及固态网络模型使我们能够为研究基于代理的模型和动态网络模型搭建起一个平台。 223

图 11-1 提供了一个网络模型技术的预览，以数学模型开始，以基于代理的模型结尾。在数学模型中，先有假设，后有结果，也就是说，结果都是经过正确性证明而得到的。Erdös-Rényi 提出的随机图模型（Erdös and Renyi, 1959, 1960, 1961）就是一个典型的面向网络的数学模型。面向网络的数学模型为网络科学提供了理论基础（Lewis 2009; Kolaczyk 2009; Newman 2010）。

面向网络的标准数学模型包括随机图、偏好连接（Barabási and Albert 1999）以及在本书前面曾经提到过的小世界模型（Watts and Strogatz 1998）。面向固态网络的离散事件模拟建立在数学模型之上，并为建模提供一个灵活的结构。通过使用离散事件模拟，我们论述结论而不是去证明结论。我们事先定义一个已知的固态网络结构，然后在这个固态网络结构中开展性能方面的研究。

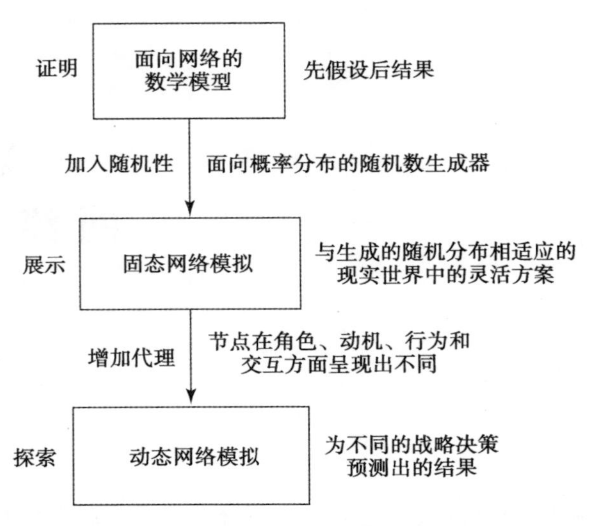

图 11-1 网络建模技术

面向固态网络的离散事件模拟对于研究网站与物理通信网络、传输以及产出问题都非常有意义。节点和连接是固定的，资源的上限也是已知的。任何通信以及对资源的请求都随着模拟的时间而发生变化。

假设我们想要为在线零售商设计一个网站，目的是让顾客从登录页面尽可能快地到达订购页面，从而降低在线购物车被抛弃的可能性。我们也许会从使用一个有向图来表达这个问题开始。节点就是订购的里程碑，比如说到达了登录页面、到达了产品信息页面，或者是到达了支付信息页面。连接代表活动，每一条连接都是从一个节点到下一个节点的动作。

在这个场景中，完全遍历这个网络所需要的总时间是我们特别关心的。我们想要寻找一种高效的方案，以做到没有购物车被丢弃。网络结构已经由于目前的网站而被固定。网络中的不同路径由不同的登录页面（源点）以及用户做出的选择来决定。我们使用具有不同概率分布的随机数生成器来模拟用户的选择、事件以及网络流量。

作为管理科学的一个重要分支，项目管理问题可以通过数学规划 (Bradley, Hax, and Magnanti 1977; Williams 2013; Vanderbei 2014) 或者离散事件模拟 (Burt and Garman 1971) 加以解决。每一个项目被定义为一个活动，其中一些活动可以并行完成，其他一些则要按照一定的顺序完成。其他的表达方式包括面向任务的网络，其中节点代表活动，以及面向事件的网络，其中连接代表活动。为了通过模拟来解决这些问题，我们构造项目网络的结构，随机设定每个活动持续的时间，并且寻找穿过网络的关键路径。目标也许就是寻找最短路径、成本最低路径、网络最大流量或者是项目完成时间的分布。

224～225

对网络进行基于代理的建模可以建立在对固态网络进行离散事件模拟的基础之上。我们都认可，节点会由于所承担的角色、动机、行为或交互而有所不同。基于代理的技术为对随着时间变化的网络（动态网络）进行建模提供了便利。

管理上的很多问题都涉及不同类型的代理以及随着时间而变化的网络。通过将代理引入进模拟模型，我们可以探索不同代理的输入及管理行为，并在此过程中预测与战略决策相关的输出（North and Macal 2007）。

假设一个消费型电子公司计划在可穿戴技术领域推出一款新产品。这个公司想尝试通过不同的方法对产品进行促销。通过大众媒体可以做到广阔的人群覆盖，但是与每一位客户的接触却非常有限。对目标群体进行直接宣传则会造成较小的人群覆盖。但正因为面向一个覆盖面狭小但却是目标非常明确的群体，可以让商家为这个目标群体量身定制宣传信息，使商家能够更为紧密地接触顾客。

大面积销售与有针对性的直接销售的成功与否取决于对人群的覆盖情况、与客户直接接触的程度以及客户对营销信息的反应。就像疾病在人群中的传播取决于网络的结构以及每个人对疾病的抵御程度一样，一个新产品是否能被人们所接受也取决于消费者的购买意愿以及消费者网络的结构。

某些消费者会比其他消费者更容易接受新的想法、信息和产品。我们可以利用基于代理的模型来探索消费者之间的差异以及这些差异对整个网络的影响。

消费者对最初的市场宣传信息的反应是整个过程的开始。部分消费者购买了可穿戴技术产品并且实际在使用。这些消费者又与其他消费者进行交流，从而影响其他消费者的购买意愿。这整个过程完全体现在模拟中。随着模拟向前推进，我们记录下受到影响的消费者以及他们对获得的信息做出的反应（即是否会选择购买）。

226

基于代理的构架使我们能够对消费者的特性进行微调，使之更加接近我们对购买市场中消费者特征的理解。我们可以测试不同的市场营销行为（大范围的或有针对性的）以及它们在消费者接受产品方面产生的效果。

基于代理模型的讨论与问题的复杂程度相挂钩。因为只有在我们对整个系统的了解程度不如我们对各个组成部分的了解程度时，基于代理的模型才更加合适。因此，我们往往更倾向于构建许多较小的模型，而不是构建一个复杂的系统模型。基于代理的模型在底层的流程同时发生而不是顺序发生时，也会有其用武之地。

对基于代理的模型进行验证十分具有挑战性。模拟出的行为与现实世界中的行为很相像是否就达到了我们的目标呢？而我们又该如何去判断相像的程度呢？

基于代理的模型目前停留在研究和实验阶段，还没有大范围地在商业研究中得到应用。要想从学术世界转为商用，即不仅仅是作为模型设计者的玩物，基于代理的模型需要做出一些有实用价值的事情，并且要让研究模型群体以外的人们去理解它。

Law (2014) 对模拟建模这个领域进行了讨论，其中包括基于代理的模型。North and Macal (2007), Miller and Page (2007) 和 Šalamon (2011) 对复杂理论以及基于代理的模型进行了讨论。这个领域中的很多工作都受生物学领域相关研究的启发，这些研究的目的是通过对各个生物体的行为进行建模来观察对整个群体的影响（Resnick 1998; Mitchell 2009）。

像其他很多模拟方面的研究一样，基于代理的研究也按照阶乘实验的思路展开。在模拟过程中，我们观察不同的网络结构和代理特性如何对模拟产生影响。其中一个例子就是信息的串联倾泻效应，也是建立在疾病传播的生物学模型上。我们考虑的因素可能会包括网络中节点的数量、网络密度以及最初受到影响或者是被病毒感染的个体的数量。我们也可能会考虑与代理相关的特性，比如风险预防能力或是对病毒的抵抗能力。实验中的每项因素都会很系统地发生变化，其最初值通过一个随机数来产生，以确保实验结果都是相互独立的。

227

我们可以像学习其他很多知识那样对基于代理的模拟进行更多了解，即通过实际动手去做、去编写代码。基于代理的模拟需要建立一个自身的环境和框架，编程也需要花费很多时间。之后在本书上网站添加范例程序后，我们可以做更多有趣的模拟游戏。

我们就像做其他实验那样对基于代理模拟的结果进行分析。对变化和线性模型的分析适用于连续响应变量。对差异和广义线性模型的分析适用于二进制响应变量、比例和计数。

代码 11-1 展示的是一个使用 Python 编写的进行差异分析的程序。代码 11-2 则是相对应的 R 语言程序。通过设计不同的因素，交互图能够很好地总结出实验结果。R 语言程序展示出如何生成一个双向或双因素交互图，此软件的开发者是 Wickham and Chang (2014)。

228

代码 11-1 基于代理的模拟分析（Python）

```
# Analysis of Agent-Based Simulation (Python)

# install necessary packages

# import packages into the workspace for this program
import numpy as np
import pandas as pd
import statsmodels.api as sm

# -----------------------------
# Simulation study background
# -----------------------------
# an agent-based simulation was run with NetLogo, a public-domain
# program available from Northwestern University (Wilensky 1999)

# added one line of code to the Virus on a Network program:

if ticks = 200 [stop]

# this line was added to stop the simulation at exactly 200 ticks
# the line was added to the <to go> code block as shown here:
#
# to go
#   if all? turtles [not infected?]
#     [ stop ]
#   ask turtles
#   [
#      set virus-check-timer virus-check-timer + 1
#      if virus-check-timer >= virus-check-frequency
#        [ set virus-check-timer 0 ]
#   ]
#   if ticks = 200 [stop]
#   spread-virus
#   do-virus-checks
#   tick
# end

# the simulation stops when no nodes/turtles are infected
# or when the simulation reaches 200 ticks

# To see the results of the simulation at 200 ticks, we route the simulation
# world to a file using the GUI File/Export/Export World
# this gives an a comma-delimited text file of the status of the network
# at 200 ticks. Specifically, we enter the following Excel command into
```

```
# cell D1 of the results spreadsheet to compute the proportion of nodes
# infected:   = COUNTIF(N14:N163, TRUE)/M10

# NetLogo turtle infected status values were given in cells N14 through N163.
# The detailed results of the simulation runs or trials are shown in the files
# <trial01.csv> through <trial20.csv> under the directory NetLogo_Results

# this particular experiment, has average connectivity or node degree
# at 3 or 5 and the susceptibility or virus spread chance to 5 or 10 percent.
# we have a completely crossed 2 x 2 design with 5 replications of each cell
# that is, we run each treatment combination 5 times, obtaining 20 independent
# observations or trials. for each trial, we note the percentage of infected
# nodes after 200 ticks---this is the response variable
# results are summarized in the comma-delimited file <virus_results.csv>.

# -----------------------------
# Analysis of Deviance
# -----------------------------

# read in summary results and code the experimental factors
virus = pd.read_csv("virus_results.csv")

# check input DataFrame
print(virus)

Intercept = np.array([1] * 20)

# use dictionary object for mapping to 0/1 binary codes
degree_to_binary = {3 : 0, 5 : 1}
Connectivity = np.array(virus['degree'].map(degree_to_binary))

# use dictionary object for mapping to 0/1 binary codes
spread_to_binary = {5 : 0, 10 : 1}
Susceptibility = np.array(virus['spread'].map(spread_to_binary))

Connectivity_Susceptibility = Connectivity * Susceptibility

Design_Matrix = np.array([Intercept, Connectivity, Susceptibility,\
     Connectivity_Susceptibility]).T

print(Design_Matrix)

Market_Share = np.array(virus['infected'])

# generalized linear model for a response variable that is a proportion
glm_binom = sm.GLM(Market_Share, Design_Matrix, family=sm.families.Binomial())
res = glm_binom.fit()
print res.summary()
```

229

230

代码 11-2　基于代理的模拟分析（R 语言）

```
# Analysis of Agent-Based Simulation (R)

# install necessary packages

library(ggplot2)  # for pretty plotting

# -------------------------------
# Simulation study background
# (same as under Python exhibit)
# -------------------------------
# note. it is possible to run NetLogo simulations within R
# using the RNetLogo package from Thiele (2014)
```

```
# ------------------------------
# Analysis of Deviance
# ------------------------------
options(warn = -1)  # drop glm() warnings about non-integer responses

# read in summary results and code the experimental factors
virus <- read.csv("virus_results.csv")

# define factor variables
virus$Connectivity <- factor(virus$degree,
  levels = c(3, 5), labels = c("LOW", "HIGH"))
virus$Susceptibility <- factor(virus$spread,
  levels = c(5, 10), labels = c("LOW", "HIGH"))

virus$Market_Share <- virus$infected

# show the mean proportions by cell in the 2x2 design
with(virus, print(by(Market_Share, Connectivity * Susceptibility, mean)))

# generalized linear model for response variable that is a proportion
virus_fit <- glm(Market_Share ~ Connectivity + Susceptibility +
  Connectivity:Susceptibility,
  data = virus, family = binomial(link = "logit"))
print(summary(virus_fit))

# analysis of deviance
print(anova(virus_fit, test="Chisq"))

# compute market share cell means for use in interaction plot
virus_means <- aggregate(Market_Share ~ Connectivity * Susceptibility,
  data = virus, mean)

# ------------------------------
# Interaction Plotting
# ------------------------------
# generate an interaction plot for market share as a percentage
interaction_plot <- ggplot(virus_means,
  aes(x = Connectivity, y = 100*Market_Share,
    group = Susceptibility, fill = Susceptibility)) +
  geom_line(linetype = "solid", size = 1, colour = "black") +
  geom_point(size = 4, shape = 21) +
  ylab("Market Share (Percentage)") +
  ggtitle("Network Interaction Effects in Innovation") +
  theme(plot.title = element_text(lineheight=.8, face="bold"))
print(interaction_plot)
```

231

232

第12章
Web的未来

[新泽西州，普林斯顿市，某酒吧里]

Nash："亚当·史密斯需要修正他的理论了。"

Hansen："你在胡说些什么？"

Nash："如果我们都去追求那个金发美女，我们会相互妨碍，最终谁也得不到她。然后，我们又去追求她的女伴们。然而，她们肯定会不理睬我们，因为没有人会愿意作为第二选择。但是，如果没有人去追求那个金发美女呢？那么我们之间会互不妨碍，我们也不会去羞辱其他那些女孩。这是我们共赢的唯一办法。这也是我们跟她们共度良宵的唯一办法。"

"亚当·史密斯说过，最好的结果来自于团队中的每个人都去做对自己最有益的事情，对吗？并不完全对，并不完全对。因为最好的结果来自于团队中的每个人都去做同时对自己和对这个团队最有益的事情。"

Hansen："Nash，如果你想通过使这套歪招好去独占美人，你就做美梦去吧。"

John Nash（Russell Crowe饰）、Hansen（Josh Lucas饰）

电影《美丽心灵》(2001年)

心理学家Jean Piaget从生物学的思考中得到启发，提出了人们学习的两种方式：同化和适应。同化指的是接收新的信息，并将它叠加在已有的认知结构之上。我们接收的信息越来越多，然而，我们却依然按照同样的方式在思考。

当然，也有另外一种学习会迫使我们做出改变，这就是被Piaget称之为适应的方法。当接收到的新的信息无法与我们已经知道的内容相适应时，我们就必须更新我们的认知结构。

教师的责任不应该是让学习变得容易。灵活变通地学习，也就是学习的本质，绝非易事。教师的责任是帮助学生学会变通，打开学生们面向新的世界和新的思考方式的眼界。

每当回想起曾经影响过我的人生的那些老师时，我都会意识到他们是教我如何使用新思维模式的老师。虽然他们都有自己的原则，但他们并不是特别严厉，不是监工头。他们所做的大部分都是给予我支持。他们使我必须要努力学习的东西变得更加容易。

对于Web，我们也需要学习。这方面的学习需要很多年的时间才能见到效果，而且这些学习没有尽头。然而，正是这种学习将会改变我们的生活、工作及思维模式。这种学习就是机器学习，它涉及语义网。

我们常常将 Web 数据形象地描述为非结构化或半结构化的文本。在本书中，我们介绍了许多实例，包括爬行、抓取、解析和分析，所有这些都是为了给事物赋予含义。

文档对象模型（Document Object Model）可以帮助我们遍历文本域。XML 和 HTML 以及赋予节点的标签为储存和显示完全不同的信息提供了一个结构。数据交换方面的标准促进了研究者之间的交流与合作，就像使用计算机支持进程之间的通信那样。但是必须承认，这些仅仅是对文本进行格式化的规则而已，它们并不能形成一个网络知识的地图。

我们经常感叹文本分析学很难。我们忍受着自然语言的奇葩，并开发出专门的工具和处理方案。然而，假如获取我们所熟悉的事物的文本，并用更容易被计算机和人们获取的方式进行存储会是一番什么情景呢？假如我们改变网络的认知结构，系统性地将信息进行组织，将它们按照知识展现出来又会是一番什么情景呢？

234

我们可以想象未来的研究世界由众多 Web 服务构成，它们将信息供应商和客户连接在一起。万维网（World Wide Web）作为一个信息仓库或语义网、一个巨大的机器可检索和机器可理解的数据储存库具有巨大的潜力。

语义网的目标是将文本结构化，以用于表达知识。对于组织文本来说，这已经不再只是一个想法了，研究者们已经有了一系列技术可以使用。Allemang and Hendler (2007) 和 Wood, Zaidman, and Ruth (2014) 对这些技术进行了概述，有关数据表达和编程技术方面的更多信息还可以通过其他渠道获得（Powers 2003; Lacy 2005; Segaran, Evans, and Taylor 2009; DuCharme 2013）。

语义数据库、自然语言处理以及机器学习结合在一起形成了智能问答领域。要想让计算机能够像人类一样智能地进行网络查询还有很长的路要走，但是智能问答的人工智能技术已经取得了很大的进步。成功的实例包括：Cyc (Lenat et al. 2010), Halo (Gunning et al. 2010) 以及 IBM 的 Watson (Ferrucci et al. 2010; Lopez et al. 2013)。从互联网中获取相互链接的信息也是一些开源项目的目标，如为 Watson 提供语义数据资源的项目 DBpedia (Bizar et al. 2009, Lehmann et al. 2014)。想象有那么一天，一个维基百科机器程序能够真正地理解事物之间如何相互关联并能够回答你提出的任何问题。实现这个目标或许只是一个很小的编程问题（我们习惯性地称之为“SMOP（small matter of programming)”）。

软件是集体努力的结果。在开源环境中共享代码是提高工作效率的关键。我们的工作建立在他人创造的工具的基础之上。Python 和 R 语言社区的发展以及 Github 故事（Bourne 2013）都展示出了协同软件环境的潜力。

Evan “Ev” Williams 和 Jack Dorsey 谈到让他们公司知名的微博的起源。Jack 说他是一个发明者，Ev 反驳道：“不，你没有发明 Twitter，我也没有发明 Twitter，Biz（Christopher “Biz” Stone）更没有。人们在互联网上并没有什么发明，他们仅仅是将一个已经存在的想法进行了拓展。”(Bilton 2013, p. 203）

235

这就是互联网方式，即相互合作、团队精神以及共享程序和想法。在这个发展过程中，只有少数人富了起来，很多人只能被列为是贡献者。数据的列车继续向前行进。

只有时间能够告诉我们还有哪些新的团队努力最终会冒出来。但是我怀疑我们是否能够看到将信息转变为知识、将公式化的推理转化为人工智能的产品和服务。

孩子们都玩过圆圈游戏，也就是讲故事游戏。他们围成一个圆圈。故事开始，每个小孩都在下个小孩的耳边轻语，讲故事就沿着这个圈往下进行，直到转回到第一个小孩。

“噢，不对。”她说，“这并不是我讲的故事，绝对不是。”

有人会认为这个故事游戏告诉我们的是信息的丢失和曲解。然而，假如反过来成立又会怎么样？假如这个故事在传递的过程中变得更好又会怎么样，也就是经过美化，经过改进，成为一个学习型团队的产出会如何？假如我们学到的是关于合作创新又会怎么样？

从我家厨房的窗户朝北望去，我看到的是圣盖博山（San Gabriel Mountains）。从远处看，它非常美丽。这就好像我看到了美好的未来。未来并不可怕，我很好奇。我很好奇接下来还要学习什么。

236

附录 A

数据科学方法

Artim：“Data，……你从来没有为了开心而玩吗？”

Data：“呃……没有……开心。”

Artim：“为什么不呢……？”

Data：“从来没有人问过我这个问题。”

Artim（Michael Welch 饰）、Data（Brent Spiner 饰）

电影《星际迷航：起义》(1998 年)

从事数据科学工作就意味着要去实现灵活、有弹性、可扩展的系统，以实现数据准备、分析、可视化以及建模。开源的发展给了我们这样一个机会。不论是建模技术还是应用程序，我们都可以找到其他人已经完成或者正在实现的相关软件包、模块或程序库。面向 Web 和网络研究数据科学就意味着使用 Python 和 R 语言，而且还要根据具体需要使用其他语言。

工作在预测分析学领域的数据科学家使用的是商业化语言，如会计、金融、市场和管理。他们具有信息技术方面的知识，包括数据结构、算法和面向对象程序设计。他们也懂得统计建模、机器学习和数学规划。

以下是数据科学家的工作内容：

- **发现问题**。这是我们首先要做的事情，即信息查询，找到前人已经做过的工作、通过阅读文献汲取知识。以很多领域中学者和实践人员的工作，以及对预测分析学和数据科学做出贡献的专家的工作为基础。
- **准备文本和数据**。文本是非结构化或者半结构化的，而数据则通常是混乱或残缺的。我们从文本中提取出特征。我们定义度量。我们为分析和建模准备文本和数据。
- **检查数据**。为了获得新的发现，我们对数据进行探索性的分析以及数据可视化。我们寻找数据组。我们找出异常值。我们确定共同的维度、模式和趋势。
- **做好预算**。经常会有人让我们对能够卖出多少数量或美元的产品进行预测，或者是对金融证券和房地产进行预测。回归技术对于这方面的预测非常有效。预测与解释不同。[⊖]我们也许不知道为什么模型会有效，但是我们需要知道它们在什么情况下能够有效工作，在什么情况下给别人展示出它们能够有效工作。我们确定模型中最关键的部分，然后把重点放在会造成影响的事物上。

- **预测是与非**。许多商业问题都是分类问题。我们使用分类方法来预测某人是否会购买某个产品，是否会违约支付贷款，是否会访问某个网页。
- **测试**。我们使用诊断图来验证模型。我们观察基于一个数据集开发的模型用于其他数据集时得到的效果。我们采用训练再测试的方式，并且使用各种数据分区，交叉验证或引导方法。
- **做出各种假设**。我们改变关键变量以观察这些变量对预测结果的影响。我们在模拟过程中使用各种假设。我们对数学规划模型进行灵敏度或压力测试。我们观察输入变量的值如何影响输出结果、收益以及预测值。我们对预测中的不确定性进行评估。
- **解释一切**。数据和模型帮助我们理解这个世界。我们将学到的知识用一种通俗易懂的语言表达出来，便于其他人理解。我们以清晰简明的方式展现我们的项目成果。 238

数据科学家都是方法折中主义者，他们根据许多科学领域的基础进行推论，并将以观察或实验为依据的研究结果转换成管理者可以理解的文字和图片进行展现。这些展现成果的方式得益于已经构建出的良好的数据可视化。在与管理者进行沟通时，数据科学家需要跳过公式、数字、术语的定义以及算法的奇妙性，而将预测模型的结果通过通俗易懂的方式表达出来，使其他人能够理解。

数据科学家都是出类拔萃的知识工作者。他们在如今的数据密集型世界中扮演着通信员这样一个至关重要的角色。他们将数据转换成模型，再将模型转变成付诸行动的计划。

关于数据科学在商业中扮演的角色很多学者讨论过（Davenport and Harris 2007; Laursen and Thorlund 2010; Davenport, Harris, and Morison 2010; Franks 2012; Siegel 2013; Maisel and Cokins 2014; Provost and Fawcett 2014）。更加深入的讨论还包括 Izenman (2008), Hastie, Tibshirani, and Friedman (2009) 和 Murphty (2012)。对具有开源工具的数据科学的讨论还可以从以下文献中获得更多的信息：Conway and White (2012), Putler and Krider (2012), James et al. (2013), Kuhn and Johnson (2013), Lantz (2013) 和 Ledoiter (2013)。

本附录对各种方法进行分类，同时对数据库与数据准备、统计学、机器学习、数据可视化以及文本分析学方面的方法进行概述。我们对这些方法进行综述，并对相关的资源进行引用，以方便读者进一步阅读。 239

A.1 数据库与数据准备

如前所述，我们能够使用的数据只是所有数据中的很小一部分。方便地收集数据和降低

⊖ 数据统计师将模型区分为解释性模型和预测性模型。解释性模型用于检验因果理论，而预测性模型则用于对新的或将来的观察进行预测。请参考 Geisser (1993), Breiman (2001b) 和 Shmueli (2010)。

存储数据的成本已经取得了很多新的进展。数据的来源有很多。它们源于在线系统的非结构化文本数据，源于传感器和摄像机的像素数据，源于世界范围内存在于空间和时间中的移动电话、平板电脑和计算机中的数据。因此，我们需要灵活、可扩展的分布式系统来容纳这些数据。

与电子表格相似，关系数据库具有一个行与列的表结构。我们使用结构化查询语言（SQL）来访问并处理这些数据。由于这些面向事务的数据具有强制数据完整性的要求，关系数据库为销售订单处理和财务会计系统提供了基础支持。

很容易理解为什么非关系数据库（NoSQL）近年来会受到如此多的关注。非关系数据库以可用性和可扩展性作为侧重点。这一类数据库可以基于键、值、面向列、面向文档或图形结构。某些此类的数据库专为在线或实时应用而设计，其关键点是能够快速响应。而其他的此类数据库则适用于大量数据的存储及线下分析，其中使用 map-reduce 作为一个关键的数据聚合工具。

很多公司都在摒弃直接拥有的集中式计算系统，而走向采用分布式云计算服务。随着机构的数据管理需求的不断增加，包括数据库系统在内的分布式硬件和软件系统可以很容易地随之不断扩展。

研究数据科学意味着要能够从各类数据库系统中获取数据，包括关系数据库、非关系数据库、商业数据源和开源数据源。我们使用数据库查询与分析工具，从分布式系统中收集信息、核对信息、创建情形分析表、为所关注的变量计算出关系指数。我们最大限度地使用信息技术和数据库系统。我们还要做更多的工作，这就是应用已有的统计推断和预测分析建模技术。

240 对于分析学来说，我们承认在数据科学领域一直有不成文的约定。我们不能只选择我们喜欢的数据。我们不能将数据改变为我们想看到或者希望看到的格式。就如同在扑克牌游戏中，我们可能只能拿到一两张梅花牌，因此必须与其他牌一起使用才能发挥出相应的效果，况且我们出牌之前都需要先整理手中的牌。科学的标志是对变异的认知、对错误来源的理解以及对数据的尊重。数据科学是一门科学。

我们经常被要求在一团混乱之上建立一个模型。管理者希望看到的是答案，但是这些数据中充满着各种错译或缺失的观察、异常值以及来历不明的值。我们使用最好的判断为分析而准备数据，同时我们意识到自己做出的许多决策都是主观且难有说服力的。

数据缺失在应用研究中一直是个问题，因为很多建模算法都需要完整的数据集。在具有大量的解释型变量时，大多数情况下都至少会有一个变量存在数据丢失。列表删除法是不可取的。若用某个值去代替缺失的数据，不论是均值、中值还是模式值，都会对变量的分布以及变量之间的关系造成扭曲。若用随机选择的值去代替缺失的数据将会加入噪音，使发现与其他变量的关系更加难上加难。统计学家更倾向于使用多重估算法。

Garcia-Molina, Ullman, and Widom (2009) 对数据库系统进行了综述。White (2011), Chodorow (2013) 和 Robinson, Webber, and Eifrem (2013) 选择了几个特定的非关系系统进行综述。关于 map-reducede 操作，请参考 Dean and Ghemawat (2004) 和 Rajaraman and Ullman

(2012)。

Osborne (2013) 对与数据准备相关的问题进行了综述，McCallum(2013) 编辑卷则对如何处理混乱的数据提出了一些建议。Rubin (1987), Little and Rubin (1987), Schafer (1997), Groves et al. (2009) 和 Lumley (2010) 讨论了应对数据缺失的方法，实现这些方法的 R 软件包来源于 Gelman et al. (2014), Honaker, King, and Blackwell (2014) 和 Lumley (2014)。

241

A.2 经典统计学与贝叶斯统计学

我们如何基于数据进行推论呢？形式化的科学方法要求先构建理论，然后再通过样本数据对这些理论进行验证。这个过程包括基于点估算的统计推论和基于区间估算的统计推论，或者是对总体假设进行检验。无论是什么形式的推论，我们都需要与感兴趣的问题相关的样本数据。为了使统计方法得到有效运用，我们希望从总体样本中进行随机抽样。

我们究竟要相信哪些统计结果呢？统计是样本数据的函数，因此，当样本对总体来说具有代表性时，统计结果就更可信。大量的随机样本、小的标准误差、狭窄的可信区间都是首选。

经典统计学和贝叶斯统计学代表着两种不同的用于进行推断以及度量不确定性的方案。经典假设检验首先对总体参数设置零假设，然后根据样本数据判断接受或废弃这些假设。典型的零假设（也就是零所隐含的）是指比例或组平均值之间没有任何差异，或者是指模型中的变量之间不存在任何关系。零假设也可以指具有多变量模型中的参数。

为了检验零假设，我们计算一个被称为检验统计量的特殊统计以及与之相关联的 *p*-value 值。假设零假设成立，我们可以得到检验统计量的理论分布。我们再根据理论分布的样本检验统计量得出 *p*-value 值。这个 *p*-value 值本身就是一个样本统计量，代表当零假设成立时，拒绝零假设的概率。

我们先假定有效推理的条件已经得到满足。然后，当我们观察到一个非常低的 *p*-value 值时（如 0.05、0.01 或 0.001），我们知道以下这两个事件中有一个肯定为真：（1）若零假设成立，则会发生一个小概率事件；（2）零假设不成立。一个小的 *p*-value 值会让我们拒绝零假设，这也代表研究结果具有统计上的重要性。有些结果在统计上是重要的和有意义的。另外一些结果在统计上是重要的但却是没有价值的。

242

在传统的应用研究中，我们希望得出 *p*-value 值很低的统计量。我们设定零假设只是为了拒绝这些假设。在检查组之间的差别时，我们将零假设定为组之间不存在任何差异。在研究变量之间的关系时，我们将零假设定为变量之间相互独立，然后再收集数据来拒绝这些假设。如果我们能够收集到足够多的数据，测试过程在统计上就具有足够的说服力。

易变性既是敌人也是朋友。如果它是由无法解释或者异变的样本造成，也就是说统计的结果随着不同的样本而变化，那么它就是我们的敌人。但是易变性又是我们的朋友，这是因

为如果没有易变性，我们将无法看清楚变量之间的关系。[⊖]

经典方法将参数视为固定的、需要估算的未知量，而贝叶斯方法却将参数视为随机变量。也就是说，我们可以将参数想象成具有概率分布，是我们对这个世界不确定性的表达。

贝叶斯方法得名于贝叶斯定理，是统计学中一个著名的定理。除了对总体分布、随机样本和样本分布做出假设，我们还可以针对总体参数提出假设。采用贝叶斯方法，首先是将我们对这个世界的不确定性用概率分布的形式表示出来，然后再通过收集关联的样本数据来降低这个不确定性。

我们要如何表达参数的不确定性呢？首先为这些参数设定先验的概率分布。然后，我们再使用样本数据和贝叶斯定理推导出这些参数的后验概率分布。最后，基于后验概率分布估计条件概率。

许多人都认为贝叶斯统计为以观察或实验为依据的研究提供了一种保持逻辑上一致的方法。暂且先抛开零假设，而把注意力集中在我们所关注的科学问题上，即科学的假设。当用一个概率区间来描述不确定性时，就没有必要考虑可信区间了。当从决策理论的角度来观察所有的科学和商业问题时，也就没有必要对零假设做出决断了（Robert 2007）。贝叶斯概率观点也可以应用在机器学习以及传统的统计模型中（Murphy 2012）。

243

为后验概率分布推导出数学公式也许很有挑战性。其实，对于研究的很多问题，我们都无法得到后验分布的公式。然而，这并不能阻止我们使用贝叶斯方法，因为可以用计算机程序来生成或估算后验分布。马尔可夫链蒙特卡洛（Monte Carlo）模拟是贝叶斯实践的核心所在（Tanner 1996; Albert 2009; Robert and Casella 2009; Suess and Trumbo 2010）。

贝叶斯统计方法由于能够帮助我们解决现实世界中的问题而盛行至今（McGrayne 2011; Flam 2014)。然而，正如 Efron(1986) 所指出的那样，并不是所有人都认可贝叶斯统计方法是有一定道理的。经典推理方面有很多不错的工作可以借鉴（Fisher 1970; Fisher 1971; Snedecor and Cochran 1989; Hinkley, Reid, and Snell 1991; Stuart, Ord, and Arnold 2010; O'Hagan 2010; Wasserman 2010）。也有很多很好的学习贝叶斯方法的资料（Geisser 1993; Gelman, Carlin, Stern, and Rubin 1995; Carlin and Louis 1996; Robert 2007）。

如果问题是两组之间的差异是否会偶然出现，我们会倾向于选择经典方法。具体做法是估算出一个 *p*-value 值作为条件概率，同时做出一个组之间不存在差异的零假设。但是，如果要估计苹果公司的股价在明年年初会高于 100 美元的概率，我们则会选择贝叶斯方法。究竟哪一种方法更好，是经典还是贝叶斯？无所谓。我们两个都需要。究竟哪一种语言更好，是 Python 还是 R 语言？无所谓。我们两个都需要。

244

A.3 回归与分类

很多数据科学方面的工作都会涉及探究变量之间有意义的关系。我们使用散点图和相关

⊖ 为了进一步认识易变性对于发现关系的重要性，我们可以先对两个高度关联的变量画出一个散点图。然后限制其中一个变量的取值范围。在大多数情况下，由此得到的散点图在被限制的取值范围内会呈现出很低的关联性。

系数来寻找一对连续变量之间的关系。我们使用情形分析表和分类数据分析方法来寻找分类变量之间的关系。我们使用多元方法和多维情形分析表来检验多个变量之间的关系。最后，我们建立预测模型。

目前主要有两类预测模型：*回归*和*分类*。回归是对有意义的量级的响应值进行预测，分类则涉及对分类或类别进行预测。使用机器学习的术语来描述，这些都是监督式学习的方法。

最常见的回归有*最小二乘回归*（也称为一般最小二乘回归）、线性回归及多元回归。当我们使用一般最小二乘回归时，要先估算回归系数，以使残差值的平方和最小化，在数理统计中，残差值是指实际观察值和估计值之间的差。对于回归问题，我们可以取实数线上的任意值作为响应值，虽然在实际情况中只可能取非常有限的不同值。很重要的一点是，响应值在回归问题中占据着举足轻重的地位。

在统计学上，*泊松回归*非常有用。响应值虽然重要但只能取到从 0 开始的离散（整数）值。面向频率的对数线性模型、分组频率和面向交叉分类观察的情形分析表都属于这个范畴。

就像在*生存分析*中那样，对于事件模型、持续模型和生存模型，我们通常必须支持截尾，因为有些观察结果测量得很精确，有些则不够精确。若采用左截尾，我们只知道测量不精确的观察结果都会小于某个值。若采用右截尾，我们只知道测量不精确的观察结果都会大于某个值。

大多数传统建模技术只会涉及*线性模型*或线性方程。响应值或者转换后的响应值置于线性模型的左边。*线性预测器*则置于右边。线性预测器涉及解释型变量，且它的参数是线性的。也就是说，它涉及通过解释型变量对系数进行相加或相乘运算。我们在线性模型中使用的系数代表对总体参数的估算值。最小二乘回归、泊松回归、逻辑回归以及生存模型都属于*广义线性模型*的范畴，它们处于传统统计推理的核心位置，包括经典统计和贝叶斯统计。 245

在本书中所举的回归实例中，我们使用 R 平方或者决定系数作为拟合度的指标。作为通过模型得出的响应方差的比率，这个指标很容易解释给管理层。另外一个统计学家倾向使用的指标是*预测的根均方差*（RMSE），这是一个衡量拟合劣度或者失拟度指标。其他有关拟合劣度的指数，如预测中误差的百分比，有时则是管理层的首选。

*逻辑回归*方法虽然被称作“回归”，但实际上是一种分类方法。它适用于预测出一个只具有两个可能值（二元值）的响应。有序和多项式分对数模型对逻辑回归进行了扩展，可以用于研究多于两个分类的问题。线性判别分析是传统统计领域中的另外一种分类方法。在情感分析那一章中对文本分类方法的基准研究就使用了逻辑回归和一些机器学习算法来进行分类。

对分类器的性能进行评价是一个挑战，因为很多问题都是低基准率问题。只有不到 5% 的客户可能会对直接使用邮件进行的宣传做出响应。发病率、贷款违约和诈骗往往也都是低基准率事件。要在低基准率条件下对分类器进行评价，我们就不能只看正确预测出事件的百分比。基于一个被称为*混淆矩阵*的四倍表，图 A-1 对二元分类器的各项指标进行了归纳。 246

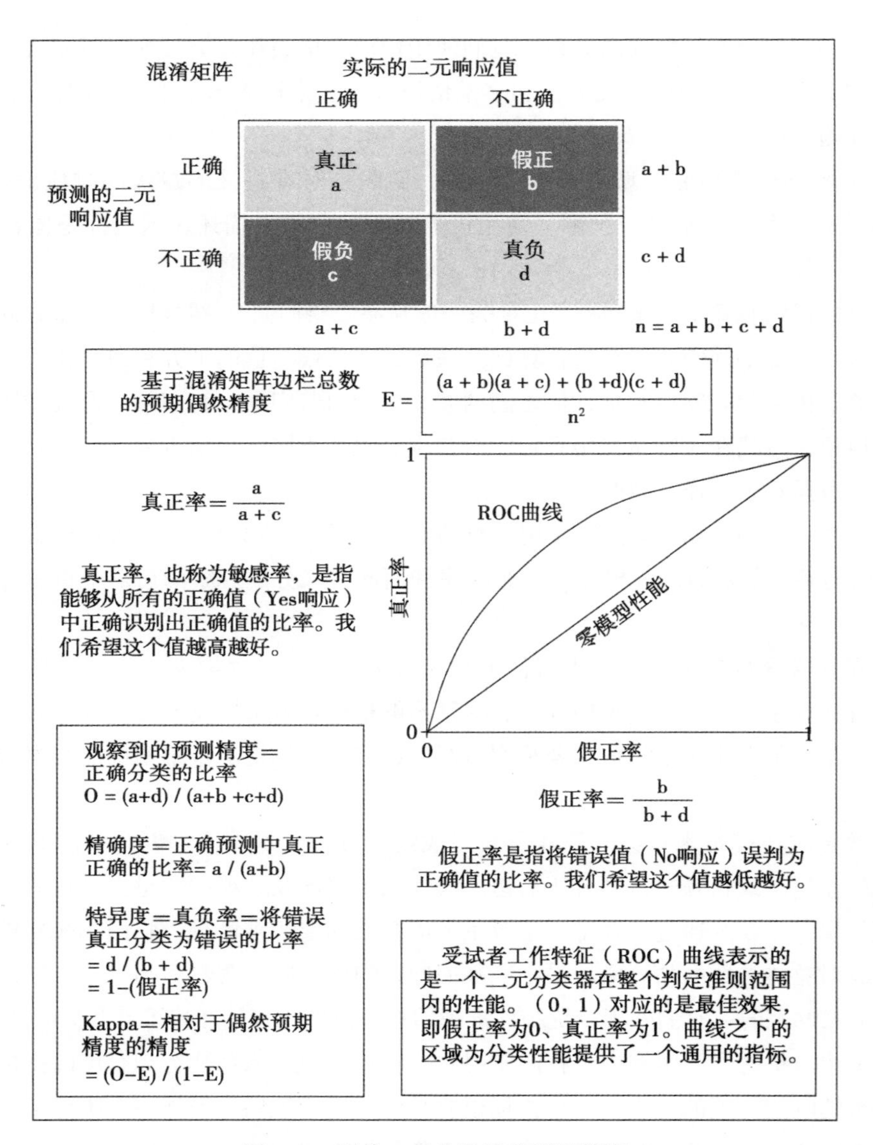

247

图 A-1 评估二元分类器的预测精度

汇总统计，如 Kappa（Cohen 1960），以及受试者工作特征（ROC）曲线之下的区域时常用来对分类器进行评价。Kappa 依赖于在分类中运用概率截断，而 ROC 曲线之下的区域则不需要这么做。[⊖]

很好的线性回归参考文献包括 Draper and Smith (1998), Harrell (2001), Chatterjee and Hadi (2012) 和 Fox and Weisberg (2011)。Berk (2008), Izenman (2008) 和 Hastie, Tibshirani,

⊖ 对于低基准率问题，ROC 曲线之下的区域是分类性能的首选指标。ROC 曲线是一个真正率对假正率的图，表达的是敏感度与特异度之间的平衡，用于度量模型将正确与错误区分开的效果。曲线之下的区域为预测精度提供了一个指标，独立于用于对事物进行分类的概率截断。最佳的预测对应于一个面积为 1.0 的区域（触及到左上角的曲线）。面积为 0.5 的区域则表示随机（零模型）预测精度，此时的曲线是从左下角到右上角的对角线，对应的面积是下三角。

and Friedman (2009) 对数据自适应回归方法以及机器学习算法进行了综述。Bates and Watts (2007) 对传统的非线性模型进行了介绍。

数据科学家特别关心的问题是回归模型的结构。在什么条件下要对响应或部分解释性变量进行变换？交互效应是否应该包含在模型中？

回归诊断是指我们通过数据可视化和索引对回归模型的适用性进行检查。文献 Belsley, Kuh, and Welsch (1980) 和 Cook (1998) 中有相关的讨论。最基本的 R 系统提供了很多种诊断，Fox and Weisberg (2011) 提供了更多的诊断。诊断会告诉我们是否需要对响应或解释性变量进行变换，以满足模型中的假设或者改进预测性能。Box and Cox (1964) 对变电理论进行了介绍，Fox and Weisberg (2011) 也有相关的综述。

当定义一个带有参数的模型时，我们希望恰当地引入一组格式适合的解释性变量。变量太少或缺少关键变量都会造成预测结果的偏差。另一方面，变量太多也可能会造成过度拟合以及较高的样本外预测错误。这个在偏差 - 方差之间的取舍或平衡（通常会这么叫）是生活中在统计上存在的一个现实。

收缩与正则化回归方法为微调、平滑或调整模型的复杂度提供了手段 (Tibshirani 1996; Hoerl and Kennard 2000)。此外，我们可以为预测模型挑选部分解释性变量。当被估算的变量太多，也许在超过了观察的数量时，就需要使用特殊的方法 (Bühlmann and van de Geer 2011)。文献 Izenman (2008) 和 Hastie, Tibshirani, and Friedman (2009) 还有更多关于偏差 - 方差平衡、正则化回归以及变量子集选择方面的讨论。 248

Graybill (1961, 2000) 和 Rencher and Schaalje (2008) 对线性模型进行了综述。McCullagh and Nelder (1989) 和 Firth (1991) 讨论了广义线性模型。Kutner, Nachtsheim, Neter, and Li (2004) 对线性和广义线性模型做了一个较全面的综述，包括这些模型在实验设计中的应用。Chambers and Hastie (1992) 和 Venables and Ripley (2002) 对 R 方法应用于评价线性和广义线性模型进行了综述。

Christensen (1997) 和 Hosmer, Lemeshow, and Sturdivant (2013) 对逻辑回归进行了讨论。Fawcett (2003) 和 Sing et al. (2005) 对受试者工作特征（ROC）曲线做了进一步讨论。Hand (1997) 和 Kuhn and Johnson (2013) 讨论了对分类器进行评价的其他方法。

在泊松回归以及路情形分析表分析方面，可以参考的资料包括 Bishop, Fienberg, and Holland (1975), Fienberg (2007), Tang, He, and Tu (2012) 和 Agresti (2013)。Andersen, Borgan, Gill, and Keiding (1993), Le (1997), Therneau and Grambsch (2000), Harrell (2001), Nelson (2003), Hosmer, Lemeshow, and May (2013) 和 Allison (2010) 对生存数据分析进行了综述，Therneau (2014) 和 Therneau and Crowson (2014) 则提供了实现解决方案的程序。

当存在会产生较大影响的异常点或极端的观测值时，我们市场就会考虑稳健的回归方法。稳健的方法是一个较为活跃的研究领域，使用的是统计模拟工具（Fox 2002; Koller and Stahel 2011; Maronna, Martin, and Yohai 2006; Maechler 2014b; Koller 2014）。Huet et al. (2004) 和 Bates and Watts (2007) 对非线性回归进行了综述，Harrell (2001) 讨论了回归问题中的样条函数。 249

A.4 机器学习

推荐系统、协同过滤、关联规则、基于试探法的优化方法以及大量的回归、分类及聚类方法都归属在机器学习的范畴内。这些都属于数据自适应方法，也称为数据挖掘。

在第8章中对文本分类方法的基准研究应用到了多个机器学习算法，包括支持向量机、分类树及随机森林。其他用于分类的机器学习工具包括朴素贝叶斯分类器和神经网络。

机器学习方法通常比传统的线性或逻辑回归方法的性能要好很多，但是要解释清楚这些方法的工作原理却并非易事。由于这个原因，机器学习模型有时会被称为*黑盒模型*。底层算法可能产生出成千上万与训练数据相匹配的公式或节点拆分。

更多关于机器学习的算法可以通过查阅 Duda, Hart, and Stork (2001), Izenman (2008), Hastie, Tibshirani, and Friedman (2009), Kuhn and Johnson (2013), Tan, Steinbach, and Kumar (2006) 和 Murphy (2012) 得到。除了已有大量的使用R语言实现的机器学习工具之外，也有很多工具使用 Python (Pedregosa et al. 2011, Demšar and Zupan 2013) 和 Java 实现（Witten, Frank, and Hall 2011）。使用 Python 和 R 语言编写使用机器学习算法的用户图形界面可以在开源解决方案 KNIME（Berthold et al. 2007）和 Orange（Demšar and Zupan 2013）以及由 Alteryx、IBM 和 SAS 提供的商业解决方案中找到。

Hothorn et al. (2005) 对基准研究设计的原理进行了介绍，Schauerhuber et al. (2008) 则对一个针对分类方法的基准研究进行了说明。Alfons (2014a) 为基准研究提供了交叉验证工具。基准研究，又称作统计模拟或统计实验，可以与专门为这一类的研究所设计的编程软件包一起使用 (Alfons 2014b; Alfons, Templ, and Filzmoser 2014)。

Duda, Hart, and Stork (2001), Tan, Steinbach, and Kumar (2006), Hastie, Tibshirani, and Friedman (2009) 和 Rajaraman and Ullman (2012) 从机器学习的角度介绍了聚类。Everitt, Landau, Leese, and Stahl (2011) 和 Kaufman and Rousseeuw (1990) 对传统的聚类方法进行了综述。Izenman (2008) 对传统聚类、自组织映射、模糊聚类、基于模型的聚类、双聚类（块聚类）进行了综述。在机器学习文献中，聚类分析被称作为*无监督式学习*，从而与分类加以区分，而分类是*监督式学习*，使用响应变量或类的已知编码值进行引导。

250

Leisch and Gruen (2014) 描述了各种聚类算法的编程软件包。Maechler (2014a) 编程实现了 Kaufman and Rousseeuw (1990) 提出的方法，包括用于确定聚类数量的轮廓模型和可视化技术。Rousseeuw (1987) 提出了轮廓模型，Kaufman and Rousseeuw (1990) 和 Izenman (2008) 提供了更多的文档和示例。

主成份分析是另外一项无监督技术，它建立在线性代数的基础上，提供了一个降低我们感兴趣的问题中度量或可量化特征的数量的新途径。长期作为度量的专用方法以及因素分析的基础，主成份分析近年来已经被应用于潜在语义分析中，潜在语义分析是一种从多个文档语料库中识别出重要主题的技术 (Blei, Ng, and Jordan 2003; Murphy 2012; Ingersoll, Morton, and Farris 2013)。

当训练集中只有部分观测值具有编码响应时，我们采用一种*半监督式学习*的方式。监督部分中的编码观测值集合可能会比非监督部分中的非编码观测值集合更小（Liu 2011）。

从更广义的角度上去思考机器学习，我们可以将机器学习看成是人工智能的一个分支（Luger 2008; Russell and Norvig 2009）。机器学习包含生物启发式方法、遗传算法和探索法，可以用于解决复杂的优化问题、调度问题和系统设计问题（Mitchell 1996; Engelbrecht 2007; Michalawicz and Fogel 2004; Brownlee 2011）。 251

A.5 数据可视化

数据可视化对数据科学工作来说非常关键。本书中的例子展示了数据可视化在发现、诊断以及设计中的重要性。我们使用探索性数据分析（发现）和统计建模（诊断）工具。在与管理层沟通结果时，我们采用展示图形（设计）。

统计式的结果无法完整地表述出数据中的故事。为了更好地理解数据，我们必须超越数据表、回归系数以及统计测试结果。可视化工具可帮助我们从数据中学习知识。我们探索数据、发现数据中的规律、找出相关联的观测值群组以及异常的观测值或异常点。我们关注变量之间的关系，有时也会发现数据中存在的维度。

Tukey (1977) 和 Tukey and Mosteller (1977) 的经典文献对面向探索性数据分析的制图算法进行了综述。Cook (1998), Cook and Weisberg (1999) 和 Fox and Weisberg (2011) 涵盖了回归制图方面的内容。文献 Tufte (1990, 1997, 2004, 2006), Few (2009) 和 Yau (2011, 2013) 中的工作阐述了统计制图和数据可视化。Wilkinson (2005) 对用户感受与制图以及用于理解统计图形的概念性结构进行了综述，Cairo (2013) 对信息化制图进行了概括性综述。Heer, Bostock, and Ogievetsky (2010) 展示了新一代面向 Web 分布的可视化技术。处理非常大的数据集时可能需要一些特殊的方法，如部分透明度和 hexbin 绘图（Unwin, Theus, and Hofmann 2006; Carr, Lewin-Koh, and Maechler 2014; Lewin-Koh 2014）。

R 语言在数据可视化中占有特别重要的地位。Murrell (2011) 对 R 语言制图进行了综述。Sarkar (2008，2014) 讨论了 R 语言点阵制图，建立在一个早期被称作为 S-Plus Trellis™ 系统的概念结构上 (Cleveland 1993; Becker and Cleveland 1996)。Wilkinson (2005) 提出的“图形语法”方式已经在 Python 的 ggplot 软件包（Lamp 2014）和 R 语言 ggplot2 软件包（Wickham and Chang 2014）中实现，其中 R 语言的编程实例由 Chang (2013) 提供。Cairo (2013) 和 Zeileis, Hornik, and Murrell (2009, 2014) 提出了统计图形的颜色方面的建议。Ihaka et al. (2014) 展示了如何使用 R 语言通过色调、色度和亮度来设定颜色。 252

A.6 文本分析学

文本分析学建立在多个学科的基础上，包括语言学、沟通与语言艺术、实验心理学、政治话语分析、新闻、计算机科学以及统计学。同时，由于机构收集和存储文本的数量非常大，文本分析学成为预测分析学中一个重要及快速发展的领域。

我们已经讨论了 Web 爬行、抓取以及解析。通过这些过程得到的输出结果是一个文档集或文本语料库。这个文档集或语料库都是用自然语言表达的。分析文本语料库的两个主要方法是词袋方法和自然语言处理。我们对这个语料库进行更进一步的解析，以常用的格式创建出表达式、索引、密钥以及更容易使用计算机进行分析的矩阵。这种额外解析有时被称为文本注释。我们从文本中提取特征并将其用于后续的分析中。

自然语言是我们每天都在说和写的语言。然而，自然语言的处理不仅仅是收集单独的词汇这么简单。自然语言传递信息。自然语言文档包含多个段落，段落又包含多个句子，而句子又由多个单词构成。自然语言有语法规则，而表达相同的想法又有很多不同的方式，因此规则又有例外以及有关例外的规则。组合在一起的单词以及语法规则构成了文本分析学的语言基础，如图 A-2 所示。

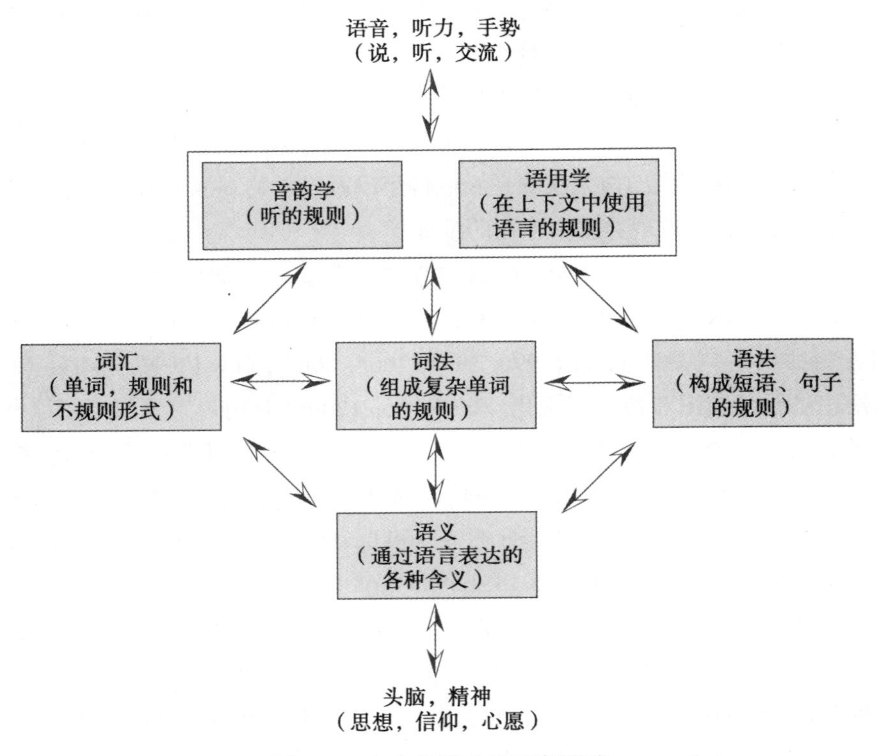

图 A-2　文本分析学的语言基础

来源：改编自 Pinker (1999)。

语言学家研究自然语言，研究组成各种言论的词汇和规则。“生成语法”是描述规则的一个通用术语；“词法”“语法”和“语义”则是更具体的术语。用于处理自然语言的计算机程序使用语言规则来模仿人类交流，并将自然语言转换为结构化文本以进行进一步分析。

自然语言处理本身就是学术研究中一个较为广泛的领域，更是计算语言学的一个重要领域。单词在句子中的位置是理解文本的关键。单词之间遵循某种顺序，先出现的单词通常比后出现的单词更重要，先出现的句子和段落也通常比后出现的句子和段落更重要。

253～254

出现在文档标题中的单词对于理解文档的内容尤为重要。某些单词出现的频率比较高，也有助于定义文档的含义。其他单词，如定冠词“the”和不定冠词“a”“an”以及很多代词和介词，出现的频率虽高但对理解文档没有什么帮助。这些停止词在分析时可以直接忽略。

文本的特征或属性通常与术语相关，而术语是由表达特定含义的一组单词构成的。构成术语的这个单词的集合与同一概念或词干相关联。词“marketer”“marketeer”和“marketing”都有一个共同的词干“market”。需要考虑句法结构，例如跟随在形容词后面的是名词，名词后面的也是名词。对于文本分析学来说，最重要的是构成术语单词的顺序。单词“New”和“York”结合在一起构成“New York”时就具有了特殊的含义。单词“financial”和“analysis”结合在一起构成“financial analysis”时也具有了特殊的含义。我们常常使用*词干提取*，也就是标识出词干，以去掉单词的后缀（有时是前缀）。一般来说，我们对自然语言文本进行解析，使其转换成为结构化文本。

在英语中，通常是主语在动词之前，宾语在动词之后。在英语中，动词的时态也很重要，例如“Daniel carries the Apple computer”与“The Apple computer is carried by Daniel”作为语句来说意思相同。“Apple computer”是主动动词“carries”的宾语，同时也是被动动词“is carried”的主语。理解这两个语句具有相同的含义是构建智能文本应用的重要组成部分。

文本分析中的一个关键步骤是创建一个术语 - 文档矩阵（有时称为词汇表）。这个数据矩阵的行对应于来源于文档集合中的单词或词干，列则对应于文档集合。在术语 - 文档矩阵的每个单元中填入的可以是一个二元标识，用于表达一个术语是否在一个文档中存在，也可以是一个频率值，用于表达一个术语在一个文档中出现的次数，还可以是一个加权频率，用于表达一个术语在一个文档中的重要性。 255

图 A-3 说明了创建一个术语 - 文档矩阵的全过程。第一个文档来自 Steven Pinker 的《Words and Rules》(1999, p.4)，第二个文档来自 Richard K. Belew 的《Finding Out About》(2000, p.73)。术语对应于出现在文档中的单词或词干。在这个实例中，每一个矩阵单元中的值都代表着一个术语在文档中出现的次数。有了词干的定义，我们将名词、动词、形容词视为类似词语。词干“combine”既代表动词“combine”，也代表名词“combination”。同样，“function”代表动词、名词以及其形容词形式“functional”。另外一种系统可以区分出讲演的不同部分，能够进行更复杂的跨文档语法搜索。在成功创建后，术语 - 文档矩阵就像是一个索引，文档标识映射到术语（关键字或词干），反之亦然。对于信息检索系统或搜索引擎，我们还可以保留术语在文档中所处特定位置的信息。

典型的文本分析学应用中的术语比文档中要多得多，造成术语 - 文档矩阵是一个长方形的稀疏矩形。为了从文本分析学应用中获取有用的结果，分析师们研究术语在文档集中的分布。出现频率很低且只出现在很少文档中的术语从术语 - 文档矩阵中直接删除，以减少矩阵的行数。

无监督式文本分析问题是指那些不对响应或者类型进行预测的问题。正如我们在电影标语实例中所展示的那样，我们的目的是为了确定数据中的共同模式或趋势。作为这项工作的一个组成部分，我们可以通过定义文本度量对语料库中的文档进行描述。

Pinker (1999)

> People do not just blurt out isolated words, but rather *combine* them into phrases and sentences, in which the meaning of the combination can be inferred from the meanings of words and the way they are arranged. We talk not merely of roses, but of the red rose, proud rose, sad rose of all my days. We can express our feelings about bread and roses, guns and roses, the War of Roses, or days of wine and roses. We can say that lovely is the rose, roses are red, or a rose is a rose is a rose When we combine words, their arrangement is crucial: *Violets are red, roses are blue*, though containing all the ingredients of the familiar verse, means something very different.

Belew (2000)

> The most frequently occurring words are not really about anything. Words like NOT, OF, THE, OR, TO, BUT, and BE obviously play an important functional role, as part of the syntactic structure of sentences, but it is hard to imagine users asking for documents about OF or about BUT. Define function words to be those that have only a syntactic function, for example, OF, THE, BUT, and distinguish them from content words, which are descriptive in the sense that we're interested in them for the indexing task.

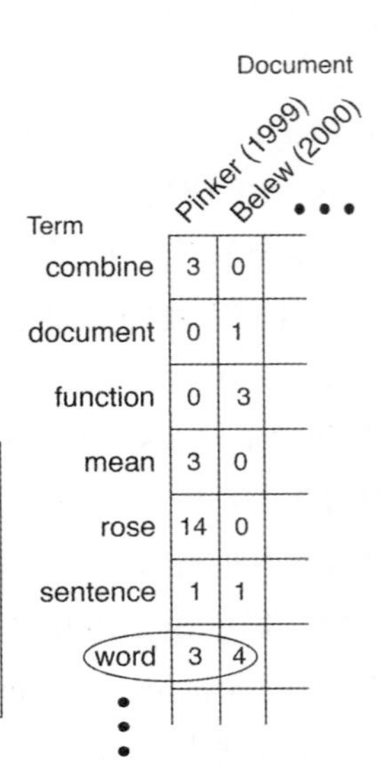

Term	Pinker (1999)	Belew (2000)	• • •
combine	3	0	
document	0	1	
function	0	3	
mean	3	0	
rose	14	0	
sentence	1	1	
word	3	4	
⋮			

图 A-3 创建一个“术语 – 文档”矩阵

来源：改编自 Miller (2005)。

监督式文本分析问题是指要对响应或者文档类型进行预测。我们在训练集上构建一个模型，然后通过测试集进行测试。文本分类问题很常见。垃圾邮件过滤作为一个分类问题一直是一个热门的话题，很多电子邮件用户都通过使用这个领域中不断演变的有效算法而受益。在信息检索场景中，搜索引擎将文档分成与搜索相关或不相关。适合于文本分类的建模技术包括逻辑回归、线性判别函数分析、分类树和支持向量机。当然，也可以采用各类组合起来的方法。

256 ~ 257

自动文本摘要是一个有助于信息管理的研究和发展领域。想象存在一个文本处理程序，它不仅具有从一个文档集中读取每个文档的功能，还可以用一两句话来概括每一个文档，也许引用的是文档本身的语句。目前的搜索引擎可以在显示文档之前就对其进行部分分析。它们为快速的信息检索自动创建摘要。它们识别出与用户请求相关联的公共的文本字符串。这些文本分析应用中使用到的信息搜索工具都被我们想当然地作为日常生活中的一部分。

具有句法处理能力的程序，如 IBM 的 Watson，让我们瞥见了文本分析智能代理时代正在到来。这些程序执行语法解析的功能，并懂得主语、动词、宾语和修饰符扮演的角色。它们还理解演讲的不同部分（名词、动词、形容词、副词）。此外，如果使用标识实体表示人、地方、事物和机构，它们还可以进行关系搜索。

读者若有兴趣了解更多的文本分析知识，可以参考 Jurafsky and Martin (2009), Weiss, Indurkhya, and Zhang (2010) 和由 Srivastava and Sahami (2009) 主编的资料。Miller (2005), Trybula (1999), Witten, Moffat, and Bell (1999), Meadow, Boyce, and Kraft (2000), Sullivan (2001), Feldman (2002) 和 Sebastiani (2002) 对文本分析进行了综述，Hausser (2001) 叙述了生成性语法及计算语言学。Charniak (1993), Manning and Schütze (1999) 和 Indurkhya and Damerau (2010) 讨论了统计语言学习和自然语言处理。

Steven Pinker (1994, 1997, 1999) 的著作深层次地剖析了语法和语言心理学。Maybury (1997) 综述了文本分析中的数据准备以及与源检测、翻译与转换、信息提取以及信息利用相关的工作。检测涉及识别相关的信息源，转换和翻译则是将一种介质或编码形式转换为另外一种。 258

Belew (2000), Meadow, Boyce, and Kraft (2000) 以及 Baeza-Yates and Ribeiro-Neto (1999) 主编的书籍叙述了信息检索方面的计算机技术，这些技术依赖于文本分类以及其他技术和算法。

作者识别多年前已经由 Mosteller and Wallace (1984) 在统计方面的文献中提出，目前仍然是一个非常活跃的研究领域（Joula 2008）。Merkl (2002) 探讨了聚类技术，也就是探讨文档的相似性并将文档组合成不同的类别。Dumais (2004) 回顾了潜在语义分析和统计方法，用于提取文档集中术语之间的关系。

计算机语言学提出的特殊专题增加了人们对与文本相关工作的深入了解。文本块（Hearst 1997）是指将文本文档自动地分割成为多个块或者单元以做进一步的分析。相邻的文本块有更大的可能具有共同的单词，因而在主题上更加相关。文本块可用于文本摘要和信息检索以及对文学和论述进行文体方面的分析（Youmans 1990; Youmans 1991）。

对于文本分析学使用 R 语言的实现，请参阅 Feinerer, Hornik, and Meyer (2008), Feinerer and Hornik (2014a) 和 Feinerer (2014)。特别值得一提的是有些书籍描述了使用 R 语言处理语言学和统计之间的重叠（Baayen 2008; Johnson 2008; Gries 2013）。在这里特别需要指出的是 Gries (2009) 的著作，它讨论的是如何使用文档语料库。Grothendieck (2014a, 2014b) 和 Wickham (2010, 2014b) 提供了文本分析的共享程序。对于有兴趣从事文本分析学的任何人，掌握 R 语言中的正则表达式编码非常有用（Friedl 2006; Wickham 2014b）。词典功能可以通过一个到 WordNet 的接口来获得（Miller 1995; Fellbaum 1998; Feinerer 2012; Feinerer and Hornik 2014b）。只是为了好玩而已，我们还有一些词云，其中包含能够将非重叠文本绘制成散点图的共享程序（Fellows 2014a; Fellows 2014b）。 259

附录 B

在线初步研究⊖

在线初步研究以网络为基础，应用都是基于 Web 浏览器的。在 Web 发展的早期，研究性应用程序都始于先使用 Web 浏览器进行访问，然后再切换到另一个应用程序或小程序（通常使用 Java 语言编写而成）。如今，应用程序在客户端主要基于浏览器和 JavaScript，同时与服务器进行交互提交数据请求。超文本传输协议（HTTP）为浏览器上的数据展示提供了格式化的编码。使用的数据模型是文档对象模型（Document Object Model，DOM）。除了 JavaScript 之外，像 Perl、PHP、Python 这样的脚本语言与 Web 服务器和数据库系统协同工作，为在线研究应用程序提供动态网页，包括在线焦点小组、焦点对话、聊天、博客、论坛、深入的访谈以及网上调查。

调查研究的方法，有时也称为初级定量研究，既容易理解也容易形成文档。Groves et al. (2009) 对调查研究进行了系统的综述，包括抽样、数据收集方式、数据质量、无应答以及分析方面的问题，Miller (2015a) 讨论了度量、项目格式以及商业研究中其他度量的尺度。

261

在本附录中，我们主要讨论在线定性研究，将其置于社会科学和商业定性研究这个更广阔的背景下。在线定性研究是一个正在发展的领域，其发展不仅仅只是涉及开展研究的一些新技术或新模式，而是会对开展定性研究的方式以及对定性研究或一般性研究进行思考的方式带来一些文化层面上的改变。

当我们在线时，往往会觉得事情发生得很快。我们能够访问网络资源。我们用计算机来收集并分析数据。我们从电子邮件和在线讨论中获取文字，将它们自动转换成用于进行文本分析的副本。原始数据就在我们的指尖上，只需要点击几下鼠标或触摸几下屏幕即可。

虽然许多研究者都认识到在线研究在时间和成本上的优势，仍有一些人在谴责该方法存在的缺陷。有些人则担心在线方法可能会取代他们认为更适合的传统方法。

传统的定性方法，如面对面访谈、焦点小组、案例分析、实地考察以及民俗，依然是研究人类和文化的重要方法。然而，在线定性研究在社会科学和商业领域中也有一定的作用。

⊖ 本附录是在获得了出版商 Research Publishers LLC 的许可后，从《Qualitative Research Online》（Miller and Walkowski 2004）这本书中摘取的。在此感谢 QualCore.com 公司的 Jeff Walkowski 在 10 多年前就向我介绍了在线定性研究。

它有什么作用呢？它又是如何更令人信服或更有效地发挥这些作用呢？通过与许多定性研究人员进行访谈，Miller and Walkowski (2004) 回答了这些问题。Dale and Abbott (2014) 则对在线定性研究进行了更多的讨论。

与传统的定性研究一样，在线定性研究也是有目的性的研究。我们观察人们的所作所为。我们通过人们输入和点击来了解他们在说什么。我们收集数据并解释这些数据。我们会尽可能地去客观评判，虽然我们也知道从数据中获得的信息往往会被我们强加给这些数据的主观范畴所影响。在线定性研究，就如同所有的研究那样，需要谨慎地去设计、分析和解析。

传统的定性研究有许多不同的类型。民族志学者是最原型的观察者，因为他们观察人类在自然和社会环境中的行为："在工作和娱乐时人们都会做些什么？他们是如何生活的？他们是如何组成不同群体的？文化的意义是什么？"通过深入访谈，民族志学者探讨人类认知的复杂性："人类如何将他们看到的和听到的转变成有意义的事物？他们相信什么？他们如何看待这个世界？标志和符号的含义是什么？"研究人员主导着焦点小组中客户感兴趣话题的讨论："选民们对当前这个州长的候选人持什么看法？有什么东西能够促使消费者使用这些产品和服务？观众对这个新的电视广告的感觉如何？" 262

究竟什么能够使研究成为定性的？定性数据出现在我们面前时是未结构化的、杂乱无章的，毫无形式可言。我们开始于发生的事件，开始于表述的词语，并且陈述出与它们相关的一个故事。当然，不同的研究者相对于同样的数据讲述出不同的故事并不奇怪。但是如果我们做好自己的工作，如果我们是称职的研究者，我们将更加接近于真实的数据，我们的故事听起来也会真实。

在线定性研究也会涉及对辅助数据进行分析，如用户组讨论或使用电子邮件通信的计算机日志。研究人员可能对人机交互和在线交流进行观察性研究。大多数在线定性研究属于初步研究，涉及为特定目的而设计定制的研究，如在线深度访谈和焦点小组。我们可以设定可用性研究，在这些研究中，用户在网站上完成被赋予的明确的任务，由观察员将他们的行为记录下来。

作为一种常用的存储与转发通信工具，电子邮件为人际沟通提供了一种便利的方法。电子邮件是异步的，也就是说发送方和接送方不需要同时在线，用户也不需要在线接收邮件。为了发送或接收电子邮件，用户需要与网络建立连接，并运行电子邮件应用程序。对于发送电子邮件，用户必须要有收件人的地址。而对于接收邮件信息，用户需要运行电子邮件程序，定位到这封邮件所在的收件区域，然后打开这封邮件，这好像是走到一个邮箱旁，伸手进去取出一封信，然后打开这封信件。许多电子邮件系统都有通知用户收到新邮件的功能。

Listserv 是一个创建自动化的多用户电子邮件组的程序。发送到一个 listserv 上的用户和特殊兴趣小组的信息会自动发送到列表中成员的邮箱中。 263

像电子邮件一样，短信也是一种人际交流工具。消息的发送者和接收者都必须有唯一的名字或地址。但是，短信提供的是同步或实时通信。

聊天室是一个组内同步或实时通信工具；它是运行在互联网中一台服务器上的一个应

用程序。为了使用这个应用程序，聊天室成员需要输入聊天室的 URL 地址。进入聊天室后，他们使用用户名或聊天室的管理员给他们建立的化名来标识自己。公共聊天室向所有能够上网的用户开放。

公告栏，也称为留言板或 Web 板，是组间异步通信工具。像聊天室一样，它们也是运行在互联网中服务器上的应用程序。访问公告栏也需要通过使用 URL，公共公告栏对所有知道相关 URL 的人开放。对于聊天室来说，组织机构可以使用公共域或商业软件工具来开发公告栏的应用程序。

与向所有能上网的人开放的公共聊天室和公告栏不同的是，大多数在线定性研究是专门为合格的受访者、招募而来以研究为目的的目标人群而设计的。招募可以通过电话、邮件、电子邮件、基于 Web 的调查表或组合起来的方式来进行。当然，也有自动的招募服务，如亚马逊的 Mechanical Turk。

多数在线定性研究都设有专业主持人。提前准备好讨论指南，了解委托客户的信息需求，主持人要时刻保持参与者的活跃度和感兴趣度，并且能够专注在选定的研究主题上。

博客是一个注有日志格式日期的在线信息网站。博客对于想要提供与特定领域相关的最新信息的个人或机构来说非常有用。博客上的文章可以被看做是个人陈述或独白，而并非是在线的交谈。作为针对具有同一个共同主题的用户社区提供的应用，让用户来张贴与这个共同主题相关的信息，博客还可以为人类分析和在线定性研究带来丰富的数据资源。

264

Lindahl and Blount (2003) 对博客的历史和技术进行了综述，其中指出使用博客的人数已从 1998 年的很少数增长为 2003 年的百万人以上，如今人数更多。

据 Wolcott (1994，1999，2001) 介绍，传统的民族志研究涉及实地考察，也就是研究现实生活中的社会和消费者行为。手机，包括融入手机的摄像机和个人数据助理，正在逐步改进传统民族志研究的面貌。传统的民族志正在慢慢变成“数字民族志”，也就是辅助以数字技术进行考察，这些技术很容易融入进这个领域。

在线民族志、虚拟民族志、或“网络志”，正如 Sherry and Kozinets (2001) 称呼的那样，是关于网络行为本身的研究。这是一种利用计算机或网络数据和物体开展的民族志研究。

已经有越来越多涉及在线交流、文化和社会理论方面的文献 (Walther 1996; Davis and Brewer 1997; Jones 1997; Johnson 1997; L′evy 1997; Holeton 1998; Jones 1999; Jonscher 1999; Smith and Kollock 1999; Herman and Swiss 2000; Mann and Stewart 2000; Miller and Slater 2000; Huberman 2001; Lévy 2001)。也有很多关于虚拟社区或通过互联网通信工具形成的个体聚合方面的研究 (Porter 1997; Markham 1998; Cherny 1999; Smith and Kollock 1999; Rheingold 2000)。Ellis, Oldridge, and Vasconcelos (2004) 对这些文献进行了有用的概述。

通常来说，一个在线社区由一群在一段较长时间内进行线上交流的人所构成。大型的在线社区，由数千名参与者组成也是非常有可能的。Werry and Mowbray (2001) 提供了众多由大型企业赞助和非盈利在线社区的例子。对于初步研究来说，在线社区通常由 50～200 人构成，这些人被招募来参加针对相关话题为期 3～12 个月的研究。

使用在线社区开展的研究可能采用很多方法，包括在线调查、实时焦点小组、公告栏等。Preece (2000), Powazak (2001), and McArthur and Bruza (2001) 对成功形成在线社区的关键因素进行了归纳，这些因素包括共享目标、知识和信念；参与者对在一段时间里能随叫随到所做出的承诺；以及进行交流、资源共享和数据存储与检索的技术。 265

采用与传统焦点小组相同的名称，在线焦点小组是由主持人引导进行的小组访谈。我们认为有两种主要类型：同步（实时）焦点小组和异步焦点小组。一个在线焦点小组就是一个在线上进行的有主持人的小组访谈，以同步或异步的方式进行。

在线和传统面对面焦点小组具有共同之处。与面对面小组一样，在线焦点小组有一个主持人、一组参与者以及一些观察员。参与者无法看到观察员。参与者在参与之前都经过资格审核，以确保他们都有相关的代表性和经验。主持人是一个训练有素的讨论领头人。主持人从自我介绍开始，向大家解释焦点小组座谈的目的，公开有观察员的事实，阐明大家都需要遵守的"基本规则"。讨论会涵盖特定的主题领域，事先由主持人和研究客户商量确定。在线焦点小组即将结束时，主持人感谢参与者付出的宝贵时间以及表达的见解，最后引导他们离开这个焦点小组。

在线与传统焦点小组之间也有不同之处。面对面讨论的主持人经常只做简短的讨论指南。很多主持人都是即兴发挥发表代表个人观点的评论。而在线工作时，讨论指南通常更加明确和完整。面对面交流时非正式或畅所欲言的表达在线上交流时需要正式化并输进电脑。由于无法亲眼见到参与者，主持人可能尝试着在引导过程中融入自己的个性。主持人最初的独白，在面对面焦点小组座谈中通常是通过口头表述，需要成为线上讨论指南中的一部分。因此，线上讨论指南通常比面对面指导要长。

面对面参与者可以看到围坐在一起的其他人的姓名牌和面孔，而在线参与者只能看到显示在计算机屏幕上的其他小组成员的名字或虚拟名字。面对面参与者可能会获得丰富的食物奖励，但是他们必须前往同一个地点才能得到这些食物。在线受访者经常在家里或者工作场所来参与。对于面对面小组来说，主持人往往会出示照片、概念板、录像带以及其他实体激励物作为焦点小组座谈会的一部分。在线焦点小组程序提供虚拟白板来显示数字激励材料，如计算机图形和图像。 266

让我们来看看在线定性研究所需要的最基本的信息系统架构。包括主持人、参与者、客户端的观察员在内的用户都要通过 Web 浏览器连接到网络。会话内容通过使用 HTML 标签捆绑进行格式化，并使用 HTTP 协议通过互联网传输。Apache 和微软 Web 服务器对互联网信息通道上动态会话内容的交换进行管理。其他信息系统的组成部分包括数据收集应用程序、数据库服务以及面向文本与数据分析的分析服务程序。

为了实现在线定性研究，系统程序员和管理员要负责对系统、网络和数据库进行管理。定性研究者（通常与主持人为同一人）分析来自于在线公告栏、同步焦点小组和调查的文字记录与数据。技术专家有时也参与主持人的工作，通过研究应用程序为他们提供软件和输入上的帮助。常见的还有一个团队的行政支持人员参与招聘、给焦点小组参与者支付报酬等工作。

我们可以对在线定性研究方法根据以下情况进行一般性的分类：（1）研究人员、访谈者、主持人是否积极地与参与者互动；（2）研究所涉及的同步或异步时间范围。

每一场访谈会都涉及一名访谈者（主持人）和一名或多名被访者（参与者）。前面几小节中所述的在线焦点小组是分组的在线访谈。也可能在网上进行个人访谈或者人们常说的“一对一”访谈。访谈者引导被访者进行与感兴趣的话题相关的讨论。在线个人访谈可以使用面向在线焦点小组的专业软件或者具有通用功能的软件，如电子邮件、即时通信软件、listservs 267 邮件列表、聊天室和公告栏。

访谈使研究人员（访谈者、主持人）直接接触研究主体（被访者、参与者）。研究人员采取积极主动的方式，提出问题、倾听和探试。同步访谈与对话实时发生，使主持人和参与者的时间完全重叠。异步访谈和对话在同一时间段发生，但是不要求主持人和参与者的时间完全重叠。Mann and Stewart (2002) 对在线访谈进行了综述。

传统实地观察也有与在线观察相似之处。在线观察对计算机和网络用户的行为进行实时观察。更为常见的是在计算机和网络用户的行为发生后再进行观察，这实际上是一种数据库研究形式。数据库里存储着与定性研究目的相关的过去的机构行为和文本文档的计算机记录。作为关于机构日常交流活动的一部分被收集，这些数据来自于同步应用，如即时通信与聊天，以及异步应用，如电子邮件、listservs 邮件列表与公告栏。

焦点对话和观察研究不要求研究人员与研究对象直接进行接触。来自于网络调查的开放式文本响应是在线定性研究的另外一个数据来源。

实时或同步焦点小组是一种广泛使用的在线方法。实时焦点小组是一个具有引导性的小组访谈，所有参与者同时登录。在一个实时焦点小组的访谈中，通常都有 6～12 名参与者和一名主持人。

实时焦点小组通常都使用专业软件。就像聊天室这样的应用程序，实时焦点小组应用允许用户使用标准的 Web 浏览器连接到同步在线讨论中。除了为在线实时讨论提供网络支持之外，实时焦点小组应用也包括为主持人和观察员准备的专用设备。

图 B-1 是一个包括一名主持人和三名参与者的实时焦点小组的时间序列线，展示了系 268 统的可用性、在线时间（连接中）以及活跃时间（读取或输入信息）。尽管参与者在这个持续 1～2 个小时的焦点小组的大部分时间里都在线，他们却并非都是积极的参与者。Andy 关注的似乎是焦点小组座谈以外的事情。Erika 一直在线而且大部分时间都很踊跃，尽管她曾因系统故障而有一段短暂时间的缺席。Nicole 频繁中断，而且提前离开。Tyler，正如我们所期望的主持人那样，除了短暂的停歇或注意力不集中外，整个活动期间都保持在线并且很活跃。

实时焦点小组的系统故障可能由计算机或网络问题导致。一个参与者有短暂的系统故障是可以接受的，就如 Erica 的时间序列线上所显示的那样。对于这个只持续 1～2 个小时的实时焦点小组，如果系统故障超过 10 或 15 分钟，或者由于服务器故障或网络故障而造成的系统故障影响了所有参与者，那将极具干扰性，会影响会话的流畅性。有时有必要忽略系统故障期间小组内传送的数据。

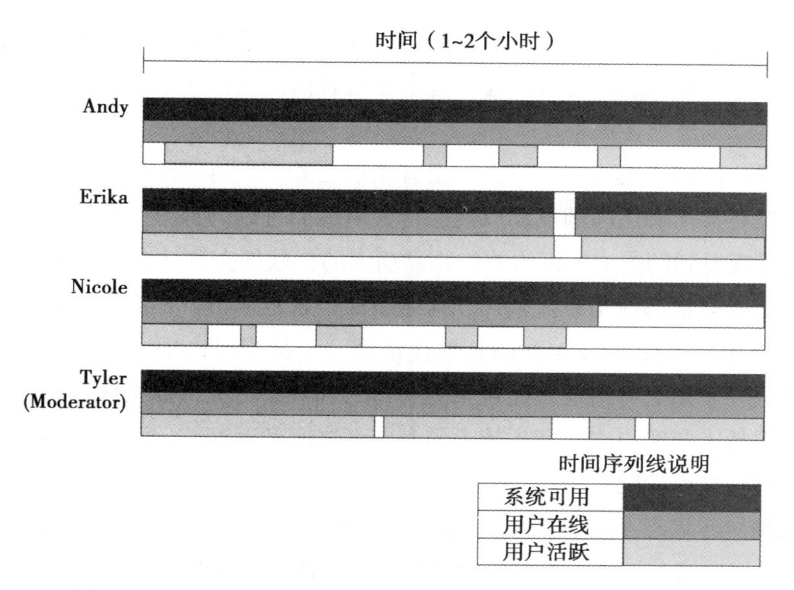

图 B-1 实时（同步）焦点小组参与者和主持人的时间序列线

基于文本的实时焦点小组的体验与面对面焦点小组有着本质上的差别。对于面对面焦点小组来说，主持人可能会鼓励参与者依次轮流发言。实时焦点小组却不是这样。在线参与者使用他们各自的计算机同时输入响应。在将响应输入后，参与者将响应发送到所有人都能看到的讨论窗口上。这样的同时响应使得讨论记录看起来比较杂乱，不像一段连续的对话。由于参与者不是轮流发言，讨论的许多条目也会出现次序混乱。实时焦点小组的批评者们觉得这点让人讨厌。

实时焦点小组的批评者将互联网看成是一个陌生的环境，在这个环境中工作以及依赖于键盘会阻碍人与人之间的交流。互联网环境是一个不属于个人的环境，会阻碍主持人深入探究受访者的真正动机和态度。研究人员所能得到的也只是简短的、随想即逝的响应，而不像面对面小组那样能获得内容丰富的响应。由于研究参与人员都是在线工作，主持人或许会错过肢体语言和面部表情表达出的线索，这些对于从讨论中得出有效的结论非常重要。

尽管受到非议，实时焦点小组的使用仍然在不断发展，这是因为面向在线研究而设计的优良的软件应用具有很多的优势。除了利用连接全球范围的互联网之外，在线焦点小组为激励显示、实验和控制这些传统焦点小组不存在的手段提供了机会。此外，由于参与者输入他们的信息，很容易就能够获得讨论的打印副本。

从主持人的角度来看，实时焦点小组包括两种并行的讨论：（1）参与者、主持人和客户都能观看的焦点小组参与者与主持人之间的讨论（2）只能由客户和主持人观看的客户与主持人之间的讨论。在线焦点小组应用使客户更加强大，因为客户可以持续不断地联系主持人，而在传统的面对面小组中，客户给主持人提出问题或建议只有有限的机会。他们可以在受访者不知道的情况在任何时候提出问题或建议。如果主持人愿意满足客户的要求，客户就会对焦点小组的方向产生巨大的影响。

269～270

设计良好的焦点小组应用程序也可以给参与者或客户发送私有信息带来很大的便利。如果某个参与者在讨论中很强势或者做了不恰当的评论，主持人可以通过发送私信来鼓励这个参与者采取更加合作和合适的行为。如果这个参与者依然我行我素，主持人可以将其从组中剔除。私信也可以发送给客户。

公告栏或异步焦点小组与实时焦点小组有相似之处。主持人使用提前准备的指南引导讨论。在讨论中存在参与者之间的互动以及主持人对他们进行的探究。参与者输入参与者、客户和主持人均可观看到的响应。因为响应是靠输入的，软件会自动生成电子副本。

异步焦点小组通常都使用专业软件。与通用的公告板或留言栏应用相似，异步焦点小组应用允许用户使用一个标准的Web浏览器连接到一个异步讨论中。但是，除了为在线讨论提供支持之外，异步焦点小组应用还包括为主持人和观察员提供的特殊工具。设计良好的应用不仅要包括主持人与客户之间和主持人与参与者之间私有交流的功能，还要有显示计算机图形和图像的白板显示功能。

主持人鼓励异步焦点小组的参与者在讨论期间每天登录一到两次。每次登录后，都希望参与者能够花上15～30分钟来阅读和回复信息。因此，异步焦点小组的参与者投入在讨论上的总时间很有可能远远超过实时焦点小组参与者投入的时间。

图B-2显示的是一个包含三名参与者和一名主持人的异步焦点小组的系统可用性、在线时间（连接）和活跃时间（阅读或输入信息）。除了一个短暂期间不可用（可能由于网络或服务器故障），焦点小组设施每一次都需要工作数天或数周。参与者登录的时间很短，并且也只是在登录期间的部分时间里很活跃（阅读和发布信息）。个人系统故障有时会导致某个参与者离线，就如Jon的时间序列线所示的那样。

271

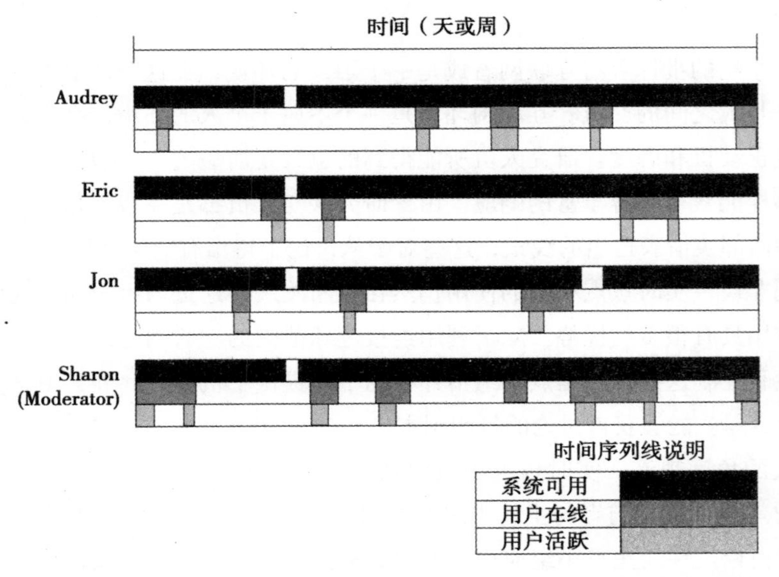

图B-2 公告发布栏（异步焦点小组）参与者和主持人的时间序列线

异步焦点小组的主持人每一天的登录次数很有可能会在两次以上，以响应参与者的信息，同时激励大家积极参与和互动。主持人还会发邮件来提醒那些不太活跃的参与者，鼓励他们主动参与到讨论中。因此，异步焦点小组中主持人所投入的总时间很有可能远远超过实时焦点小组主持人所所投入的时间。

设计良好的异步焦点小组应用程序允许主持人在预定时间发布信息。在讨论异步焦点小组时，我们使用“发布”这个词而不是“发送”，因为只有当参与者登录到这个讨论中并且提出查看的请求才能看到主持人的消息。除了讨论指南外，异步焦点小组软件还要提供了一个主持人发布消息的时间表以及白板激励。 272

异步焦点小组具有与书信交流相同的效果。当主持人发出一条消息后，便开始有了一系列相关的陈述或讨论线索。随着时间的流逝，参与者贴出他们的回复。又过了一段时间，出现了更多的回复帖。再过了一段时间，主持人发出另一条消息，开始第二个话题的讨论。就像远方朋友或恋人之间的书信，异步焦点小组的副本随着时间的推移逐渐积累，记录着一段时间内一个虚拟的关系、一次具有灵魂的会议或不同意见的共享。

由于参与者可以在任何地点或任何时间加入或继续讨论，需要一种机制来确保讨论的秩序。为异步焦点小组提供秩序保障的是围绕着讨论话题这样一种组织方式。主持人的每一条信息都会开始一个新的讨论话题。讨论话题在异步小组持续期间都是活跃的。正如我们经常做的那样，我们需要依赖参与者的智慧和善意，希望在每一个讨论话题下的消息都与该话题相关。

一般而言，异步焦点小组、实时焦点小组和在线研究具有许多共同的优点和缺点。此外，异步焦点小组对于专业或难以接近的参与小组来说有着特别的价值。忙碌的人们发现要参与实时焦点小组并非易事，因为这需要所有的参与者同时登录。异步焦点小组却使参与者能够在研究持续期间有机会在白天或晚上、工作日或周末不同的时间登录。

异步焦点小组也有利于讨论的困难话题。由于讨论可以持续数天或数周的时间，参与者不会觉得很匆忙，也不需要等其他人输入他们的响应。异步焦点小组给予参与者足够的时间，不仅能让他们描述他们的想法，也能让他们阐明是如何得出的结论。在人们讨论他们的论证过程时，我们经常会看到很长、很有深度的回复响应。

主持人和研究客户希望听到参与者用他们自己的话来陈述讨论的话题，目的是为了让响应者能够自由发挥，而不需要他们对主题的理解做出任何假设，也不用引入反映主持人或客户偏见的新的概念或词语。 273

焦点小组主持人通常都会尽量减少对焦点小组交流的干扰。他们尽量只提出一个“有魔力”的问题，就能使参与者（也许两个也许三个）进行长时间的讨论，并触及讨论中的重要话题。在接受培训的过程中，主持人会被训练去提升小组的“滚动”能力，即一个小组在不知觉的情况下能够对讨论内容和思路进行自我调节的次数。很多时候，记录的所有页中都没有显示出由主持人造成的中断。

主持人也许会说，“大家可以把我想象成是一只趴在墙上的苍蝇，只在听。你们之间要相互交谈，而不是跟我谈。但是，也就像苍蝇那样，我可能偶尔也会‘登录’到桌子上，提一个问题，或者将你们的讨论引向正确的方向。请记住，你们才是被关注的焦点，而不是我。”

尽管意图非常好，但是好的主持人绝不应该是“墙上的苍蝇”。只要他们不在讨论室里或不在计算机屏幕上标识出名字，参与者就看不见他们。实验者效应依然存在并且在定性研究中很明显，因为参与者会设法取悦主持人、访谈者和观察员。究竟有没有一种方法能够使参与者自然进行交谈、不需要主持而进行交谈呢？

没有主持人的在线交谈也司空见惯。这些交谈存在于朋友和熟人之间利用电子邮件和即时消息进行的通信。这些交谈存在于公共聊天室、listservs 邮件列表和公告栏的通信方式中。人与人进行交流，分享他们对于社会和商业研究人员感兴趣话题的意见和信息。这些自由流动的交流缺少的是结构和焦点，也缺少交流指南能够提供的质量保证。

虽然"焦点讨论"这个术语可以很容易地用于描述很多定性研究方法，包括深入面谈和具有主持人的焦点小组，同时，我们也用"焦点讨论"描述那些没有访谈者或主持人参与的在线交谈。焦点讨论是参与者之间通过一个对话指南来确定焦点的开放式对话。由于没有主持人或访谈者进行引导，焦点讨论需要参与者自我指导或自我调节，并使用会话指南对访谈进行引导。 274

焦点讨论首先最起码是一场对话，目的是通过个人表达来了解参与者。Kvale (1996) 编著的一本关于面谈的书里主张对话在定性研究中的主导地位。用他的话说，"对话是一种基本的人与人交流方式。人与人进行交谈，他们会互动，会提出问题，也会回答问题。我们通过了解他人，了解他们的经历、感受和愿望"（Kvale 1996, p.5）。

想象一下两个人进行焦点讨论，很像一段时间内相互交换的信件和信息。这个时间持续期可能很短（对于一个实时焦点小组来说，可能只持续 20～60 分钟），也可能很长（对于一个异步焦点小组来说，可能会持续数天或数周）。时间长短并不是主要的，对话的深度才是重点，也就是参与者在多大程度上互相挑战对方，以详细阐述自己的观点。

两个人之间的一段焦点讨论就好像是一场没有访谈者的深度开放式访谈。我们可以把这种访谈看成是两个并发的个人访谈，两人轮流充当访谈者和被访者。我们希望从焦点讨论得到的是一种让参与者能够畅所欲言的媒介。参与者之间会互动，然而这种互动并不受主持人的直接影响。

为了进行一段焦点讨论，研究人员需要准备一个访谈指南，其中包含一系列的激励事件和出场时间表。激励事件包括参与者说明、讨论的问题和需要看的材料。激励事件的发生可以是在预先设定的时间或者通过参与者提出请求或访谈达到的条件进行触发，如对前一个问题已经输入的响应的字数、参与者已经发送的消息数或在实时访谈中出现了冷场时。

在线技术对实现焦点讨论非常有帮助。电子邮件、即时通信、聊天室或在线焦点小组软件都可能会用到。异步焦点小组应用也可能会用到，如果这种应用具有支持按照主线组织讨论、发布文本消息、在规定的时间点上具有白板激励材料的功能的话。更好的则是使用为焦点讨论设计的专业软件。

在一场焦点讨论中，研究的参与者首先阅读访谈指南，然后再输入他们的响应。他们根据指南中列出的话题进行讨论。就像在线焦点小组那样，软件应用程序会收集并储存参与者输入的响应，从而自动生成会话记录。但是，与在线焦点小组不同的是，访谈本身没有主持人。 275

一场两个人之间的焦点讨论很可能会出现明显的角色互换情况，这是两人对话中的自然的相互作用。在每一个新的讨论话题开始时，所有参与者都有同时回答的机会。但是随着讨论的持续，轮流发言的情景就出现了。假如 Jennifer 和 Shawn 回答访谈指南上的一个问题。Jennifer 在读了 Shawn 的消息后进行回复。然后 Shawn 对 Jenifer 的消息进行回复。Jennifer 又回复 Shawn 的消息，如此循环往复，直到 Jennifer 和 Shawn 开始访谈指南上第 2 个话题的讨论。焦点讨论中的轮流发言使得对话记录读起来就像是在看一场自然的对话，井然有序而且流畅通顺。

图 B-3 显示的是同步焦点讨论的时间序列线，比如通过实时焦点小组应用、聊天室或即时通信软件方式进行的讨论。对于一个持续 20～60 分钟的同步焦点讨论，我们期望参与者大部分时间都在线并且处于活跃状态。短暂的不活跃状态有时可以进行调节，让参与者们同意做短暂的休息，就如图中 Jennifer 和 Shawn 所做的那样。图 B-4 显示的是异步焦点讨论的时间序列线，比如通过异步焦点小组、电子邮件或公告栏软件进行的讨论。

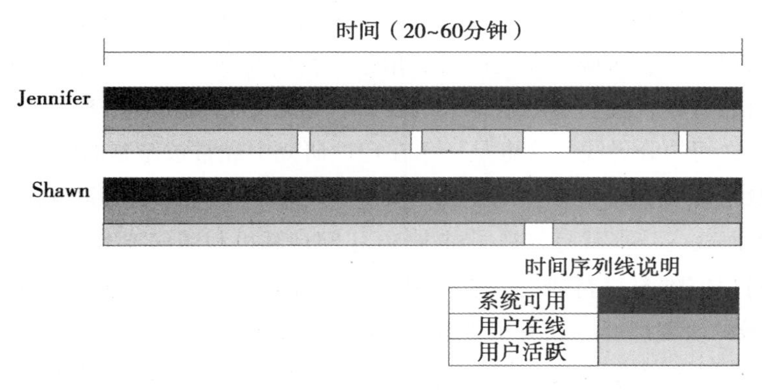

图 B-3　同步焦点讨论中参与者的时间序列线

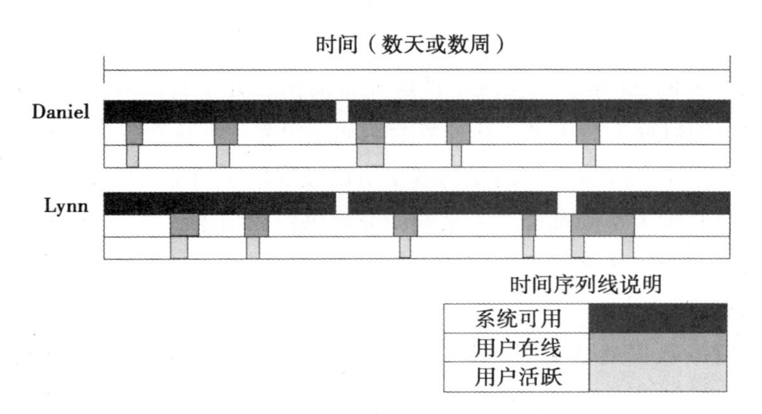

图 B-4　异步焦点讨论中参与者的时间序列线

异步焦点讨论中的参与者在研究期间可以相互协调时间表，共享自己在活动期间的在线活动计划。参与者有时也会同时在线，也可以体验同步访谈的即时性。

在线定性研究工具也可以作为通用的协作工具。它们有着在更广泛的应用领域中促进高效沟通的潜力。

我们从在线定性研究中学到的知识对一般的定性研究也有帮助。例如，一些比较成功的在线焦点讨论实验就对离线研究提出了新的方式。让我们试想一下这种可能性：两个参与者在同一个房间里打开一连串的信封，信封里面有要讨论的问题，然后他们自我调节，开始一场离线焦点讨论。

276~277

我们在此所说的定性研究是什么意思？我们从观察开始，观察在自然和社会环境中的行为，观察在工作或者娱乐中的团体，再根据我们的观察进行学习。如果与研究对象并肩站在一起，我们要尽可能地保持低调，参与到观察中但不直接牵扯观察对象。在完成观察后，我们采访，我们倾听，我们互动。我们的目标，即倾听人们用自己的言语表达出的观点，与推动者的角色相一致，即鼓励大家积极讨论但又不去影响讨论的方向。

如果我们是研究蚂蚁的生物学家，要做到低调、客观和公平很容易。但是，我们的研究对象是人。参与者可以看到我们，因此他们的表现会有所不同。多数的定性研究都涉及研究者和参与者。被访谈者知道他们正在接受访谈。焦点小组志愿者看到的只是单向的镜子。他们所期望的就是为做研究而被观察、被录音、被录像。在接受了这种社会研究的道理后，我们扮演参与式的观察者、访谈者和主持人的角色。我们互动，我们对自己的所见所闻做出各种反应，无论情愿与否，我们会影响他人的言行。这是一条被动观察与主动参与、客观与主观、科学与艺术之间的分界线。这也是一条在定性研究中经常被逾越的线。

我们既有参与者又有研究者，现在我们又有了计算机和网络。如果说定性研究人员的生活没有随着技术而改变，那就是忽视了过去50年间计算机和通信技术发展造成的影响。

我们需要对研究本身进行研究，也需要研究在线和离线模式之间的不同，就像Esipova et al. (2004) 所做的研究那样。还有很多其他问题需要我们去研究。在线定性研究的支持者们称，受访者在线时比在传统环境中更加诚实。批评者们则怀疑网络环境中的受访者可能会在事实上呈现出不同的形象，不会像在面对面环境中表现的那么诚实。我们需要去发现支持以上每一种观点的证据。

在人们开始研究的25年后，在线定性研究目前仍然处于起步阶段。还有很多工作需要愿意拥抱技术创新的研究人员去做。研究界与沟通交流界一起正在发生快速的变化。

278

在很多研究设置中，经验丰富的受访者小组已经取代了招募来的受访者。建立在自主自发的在线社区和公共聊天室的基础上，某些研究提供商提出了要发展研究受访者的网络社区以及相关的付费消费者团体来讨论确定的主题，例如他们在很长的一段时期内对某个品牌或某个产品种类的感受。

在线定性研究的数据就是文字，即文本。文字来源于消费者与研究对象、用户组讨论、在线聊天日志、电子邮件、实时焦点小组以及公告栏副本，文本来源更加广阔，它们都使用电子方式进行存储，可用于未来的研究与分析。本书中介绍的文本分析方法一般适用于对定性研究数据进行分析。

其他一些研究人员对定性数据分析方法与解释的方法进行了综述（Silverman 1993; Georgakopoulou and Goutsos 1997; Roberts 1997; Ryan and Bernard 2000; Silverman 2000a）。对社会科学家和营销研究人员来说，特别有用的还有扎根理论构建（Glaser and Strauss 1967;

Strauss and Corbin 1998; Dey 1999; Charmaz 2002）。

单词统计、自动文本摘要和文本分析的过程都在寻找单词和短语用法的规律，将单词转换成数值摘要。当我们使用数值方法来进行定性数据分析时，定性和定量研究之间的区别会变得模糊起来。Schrott and Lanoue (1994), Silverman (2000b, 2001) 和 Neuendorf (2002) 对内容分析方法进行了综述，其他研究人员讨论了文本分析的计算机辅助方法（Kelle 1995; Weitzman and Miles 1995; Roberts 1997; Stone 1997; Fielding and Lee 1998; Popping 2000; Weitzman 2000; West 2001; Seale 2002）。我们在数据科学附录 A 中介绍的文本分析学适用于初级及二次研究。

标准的研究问题依然存在，抽样就是其中之一。当我们进行初步研究时，通常要首先定义一个目标人群和一个抽样方法。我们设法获取能代表总体的样本。对于面向 Web 的二次研究，一个很不幸的事实就是抽样问题基本上不再被提起。很多二次研究使用的数据恰恰都是可用且方便的样本。我们很难基于这些样本来进行概括，因为我们不知道需要概括的总体是什么。

279

我们正处于一个前沿性研究的新时代，姑且称它为超现代的而不是落后现代的。不能一味地去责怪科学的局限性，我们应当看到涌现出的由技术支撑的科学，使网络方法渗透到不同形式的研究中，不论是定性研究还是定量研究。

有些在线研究系统依赖于文本阅读和输入。其他的有些研究方式则将文本、流式音频和媒体、电话会议功能结合起来。作为一种传输媒介，网络随着越来越宽的带宽和越来越广的分布而持续扩大。计算机系统的性能在不断提高，很多在线系统都建立在具有高度响应的客户端应用程序的基础上。

面对研究模式和方法的变化，我们应该做什么？答案是调整并适应它们。我们应该如何去应对新媒体？答案是创造性地使用它，帮助我们提高交流与研究。将前沿技术接受到我们所做的工作中去。不论是在线的还是离线的，我们都要对各种方法秉持开放接受的态度。理解了这一点，尽管已经有了很多交流与合作的方式，我们还必须持续去寻找了解彼此的更好方法。

280

附录 C

案例分析

本附录介绍预测分析学中 Web 与网络数据科学和建模技术的一些案例，就像我们在本书各章中所做的那样。案例中的数据可以从公共域或者本书的网站 http://www.ftpress.com/miller/ 获得。

C.1 电子邮件还是垃圾邮件

这个案例是关于垃圾邮件的自动检测的。垃圾邮件是未经用户许可就强行发送到用户邮箱的商业邮件和群发邮件，如产品广告、致富计划、连锁信、成人情色读物。垃圾邮件通常发送给邮件列表和新闻组里的用户。这一类型的电子邮件发送行为被认为是缺乏职业道德的，因为邮件的发送者不用承担发送消息的全部成本而且收件人并没有同意接收这种信息。

281

对电子邮件进行文本分类对商业来说非常重要。根据《信息周刊》（InformationWeek）2003 年 8 月的一篇报道，垃圾邮件泛滥的程度就像流行病蔓延一样疯狂，因为由垃圾邮件携带的病毒和蠕虫攻击了互联网上的电子邮件应用程序和服务器。虚假信息的泛滥导致许多机构被迫关闭他们的电子邮件系统。编写用于过滤或屏蔽垃圾邮件软件以及对服务器系统进行软件补丁以防止进一步安全泄露的兴趣在不断增长。

我们可以使用电子邮件的文本特性来识别垃圾邮件和正常邮件。垃圾邮件的指示符中可能含有与金钱相关的词语，也可能会使用很多的大写字母或者特殊字符来吸引读者的注意力。而正常电子邮件的指示符包含个人姓名、与工作相关的条目和与工作相关的数字，如用于业务联系的电话区号。

这个案例中的数据包括 1999 年惠普实验室收集到的 4 601 封电子邮件的描述符。在这些邮件中，有 1 813 封（39.4%）被分类为垃圾邮件。一共有 57 个潜在的解释型变量，所有的都是连续的，如表 C-1 所示。响应变量是将邮件分成垃圾邮件或正常邮件的一个二进制分类值。

改编自 Miller (2005)。据荷美尔食品公司（Hormel Foods Corporation, http://www.spam.com）称，用垃圾邮件（spam）这个词描述来路不明的商业邮件源于 Monty Python 短剧中讽刺荷美尔的肉类罐头产品 SPAM（全部字母都为大写）。在短剧中，一群维京人合唱“spam,

spam, spam”的声音完全淹没了其他的声音。用同样的类比，来路不明的商业邮件也会造成有用的电子邮件被忽略。想要了解更多的有关垃圾邮件的背景信息，请查阅 Cranor and LaMacchia (1998) 以及《信息周刊》(InformationWeek) 的在线网站 (http://information.week.com/spam)。互动营销研究组织 (http://www.imro.org) 和美国调查研究组织委员会 (http://www.casro.org) 的网站上讨论了垃圾邮件的道德问题以及将垃圾邮件应用于研究中的问题。本案例的原始数据是由 Mark Hopkins, Erik Reeber, George Forman 和 Jaap Suermondtat 在惠普 (Hewlett-Packard) 实验室形成的。George Forman 将这些数据赠给了加利福尼亚大学尔湾分校的机器学习库 (Machine Learning Repository)，并将这些数据置于公共域中。对这些数据的示例分析可以从 Izenman (2008) 和 Hastie, Tibshirani, and Friedman (2009) 获得。

表 C-1 数据编码：电子邮件还是垃圾邮件

变量名	说明
make	邮件中与 make 相对应的词或字符串所占的百分比
address	邮件中与 address 相对应的词或字符串所占的百分比
all	邮件中与 all 相对应的词或字符串所占的百分比
…	其他 45% 的变量遵循以上的结构，且变量名与单词和字符串相关联（变量名应以字母 x 开头，随后是数字） 3rd our over remove internet report addresses free business email you credit your font 000 money hp hpl george 650 lab labs telnet 857 data 415 85 technology 1999 parts pm direct cs meeting original project re edu table conference
;	邮件中与分号；相匹配的字符的百分比
(	邮件中与左括号 (相匹配的字符的百分比
[	邮件中与左方括号 [相匹配的字符的百分比
!	邮件中与感叹号！相匹配的字符的百分比
$	邮件中与美元符号 $ 相匹配的字符的百分比
#	邮件中与英镑或数字符号 # 相匹配的字的百分比
avecaprun	连续大写字母字符串的平均长度
maxcaprun	连续大写字母字符串的最大长度
totcap	邮件中大写字母的总数
class	邮件的分类（“电子邮件”或“垃圾邮件”）

这个案例中的基本问题就是建立对电子邮件进行分类的模型。很多传统的和数据自适应的方法都可以得到应用。但是哪种方法的效果最好？我们应该如何选择一个适用的模型？我们如何对性能进行评价？

282～283

C.2 开启 ToutBay 之旅

2014 年 10 月，数据科学应用的出版商和经销商 ToutBay 有限责任公司处于启动阶段，在等待它首个产品的发布。在这个日益变得数据驱动的世界里，ToutBay 的创始人 Greg Blence 和 Tom Miller 看到了数据科学即服务（DSaaS）的发展前景，数据科学即服务这个术语用于描述 ToutBay 的业务。ToutBay 的目标是要在数据科学领域创造出一个市场，出版和发行最前沿信息与竞争性情报。

本案例源于 ToutBay 网站 http://www.toutbay.com 上的信息以及 Google Analytics 报告，包括对 Scroll Depth plug-in data 进行的总结。

ToutBay 的网站 www.toutbay.com 讲述了一个在 2013 年 12 月成立的公司提供对全球的分析师、研究人员和数据科学家开发的应用进行访问的事例。这些领域专家或推销者对数据进行分析，并开发出对很多人都有用的模型。就像这个网站上一段 2 分钟的视频所称的那样，ToutBay 将提供答案的人和需要答案的人聚集在一起。这段视频还对 ToutBay 公司进行了详细介绍，解释为什么从 ToutBay 获取的信息会比从搜索引擎获取的免费信息更有价值。

目前有四个应用领域：体育、金融、市场营销和通用信息。体育推销者基于运动员和运动队的原始数据建立预测未来表现的模型。ToutBay 与这些体育推销者们进行合作，向运动员、运动队的拥有人、管理者和体育爱好者推荐这些预测模型。

金融推销者帮助个人和企业做出在何时向何处进行投资的明智选择。这些推销者具有计量经济学和时间序列分析方面的技能。他们了解市场以及预测模型。他们发现过去的趋势并对未来进行预测。

ToutBay 最早推出的金融产品之一股票投资组合构造（Stock Portfolio Constructor）源自 Ernest P. Chan 博士的工作，Ernest P. Chan 博士是定量金融领域公认的专家，已经出版过两部定量金融方面的著作（Chan 2009, 2013）。该产品的想法是让股票投资者指定他的投资目标、投资期以及投资股票的行业和在证券投资组合中所希望持有的股票数量。然后，利用当前股票价格和表现方面的信息以及特定的经济因素，Stock Portfolio Constructor 为投资者构建一个个性化的股票投资组合。它会一一列出所选择的股票以及假设每个股票的投资额相同时在投资期间的预期收益。Stock Portfolio Constructor 还显示这一股票投资组合在近年里的表现。

284

市场营销推销者扮演着类似的专家角色，在原始销售数据基础上提供关于消费者和市场的深度信息。他们不仅受过度量、统计、机器学习方面的专业培训，而且具有丰富的商业咨询经验。他们的模型在选址、产品定位、市场分割或目标营销上的结果对商业经理来说特别有吸引力。

通用科学，即第四个产品领域，涉及各行各业的科学家，目的是建立人类感兴趣的模型。ToutBay 公司致力于将科学思维和模型进行推广，而不是仅仅只发表在学术期刊上。

ToutBay 公司迄今为止最主要的公众活动就是 2014 年 6 月 30 日至 7 月 3 日举行的 R 语

言用户大会，也被称为 UseR！。这个大会在加利福尼亚大学洛杉矶分校（UCLA）举行，吸引了约 700 名科学家和软件工程师以及使用开源 R 语言（一种在统计和数据科学领域中广泛使用的语言）编写程序（脚本）的人员。ToutBay 是 UseR! 会议的赞助商之一，同时也是主要的软件开发商和出版商。

ToutBay 举办 UseR! 的目的是把公司介绍给潜在的推销者。公司的口号很简单：*你专心做研究和建模，剩下的就交给我们。我们将脚本变成产品。* ToutBay 公司的理念是数据科学家通过与 ToutBay 公司合作就可以完全专注于数据科学，ToutBay 则负责市场营销、沟通合作、销售、订单处理、物流和售后方面的事物。ToutBay 主页上有一个 *For Touts* 的页面提供详细的信息。

由于 ToutBay 完全是在线运作，它的业务依赖于一个能给访客或顾客传递明确信息的网站。成功取决于将网站访客变成在 ToutBay 建立账号的人。在信息产品上线运营之后，成功则意味着把有账号的人变成该产品的用户。 285

销售额来自于客户的订阅，由推销者设定信息产品的价格以及由 ToutBay 对产品进行在线销售并发布收取费用。在招募未来的推销者时，ToutBay 使用一则简短的广告语：*如果你是一本书的作者，你就要找到一个出版商，你还会希望这个出版商与书店合作出售你的书。但是，如果你是一个预测模型的作者呢？你会去哪里发布你的模型呢？你会去哪里出售你的模型产生的结果呢？ ToutBay 就是你应该去的地方。*

自从 2014 年 4 月网站对外开放以来，ToutBay 公司一直使用 Google Analytics 对用户流量进行跟踪。最近，该公司正在对网站访问量、页面浏览量和网站停留时间的数据进行评估。网站访问量在 UserR! 会议期间出现了微量的增长。除此之外，流量一直很有限，这是公司关注的问题的一个来源。

ToutBay 网站采用的是单页设计，主页面上的信息特别丰富，包括对公司进行介绍的一段 2 分钟的视频。单页面的网站设计比多页面设计能产生更好的性能，因为单页面方法在客户端浏览器和网站服务器之间的数据传输较少。

然而，采用单页面方法有一个难点，那就是标准的页面浏览统计提供的是一个不全面的网站使用情况。认识到这一点后，ToutBay 网站开发者使用 JavaScript 代码来检测用户在首页面上会往下深入多少。这些深入的数据都包含在网站的用户流量信息中。表 C-2 展示了评估的变量以及应用于网站数据时这些变量的定义。

表 C-2 ToutBay 开启之旅：网站数据

变　量	说　明
date	编码格式为 mm/dd/yy 的日期
sessions	会话数量（一个会话定义为一个用户停留并活跃在网站上的时间长短，如 30 分钟）
users	至少有一个会话的用户数量
new_sessions	首次访问百分比的估计值
pageviews	浏览的网页数量（包括对同一个网页的重复浏览数）
pages_per_session	一段会话期间浏览的平均网页数（包括对同一个网页的重复浏览数）

（续）

变　量	说　明
ave_session_duration	一个会话以时、分、秒记录的平均持续时间（时时 : 分分 : 秒秒）
bounce_rate	单页面的访问量（用户进入、离开相同的页面，通常是主页）
scroll_videopromo	点击 ToutBay 首页上视频介绍的用户数量
scroll_whatstoutbay	点击 ToutBay 首页上 What's ToutBay 的用户数量
scroll_howitworks	点击 ToutBay 首页上 How It Works 的用户数量
scroll_faq	点击 ToutBay 首页上 FAQ 的用户数量
scroll_latestfeeds	点击 ToutBay 首页上 Press Releases 的用户数量
chrome	源自 Google Chrome 浏览器的会话数
safari	源自 Apple Safari 浏览器的会话数
firefox	源自 Mozilla Firefox 浏览器的会话数
internet_explorer	源自微软 Internet Explorer 浏览器的会话数
windows	源自微软 Windows 操作系统的会话数
macintosh	源自苹果 Macintosh 操作系统的会话数
ios	源自苹果 iOS 操作系统的会话数
android	源自 Android 操作系统的会话数

在准备这些数据时，我们首先创建了一个将网站开发者和 ToutBay 主要人员的流量过滤后的外部流量报告段。数据集里的变量包含 Google Analytics 报告对 www.toutbay.com 从 2014 年 4 月 12 日～2014 年 9 月 19 日收集的数据，还包括来自 Scroll Depth 的统计数据，这是 Google Analytics 的一个插件，用于追踪用户向下点击的深度。Scroll Depth 特别适用于在独自的页面上放置很多信息的网站，如首页（单页面方法）。Scroll Depth 的文档可以从 http://scrolldepth.parsnip.io/ 获得。当使用 Google Analytics 时，我们无法对已经收集到的原始数据进行访问，我们只能对 Google Analytics 定义的变量和得出的统计进行访问。Google Analytics 度量（维度和指标）方面的文档可以从 https://developers.google.com/analytics/devguides/reporting/core/dimsmets 获得。

286～287

ToutBay 的创始者们希望对网站内容、结构以及网站使用情况数据的详细分析能够给开发出网站的新版本提供指南，这与公司首个产品的介绍不谋而合。

C.3 关键字游戏：道奇队与天使队

Google 为希望在 Google 网站上付费刊登广告的网站所有者提供搜索流量的估计。流量估计基于用户过去的查询行为。热门关键字会产生较高的流量估计值以及较高的报价。这些都与检测到的较高搜索频率相关。在线广告商要想使他们的广告对热门关键字的搜索进行响应，就必须支付更多的费用。关键字是一个或一组搜索查询中使用的词语，搜索引擎提供商为这些词语建立索引。

本案例中的数据是 2014 年 11 月 1 日通过 GoogleAdWords Keyword Planner 获取的。GoogleAdWords Keyword Planner 位于 https://adwords.google.com/KeywordPlanner。Google 为此工具提供了在线文档和培训，以帮助市场营销专业人员使用 Google AdWords 开展在线广告活动。在此报告的数据源自 Google AdWords Keyword Planner 的公开版。任何有 Google 电子邮件账号的人都可以免费使用这些数据和其他相关数据。

付费的搜索结果与有机搜索结果显示在同样的屏幕上，通常前者显示在屏幕的顶端或者有机搜索结果的右边。Google AdWords 采用按点击次数付费（cost-per-click, CPC）的定价模式。广告商所支付在线广告的费用取决于通过在线竞标过程确定的每次点击成本和广告获得的实际点击数（点击率）。Google AdWords Keyword Planner 提供的数据帮助广告商决定对关键字的出价并挑选广告的放置位置。

关键字流量估计反映出用户在搜索中的行为。它们表示用户在查询过程中所使用的词语和词语组合所处的相对排名。这些数据对于在线广告的规划非常有用，对网站内容设计以及改进有机搜索性能也非常有用。数据集中的关键字就是在搜索引擎提供商索引中使用的编码关键字。

我们应该期望流量估计值高的关键字会产生更激烈的竞争力，在这里，竞争力是介于0～1的一个级别值，可以通过每个关键字所收到的广告商竞标数得出。我们由此得出高的建议按点击付费的关键字对应于高的竞争力。也就是说，流量、竞争力和建议竞标价格之间成正比关系。

288

关键字数据具有很大的可变性，因此，具有相同流量估计值的关键字在竞争上存在有很大的差异以及截然不同的建议竞标价。为了从投入的广告预算中获取尽可能多的回报，在线广告商寻找价值被低估的关键字，这些关键字具有较高的流量估计值且较低的建议竞标价，因为他们意识到 Google 提供的数据不是来自于搜索中的观测数据，而是 Google 定价模型估计出的结果。

为了收集这个案例的数据，我们做了 12 项针对关键字的研究。在每一项研究中，我们填写了在线表单中的三个字段：（1）您的产品或服务；（2）您的登录页面；（3）您的产品类别。在每一项研究中，即将出售的产品或服务指定为“门票”。在一半的研究中，我们使用洛杉矶道奇队（Los Angeles Dodgers）的登录界面：www.dodgers.com。在另外一半的研究中，我们则使用了洛杉矶天使队（Los Angeles Angels）的登录页面：losangeles.angels.mlb.com。

产品或服务类型在每一次的搜索中都会变化，跨越 6 个不同的值：“艺术与娱乐”“体育娱乐”“体育赛事门票与票务服务”“体育与健身”“体育”和“棒球”。产品和服务类型按照一个层次结构来组织。“体育赛事门票与票务服务”是“体育娱乐”的一个子集，而后者又是“艺术与娱乐”的一个子集。“棒球”是“体育”的一个子集，而后者又是“体育与健身”的一个子集。在所有其他条件都相同的情况下，宽泛的产品类别与流量估计值高的关键字及高的每次点击价格相关。

美国职业棒球大联盟赛季从 4 月开始，一直延续到 9 月，但全年都在出售门票。Google 提供月流量估计和平均月流量估计。表 C-3 显示的是这 12 项研究得出的综合数据的结构。

表 C-3 面向关键字游戏的数据字典：道奇队对阵天使队

变量名	说明
group	广告关键字组名
keyword	关键字（用于查询和索引的编码字或字集）

（续）

变 量 名	说 明
traffic	平均月流量估计（点击量）
october	十月份的流量估计
november	十一月份的流量估计
december	十二月份的流量估计
january	一月份的流量估计
february	二月份的流量估计
march	三月份的流量估计
april	四月份的流量估计
may	五月份的流量估计
june	六月份的流量估计
july	七月份的流量估计
august	八月份的流量估计
september	九月份的流量估计
competition	竞争数量的级别值（范围 0～1）
cpcbid	建议的按点击付费关键字竞标价格（美元）
study	研究中使用的产品或服务类名称 艺术与娱乐 体育娱乐 体育赛事票务 体育与健身 体育 棒球
team	与登录页面网址对应的运动队名称 道奇队：www.dodgers.com 天使队：losangeles.angels.mlb.com

289～290

数据来源：https://adwords.google.com/KeywordPlanner。

C.4 安然邮件语料库与网络

安然公司（Enron Corporation）是一家总部设在得克萨斯州休斯顿的能源与期货商品服务公司。公司 1985 年开张，并迅速成长为美国十大公司之一，2000 年年中的股价高达每股 90.75 美元。然而，到了 2001 年 11 月，该公司的股价跌至每股不到 1 美元。安然公司在 2001 年 12 月 2 日申请破产，迫使美国证券交易委员会和联邦能源监管委员会对其进行调查。

安然公司的欺诈行为已经载入公共记录，使人们开始慎重思考企业会计和审计实践。安然公司的丑闻给大家上了一堂商业上的伦理课，并记载在畅销的书籍和文章中。

安然公司的大多数高管和丑闻无关，但是他们的电子邮件记录在联邦能源监管委员会公布调查结果时成为了公共记录。安然公司的案例数据以安然为中心却面向安然以外的人，我们仅仅能够看到他们与安然高管之间的通信。

McLean and Elkind (2003) 和 Eichenwald (2005) 提供了安然公司丑闻方面畅销的书籍。Tim Grieve 的沙龙（Tim Grieve's Salon）文章引用了电子邮件档案（2003）中的部分邮件，重点是 Ken Lay (1942-2006)，此人在 1985 年～2002 年大部分时间担任安然公司的首席执行官。与安然公司案件相关的原始资料可以从联邦能源监管委员会 http://www.ferc.gov/industries/electric/indus-act/wec/enron/info-release.asp 获得。由卡内基·梅隆大学 William W. Cohen 维护的安然电子邮件语料库可以通过 http://www.cs.cmu.edu/~enron/ 获得。Leskovec et al. (2009) 的显示出电子邮件网络数据的源节点和目的节点结构的数据作为斯坦福大型网络数据集的一部分可以从 http://snap.stanford.edu/data/email-Enron.html 获得。Klimmt and Yang (2004) 对安然数据进行了综述。在过去的十年中，安然数据一直是许多研究中的数据来源 (Leber 2013)，《Computational and Mathematical Organization Theory》期刊出版过一期对这些数据进行分析的专刊（Carley and Skillicorn 2005）。

作为公共域中极少的真实电子邮件数据之一，安然电子邮件档案与网络为研究文本分析学、在线通信和社交网络提供了一个重大的机遇。当前的安然邮件语料库占据了超过 2GB 的存储空间和 500 000 份文件，包含 158 位高管的文件夹以及超过 200 000 封电子邮件。电子邮件网络由超过 36 000 个节点和 183 000 条连接构成。 291

C.5 维基百科选举

维基百科的在线百科全书是面向所有人的协作式写作项目。Jimmy Wales 和 Larry Sanger 于 2001 年 1 月 15 日使用 Ward Cunningham 的 wiki 软件创办了维基百科。

维基百科网站在最初的 4 年里成长缓慢。然而，在第 5～6 年运营期间，网站规模翻了一番，条目数量从 500 000 增长到超过 1 000 000。截至 2014 年 9 月，维基百科包含 287 种语言超过 3 300 万个条目，有 4 800 多万个贡献者。

维基百科由一组选出的管理员进行维护。现有的管理员和非管理员用户进行投票选举。在任意选定时间段内的一组选票都可以用来定义一个社交网络。投票行为定义了一个有向网络中的连接，即用户 / 投票人连接到另一个用户 / 候选人。

维基百科选举是一个由 7 115 个节点（用户、投票人、候选人）和 103 689 条连接（选举）构成的网络数据集。数据囊括了维基百科的最初 7 年，即 2001 年 1 月至 2008 年 1 月，记录了网站早期的成长以及建设网站期间的用户协作。

表 C-4 列出了根据亚马逊的子公司 Alexa Internet, Inc. 给出的 2014 年 9 月世界上排名前 10 的网站。网站的排名基于页面浏览量和日访客量。维基百科排名第 6，是唯一一个不是由公司维护的网站。

显示这个社交网络中出和进结构的数据来自 Lestovec, Huttenlocher, and Kleinbert (2010a, 2010b)，目前已经成为斯坦福大型网络数据集的一部分，可以从 https://snap.stanford.edu/data/wiki-Vote.html 获得。维基百科的背景信息来源于 Wikipedia (2014c)。 292

表 C-4 2014 年 9 月排名前 10 的网站

排 名	名 称	说 明
1	Google.com	搜索引擎
2	Facebook.com	社交网络，照片共享
3	Youtube.com	在线视频
4	Yahoo.com	网络门户
5	Baidu.com	中国搜索引擎
6	Wikipedia.org	在线百科全书
7	Twitter.com	社交网络，微博
8	Amazon.com	在线零售商
9	Qq.com	中国网络门户
10	Linkedin.com	职业社交网

293 改编自 Alexa 网（2014）。

C.6 地震谈话

2011 年 8 月 23 日美国东部夏令时间下午 13:51，弗吉尼亚州路易莎县发生了 5.8 级地震。这是自 1944 年在纽约 – 安大略边界发生 5.8 级地震以来，在落基山脉（Rocky Mountains）以东发生的震级最大地震。对弗吉尼亚地震进行的地质研究表明，距震中 150 英里处（纬度 37.936111，经度 –77.933056）发生了滑坡。

鉴于地震的大小、位置和发生的时间，与美国历史上其他任何一次地震相比，更多的人对弗吉尼亚的地震有震感。大量的社会媒体活动报道此次事件，根据 Twitter 报道，仅仅在地震发生的一秒后，用户就发送了超过 5000 条的消息。

Twitter 推文的公共信息库让我们有机会分析地震当天的一个社会媒体数据样本。这些数据使用人和机器都可读的 JavaScript 对象表示法（JSON）提供。许多推文都允许我们对时间和地点进行分析。正如我们对于社交媒体文本所预料的那样，这些推文都充斥着美国口语中常见的街谈巷语、方言和各种粗暴语言。

这些地震谈话数据可从 https://github.com/maksim2042/earthquake 获得。Tsvetovat and Kouznetsov (2011) 对这些数据进行了综述。关于 2011 年 8 月 23 日地震的其他信息可以从 Wikipedia (2014a) 中获得。294

C.7 POTUS 演讲

美国总统项目（American Presidency Project）在加利福尼亚大学圣塔芭芭拉分校实施。参与该项目的研究人员维护着一个超过 100 000 份公开文档的与美国总统相关的档案。这个档案对于任何一个学习政府学或历史学的学生以及普通公民来说都是一个重要的资源。

对于这个案例，我们选择了美国总统发表的国情咨文演讲（POTUS）。这组演讲对于文本分析来说是一个丰富的语料库。

我们之所以选择国情咨文演讲有很多原因。它们与我们希望更多地了解政治进程以及美国政府在国内外事务中所扮演的角色的兴趣相关。它们跨越了若干年，有助于研究发展趋势。它们都是公开文档，不受版权限制，任何人都可以使用。最后，在美国总统项目的帮助下，这些文档的文本可以很容易地在线获取。

我们选择将国情咨文演讲用明文的方式存储，其中每一个文档对应着每一次演讲。文档名的编码使访问文档的元数据更加容易。文件名的首字母用于标注总统所属的政党（“D”代表民主党，“R”代表共和党）。元数据项之间使用下划线分隔符，用于区分发表演讲的方式（“O”代表“口头演讲”，“W”代表书面演讲），每个总统都使用唯一的名称进行编码，发表演讲的年份也如此。例如，文档 R_O_BushGW_2008.txt 表示共和党总统乔治 W. 布什在 2008 年发表的一个口头演讲。

美国宪法规定总统每年都必须向国会提供一篇演讲。因此，国情咨文演讲代表了从 1790 年联邦成立到现在的政治记录。这个案例的目的是为了用一种易于输入到文本分析程序中的方式提供一个完整的语料库。

本案例中的数据来自加利福尼亚大学圣塔芭芭拉分校的美国总统项目（American Presidency Project）。http://www.presidency.ucsb.edu/ 网站提供这些数据。关于这些文件的其他信息，可以参考 Gerhard Peters and John T. Woolley（Peters and Woolley 2014）撰写的报告。

295

C.8 微软网站的匿名数据

1998 年 2 月的某一周，研究人员对微软网站 www.microsoft.com 上的服务器日志进行采样与解析。样本中有 37 711 个用户，这个案例称为“匿名”是因为这些数据文件中没有任何用户的个人身份信息。

通常，我们是将网站节点认为是网页，但在这项研究中却是将网站区域作为节点。对用户浏览页面的请求进行的分类要反映出对微软网站中访问的区域。根据名称和数字确定了 294 个不同的区域。

为了对预测模型进行诚实的评估，我们将微软用户数据划分成训练集和测试集，用于训练的文件中有 32 711 个用户，而用于测试的文件中有 5 000 个用户。这些文件中的数据行用于显示用户识别号以及在这个星期的研究时间内访问的网站区域。另外一份文件用于显示区域识别号、网站区域的目录名以及对区域的说明。

对微软数据的分析始于查看被访问的区域以及对用户行为特征的分析。而交界区域作为区域节点之间的链接，通过对共享区域构建一条连接区域节点的链接，网络区域结构可以通过用户行为来收集。

与基于这些数据所发布的其他公开研究成果相一致，一个更宏伟的目标是基于过去访问过的区域来预测用户将要访问网站的哪些区域。对于这种类型的预测模型，我们使用关联规则分析方法和推荐系统。

本案例中的数据来自加利福尼亚大学尔湾分校机器学习与智能系统的机器学习数据库（Bache and Lichman 2013）。数据集和文档可以从 http://archive.ics.uci.edu/ml/datasets/Anonymous+Microsoft+Web+Data 获得。原始数据是由位于华盛顿州雷德蒙德市微软研究院（Microsoft Research）的 Jack S. Breese, David Heckerman 和 Carl M. Kadie 创建，已经被用于对模型进行测试，目的是基于用户访问过的其他区域的数据来预测用户将会访问的网站区域（Breese, Heckerman, and Kadie 1998）。

296

附录 D
代码与共享程序

“愿力量与你同在。”

Han Solo（Harrison Ford 饰）
电影《星球大战》(1977 年)

本附录提供了本书和预测分析学中的建模技术使用的代码与共享程序。这些代码和相应的数据集均为开源资源，可以从本书出版商的网站 http://www.ftpress.com/miller 下载。

对于实践数据科学家来说，掌握多种语言将非常具有优势。作为一种通用语言，Python 为网络和文本处理提供了丰富的工具选择。R 语言则在建模和统计图形方面非常强大。如果既懂 Python 也懂 R 语言，我们就可以对使用这两种语言开发的脚本进行对比性测试。

使用 Python 研究数据科学意味着从 GitHub 中收集程序和文档，并与 PyCon、SciPy 及 PyData 这些机构保持接触。在撰写本书时，Python 编程环境已经包含了超过 15 000 个软件包。目前还有很多大型的开源开发者社区正在开发科学编程软件包，如 NumPy、SciPy 和 SciKit-Learn。还有 Python 软件基金会在对代码开发和培训提供支持。有益于学习 Python 的参考资料包括 Chun (2007), Beazley (2009) 和 Beazley and Jones (2013)。

297

根据 IEEE Spectrum 的统计 (Cass 2014)，在排名前五的编程语言中，有四种语言需要编译周期，两种语言由大公司控制，另外三种语言以字母“C”开头。这就使 Python 稳操胜券。Python 在本书中得分为“A”。作为一种通用脚本语言，与其他专用语言相比，我能用 Python 完成更多的任务，而且能够完成得更快。Python 提供了一组丰富的数据结构。它是一种面向对象、自动形成文档、开源且免费的全球最流行的脚本语言。作为另外两种开源的脚本语言，JavaScript 在 IEEE 列表中排名第六，而 R 语言则排名第十三。

使用 R 语言研究数据科学意味着要去寻找由 Comprehensive R Archive Network(CRAN) 发布的任务视图。我们首先去 RForge 和 GitHub。我们阅读《The R Journal》和《Journal of Statistical Software》上的软件包插图及论文。在撰写本书时，R 语言的编程环境已经包含超过 5 000 个软件包，其中很多包侧重于建模方法。学习 R 语言的参考资料包括 Matloff (2011) 和 Lander (2014)。尽管 Venables and Ripley (2002) 考虑用 S/S+ 语言编写，它仍然是统计编程社区中一部非常关键的参考文献。

Web 与网络数据科学领域的研究人员有足够的理由要学习三种语言，他们也应该这么去做。首先要掌握的是 Python，其次是 R 语言，如果时间允许的话，最后还要学一点 JavaScript。如果还有一件我们已经在本书中陈述的事，那么就是：信息世界属于那些会编程的人。因此，持续编写代码吧。

代码 D-1 是一个评估二元分类器的 Python 共享程序。代码 D-2 和代码 D-3 是编写情感分析的 Python 程序时使用到的共享程序。相应的 R 语言共享程序则在代码 D-7 中。

代码 D-4 和代码 D-5 是 grid 和 ggplot2 图形中使用的 R 语言共享程序。代码 D-4 中的分离 - 绘制共享程序在很多章都应用到了，以对 R 语言中的 ggplot2 图形对象的边缘和多点布局进行渲染。

298

用于计算用经度和纬度表示的点与点之间距离的 R 语言共享程序在空间数据分析中以及在显示在线流量之源中非常有用，在这里 IP 地址用于提供地理位置信息。这些共享程序如代码 D-6 所示。

图形用户界面和集成开发环境在开发终端用户应用程序时对编程人员非常有帮助。IPython、KNIME (Berthold et al. 2007) 及 Orange (Demšar and Zupan 2013) 这些开源系统或者由 Continuum Analytics、Enthought、微软及 IBM 开发的商业系统提供使用 Python 构建解决方案的工具。KNIME 及 RStudio 这些开源系统或者由 Alteryx、IBM、微软及 SAS 开发的商业系统提供使用 R 语言构建解决方案的工具。

299

代码 D-1　评估一个二进制分类器的预测精度（Python）

```
# Evaluating Predictive Accuracy of a Binary Classifier (Python)

def evaluate_classifier(predicted, observed):
    import pandas as pd
    if(len(predicted) != len(observed)):
        print('\nevaluate_classifier error:',\
             ' predicted and observed must be the same length\n')
        return(None)
    if(len(set(predicted)) != 2):
        print('\nevaluate_classifier error:',\
              ' predicted must be binary\n')
        return(None)
    if(len(set(observed)) != 2):
        print('\nevaluate_classifier error:',\
              ' observed must be binary\n')
        return(None)

    predicted_data = predicted
    observed_data = observed
    input_data = {'predicted': predicted_data,'observed':observed_data}
    input_data_frame = pd.DataFrame(input_data)

    cmat = pd.crosstab(input_data_frame['predicted'],\
        input_data_frame['observed'])
    a = float(cmat.ix[0,0])
    b = float(cmat.ix[0,1])
    c = float(cmat.ix[1,0])
    d = float(cmat.ix[1,1])
    n = a + b + c + d
```

```
    predictive_accuracy = (a + d)/n
    true_positive_rate = a / (a + c)
    false_positive_rate = b / (b + d)
    precision = a / (a + b)
    specificity = 1 - false_positive_rate
    expected_accuracy = (((a + b)*(a + c)) + ((b + d)*(c + d)))/(n * n)
    kappa = (predictive_accuracy - expected_accuracy)\
      /(1 - expected_accuracy)
    return(a, b, c, d, predictive_accuracy, true_positive_rate, specificity,\
        false_positive_rate, precision, expected_accuracy, kappa)
```

300

代码 D-2 面向情感分析的文本度量 (Python)

```
# Text Measures for Sentiment Analysis (Python)

def get_text_measures(corpus):
    # individually score each of the twenty-five selected positive words
    # for each document in the working corpus... providing new text measures
    # initialize the list structures for each positive word
    beautiful = []; best =  []; better =  []; classic = [];
    enjoy = []; enough = []; entertaining = []; excellent = [];
    fans =  []; fun =  []; good =  []; great = []; interesting =  [];
    like =  []; love =  []; nice = []; perfect =  []; pretty =  [];
    right =  []; top = []; well = [];
    won = []; wonderful = [];  work = []; worth = []
    # initialize the list structures for each negative word
    bad = []; boring = [];    creepy = [];    dark = [];
    dead = []; death =  []; evil = []; fear = [];
    funny = []; hard = []; kill = []; killed = [];
    lack =  []; lost =  []; mystery = []; plot = [];
    poor = []; problem = []; sad = []; scary = [];
    slow = []; terrible = []; waste = []; worst = []; wrong  = []

    for text in corpus:
        beautiful.append(len([w for w in text.split() if w == 'beautiful']))
        best.append(len([w for w in text.split() if w == 'best']))
        better.append(len([w for w in text.split() if w == 'better']))
        classic.append(len([w for w in text.split() if w == 'classic']))

        enjoy.append(len([w for w in text.split() if w == 'enjoy']))
        enough.append(len([w for w in text.split() if w == 'enough']))
        entertaining.append(len([w for w in text.split() if w == 'entertaining']))
        excellent.append(len([w for w in text.split() if w == 'excellent']))

        fans.append(len([w for w in text.split() if w == 'fans']))
        fun.append(len([w for w in text.split() if w == 'fun']))
        good.append(len([w for w in text.split() if w == 'good']))
        great.append(len([w for w in text.split() if w == 'great']))

        interesting.append(len([w for w in text.split() if w == 'interesting']))
        like.append(len([w for w in text.split() if w == 'like']))
        love.append(len([w for w in text.split() if w == 'love']))
        nice.append(len([w for w in text.split() if w == 'nice']))

        perfect.append(len([w for w in text.split() if w == 'perfect']))
        pretty.append(len([w for w in text.split() if w == 'pretty']))
        right.append(len([w for w in text.split() if w == 'right']))
        top.append(len([w for w in text.split() if w == 'top']))

        well.append(len([w for w in text.split() if w == 'well']))
        won.append(len([w for w in text.split() if w == 'won']))
        wonderful.append(len([w for w in text.split() if w == 'wonderful']))
        work.append(len([w for w in text.split() if w == 'work']))
        worth.append(len([w for w in text.split() if w == 'worth']))
```

301

```
    # individually score each of the twenty-five selected negative words
    # for each document in the working corpus... poviding new text measures

        bad.append(len([w for w in text.split() if w == 'bad']))
        boring.append(len([w for w in text.split() if w == 'boring']))
        creepy.append(len([w for w in text.split() if w == 'creepy']))
        dark.append(len([w for w in text.split() if w == 'dark']))

        dead.append(len([w for w in text.split() if w == 'dead']))
        death.append(len([w for w in text.split() if w == 'death']))
        evil.append(len([w for w in text.split() if w == 'evil']))
        fear.append(len([w for w in text.split() if w == 'fear']))

        funny.append(len([w for w in text.split() if w == 'funny']))
        hard.append(len([w for w in text.split() if w == 'hard']))
        kill.append(len([w for w in text.split() if w == 'kill']))
        killed.append(len([w for w in text.split() if w == 'killed']))

        lack.append(len([w for w in text.split() if w == 'lack']))
        lost.append(len([w for w in text.split() if w == 'lost']))
        mystery.append(len([w for w in text.split() if w == 'mystery']))
        plot.append(len([w for w in text.split() if w == 'plot']))

        poor.append(len([w for w in text.split() if w == 'poor']))
        problem.append(len([w for w in text.split() if w == 'problem']))
        sad.append(len([w for w in text.split() if w == 'sad']))
        scary.append(len([w for w in text.split() if w == 'scary']))

        slow.append(len([w for w in text.split() if w == 'slow']))
        terrible.append(len([w for w in text.split() if w == 'terrible']))
        waste.append(len([w for w in text.split() if w == 'waste']))
        worst.append(len([w for w in text.split() if w == 'worst']))
        wrong.append(len([w for w in text.split() if w == 'wrong']))

    # creat dictionary data structure as a preliminary
    # to creating the data frame for the fifty text measures
    add_corpus_data = {'beautiful':beautiful,'best':best,'better':better,\
        'classic':classic, 'enjoy':enjoy, 'enough':enough,\
        'entertaining':entertaining, 'excellent':excellent,\
        'fans':fans, 'fun':fun, 'good':good, 'great':great,\
        'interesting':interesting, 'like':like, 'love':love, 'nice':nice,\
        'perfect':perfect, 'pretty':pretty, 'right':right, 'top':top,\
        'well':well, 'won':won, 'wonderful':wonderful, 'work':work,\
        'worth':worth,'bad':bad, 'boring':boring, 'creepy':creepy,\
        'dark':dark, 'dead':dead, 'death':death, 'evil':evil, 'fear':fear,\
        'funny':funny,'hard':hard, 'kill':kill, 'killed':killed, 'lack':lack,\
        'lost':lost, 'mystery':mystery, 'plot':plot,'poor':poor,\
        'problem':problem, 'sad':sad, 'scary':scary, 'slow':slow,\
        'terrible':terrible, 'waste':waste, 'worst':worst, 'wrong':wrong}

    return(add_corpus_data)
```

302

代码 D-3　情感的总体评分（Python）

```
# Summative Scoring of Sentiment (Python)

def get_summative_scores(corpus):
    # individually score each of the positive and negative words/items
    # for each document in the working corpus... providing a summative score
    summative_score = []  # intialize list for summative scores
    for text in corpus:
        score = 0  # initialize for individual document
        # for each document in the working corpus...
        # individually score each of the eight selected positive words
```

```
        if (len([w for w in text.split() if w == 'beautiful']) > 0):
            score = score +1
        if (len([w for w in text.split() if w == 'best']) > 0):
            score = score +1
        if (len([w for w in text.split() if w == 'classic']) > 0):
            score = score +1
        if (len([w for w in text.split() if w == 'excellent']) > 0):
            score = score +1
        if (len([w for w in text.split() if w == 'great']) > 0):
            score = score +1
        if (len([w for w in text.split() if w == 'perfect']) > 0):
            score = score +1
        if (len([w for w in text.split() if w == 'well']) > 0):
            score = score +1
        if (len([w for w in text.split() if w == 'wonderful']) > 0):
            score = score +1
    # individually score each of the ten selected negative words
        if (len([w for w in text.split() if w == 'bad']) > 0):
            score = score -1
        if (len([w for w in text.split() if w == 'boring']) > 0):
            score = score -1
        if (len([w for w in text.split() if w == 'funny']) > 0):
            score = score -1
        if (len([w for w in text.split() if w == 'lack']) > 0):
            score = score -1
        if (len([w for w in text.split() if w == 'plot']) > 0):
            score = score -1
        if (len([w for w in text.split() if w == 'poor']) > 0):
            score = score -1
        if (len([w for w in text.split() if w == 'problem']) > 0):
            score = score -1
        if (len([w for w in text.split() if w == 'terrible']) > 0):
            score = score -1
        if (len([w for w in text.split() if w == 'waste']) > 0):
            score = score -1
        if (len([w for w in text.split() if w == 'worst']) > 0):
            score = score -1
        summative_score.append(score)
    summative_score_data = {'summative_score': summative_score}
    return(summative_score_data)
```

303

代码 D-4 分割绘制共享程序（R 语言）

```
# Split-Plotting Utilities with grid Graphics (R)

library(grid)  # grid graphics foundation of split-plotting utilities

# functions used with ggplot2 graphics to split the plotting region
# to set margins and to plot more than one ggplot object on one page/screen

vplayout <- function(x, y)
viewport(layout.pos.row=x, layout.pos.col=y)


# grid graphics utility plots one plot with margins
ggplot.print.with.margins <- function(ggplot.object.name,left.margin.pct=10,
  right.margin.pct=10,top.margin.pct=10,bottom.margin.pct=10)
{ # begin function for printing ggplot objects with margins
  # margins expressed as percentages of total... use integers
 grid.newpage()
pushViewport(viewport(layout=grid.layout(100,100)))
print(ggplot.object.name,
  vp=vplayout((0 + top.margin.pct):(100 - bottom.margin.pct),
  (0 + left.margin.pct):(100 - right.margin.pct)))
} # end function for printing ggplot objects with margins
```

```
# grid graphics utility plots two ggplot plotting objects in one column
special.top.bottom.ggplot.print.with.margins <-
  function(ggplot.object.name,ggplot.text.tagging.object.name,
  left.margin.pct=5,right.margin.pct=5,top.margin.pct=5,
  bottom.margin.pct=5,plot.pct=80,text.tagging.pct=10) {
# begin function for printing ggplot objects with margins
# and text tagging at bottom of plot
# margins expressed as percentages of total... use integers
  if((top.margin.pct + bottom.margin.pct + plot.pct + text.tagging.pct) != 100)
    stop(paste("function special.top.bottom.ggplot.print.with.margins()",
    "execution terminated:\n   top.margin.pct + bottom.margin.pct + ",
    "plot.pct + text.tagging.pct not equal to 100 percent",sep=""))
  grid.newpage()
  pushViewport(viewport(layout=grid.layout(100,100)))
  print(ggplot.object.name,
  vp=vplayout((0 + top.margin.pct):
    (100 - (bottom.margin.pct + text.tagging.pct)),
  (0 + left.margin.pct):(100 - right.margin.pct)))

  print(ggplot.text.tagging.object.name,
    vp=vplayout((0 + (top.margin.pct + plot.pct)):(100 - bottom.margin.pct),
    (0 + left.margin.pct):(100 - right.margin.pct)))
} # end function for printing ggplot objects with margins and text tagging
```

[304]

```
# grid graphics utility plots three ggplot plotting objects in one column
three.part.ggplot.print.with.margins <- function(ggfirstplot.object.name,
ggsecondplot.object.name,
ggthirdplot.object.name,
left.margin.pct=5,right.margin.pct=5,
top.margin.pct=10,bottom.margin.pct=10,
first.plot.pct=25,second.plot.pct=25,
third.plot.pct=30) {
# function for printing ggplot objects with margins and top and bottom plots
# margins expressed as percentages of total... use integers
if((top.margin.pct + bottom.margin.pct + first.plot.pct +
  second.plot.pct  + third.plot.pct) != 100)
    stop(paste("function special.top.bottom.ggplot.print.with.margins()",
         "execution terminated:\n   top.margin.pct + bottom.margin.pct",
         "+ first.plot.pct + second.plot.pct  + third.plot.pct not equal",
         "to 100 percent",sep=""))
grid.newpage()
pushViewport(viewport(layout=grid.layout(100,100)))

print(ggfirstplot.object.name, vp=vplayout((0 + top.margin.pct):
  (100 - (second.plot.pct  + third.plot.pct + bottom.margin.pct)),
  (0 + left.margin.pct):(100 - right.margin.pct)))

print(ggsecondplot.object.name,
  vp=vplayout((0 + top.margin.pct + first.plot.pct):
  (100 - (third.plot.pct + bottom.margin.pct)),
  (0 + left.margin.pct):(100 - right.margin.pct)))

print(ggthirdplot.object.name,
  vp=vplayout((0 + top.margin.pct + first.plot.pct + second.plot.pct):
  (100 - (bottom.margin.pct)),(0 + left.margin.pct):
  (100 - right.margin.pct)))
}

# grid graphics utility plots two ggplot plotting objects in one row
# primary plot graph at left... legend at right
special.left.right.ggplot.print.with.margins <-
  function(ggplot.object.name, ggplot.text.legend.object.name,
```

```
  left.margin.pct=5, right.margin.pct=5, top.margin.pct=5,
  bottom.margin.pct=5, plot.pct=85, text.legend.pct=5) {
# begin function for printing ggplot objects with margins
# and text legend at bottom of plot
# margins expressed as percentages of total... use integers
  if((left.margin.pct + right.margin.pct + plot.pct + text.legend.pct) != 100)
    stop(paste("function special.left.right.ggplot.print.with.margins()",
    "execution terminated:\n   left.margin.pct + right.margin.pct + ",
    "plot.pct + text.legend.pct not equal to 100 percent",sep=""))
  grid.newpage()
  pushViewport(viewport(layout=grid.layout(100,100)))
  print(ggplot.object.name,
  vp=vplayout((0 + top.margin.pct):(100 - (bottom.margin.pct)),
  (0 + left.margin.pct + text.legend.pct):(100 - right.margin.pct)))
  print(ggplot.text.legend.object.name,
    vp=vplayout((0 + (top.margin.pct)):(100 - bottom.margin.pct),
    (0 + left.margin.pct + plot.pct):(100 - right.margin.pct)))
} # end function for printing ggplot objects with margins and text legend

# save split-plotting utilities for future work
save(vplayout,
  ggplot.print.with.margins,
  special.top.bottom.ggplot.print.with.margins,
  three.part.ggplot.print.with.margins,
  special.left.right.ggplot.print.with.margins,
  file="mtpa_split_plotting_utilities.Rdata")
```

305 306

代码 D-5 关联热度图共享程序（R 语言）

```
# Correlation Heat Map Utility (R)
#
# Input correlation matrix. Output heat map of correlation matrix.
# Requires R lattice package.

correlation_heat_map <- function(cormat, order_variable = NULL) {
    if (is.null(order_variable)) order_variable = rownames(cormat)[1]
    cormat_line <- cormat[order_variable, ]
    ordered_cormat <-
        cormat[names(sort(cormat_line, decreasing=TRUE)),
            names(sort(cormat_line, decreasing=FALSE))]
    x <- rep(1:nrow(ordered_cormat), times=ncol(ordered_cormat))
    y <- NULL
    for (i in 1:ncol(ordered_cormat))
        y <- c(y,rep(i,times=nrow(ordered_cormat)))
    # use fixed format 0.XXX in cells of correlation matrix
    cortext <- sprintf("%0.3f", as.numeric(ordered_cormat))
    text.data.frame <- data.frame(x, y, cortext)
    text.data.frame$cortext <- as.character(text.data.frame$cortext)
    text.data.frame$cortext <- ifelse((text.data.frame$cortext == "1.000"),
    NA,text.data.frame$cortext)  # define diagonal cells as missing
    text.data.frame <- na.omit(text.data.frame)  # diagonal cells have no text
    # determine range of correlations all positive or positive and negative
    if (min(cormat) > 0)
        setcolor_palette <- colorRampPalette(c("white", "#00BFC4"))
    if (min(cormat) < 0)
        setcolor_palette <- colorRampPalette(c("#F8766D", "white", "#00BFC4"))
    # use larger sized type for small matrices
    set_cex = 1.0
    if (nrow(ordered_cormat) <= 4) set_cex = 1.5
    print(levelplot(ordered_cormat, cuts = 25, tick.number = 9,
        col.regions = setcolor_palette,
        scales=list(tck = 0, x = list(rot=45), cex = set_cex),
        xlab = "",
        ylab = "",
```

```
        panel = function(...) {
            panel.levelplot(...)
            panel.text(text.data.frame$x, text.data.frame$y,
            labels = text.data.frame$cortext, cex = set_cex)
            }))
    }
```

307

```
save(correlation_heat_map, file = "correlation_heat_map.RData")
```

代码 D-6　空间数据分析共享程序（R 语言）

```
# Utilities for Spatial Data Analysis (R)

# user-defined function to convert degrees to radians
# needed for lat.long.distance function
degrees.to.radians <- function(x) {
  (pi/180)*x
  } # end degrees.to.radians function

# user-defined function to convert distance between two points in miles
# when the two points (a and b) are defined by longitude and latitude
lat.long.distance <- function(longitude.a,latitude.a,longitude.b,latitude.b) {
  radius.of.earth <- 24872/(2*pi)
  c <- sin((degrees.to.radians(latitude.a) -
    degrees.to.radians(latitude.b))/2)^2 +
    cos(degrees.to.radians(latitude.a)) *
    cos(degrees.to.radians(latitude.b)) *
    sin((degrees.to.radians(longitude.a) -
    degrees.to.radians(longitude.b))/2)^2
  2 * radius.of.earth * (asin(sqrt(c)))
  } # end lat.long.distance function

save(degrees.to.radians,
  lat.long.distance,
  file = "mtpa_spatial_distance_utilities.R")
```

308

代码 D-7　面向情感分析的单词评分代码（R 语言）

```
# Text Scoring Script for Sentiment Analysis (R)
# ----------------------------------------
# Word/item analysis method
# ----------------------------------------
# develop simple counts for working.corpus
# for each of the words in the sentiment list
# these new variables will be given the names of the words
# compute the number of words that match each word
amazing <- integer(length(names(working.corpus)))
beautiful <- integer(length(names(working.corpus)))
classic <- integer(length(names(working.corpus)))
enjoy <- integer(length(names(working.corpus)))
enjoyed <- integer(length(names(working.corpus)))
entertaining <- integer(length(names(working.corpus)))
excellent <- integer(length(names(working.corpus)))
fans <- integer(length(names(working.corpus)))
favorite <- integer(length(names(working.corpus)))
fine <- integer(length(names(working.corpus)))
fun <- integer(length(names(working.corpus)))
humor <- integer(length(names(working.corpus)))
lead <- integer(length(names(working.corpus)))
liked <- integer(length(names(working.corpus)))
love <- integer(length(names(working.corpus)))
loved <- integer(length(names(working.corpus)))
modern <- integer(length(names(working.corpus)))
```

```
nice <- integer(length(names(working.corpus)))
perfect <- integer(length(names(working.corpus)))
pretty <- integer(length(names(working.corpus)))
recommend <- integer(length(names(working.corpus)))
strong <- integer(length(names(working.corpus)))
top <- integer(length(names(working.corpus)))
wonderful <- integer(length(names(working.corpus)))
worth <- integer(length(names(working.corpus)))
bad <- integer(length(names(working.corpus)))
boring <- integer(length(names(working.corpus)))
cheap <- integer(length(names(working.corpus)))
creepy <- integer(length(names(working.corpus)))
dark <- integer(length(names(working.corpus)))
dead <- integer(length(names(working.corpus)))
death <- integer(length(names(working.corpus)))
evil <- integer(length(names(working.corpus)))
hard <- integer(length(names(working.corpus)))
kill <- integer(length(names(working.corpus)))
killed <- integer(length(names(working.corpus)))
lack <- integer(length(names(working.corpus)))
lost <- integer(length(names(working.corpus)))
miss <- integer(length(names(working.corpus)))
murder <- integer(length(names(working.corpus)))
mystery <- integer(length(names(working.corpus)))
plot <- integer(length(names(working.corpus)))
poor <- integer(length(names(working.corpus)))
sad <- integer(length(names(working.corpus)))
scary <- integer(length(names(working.corpus)))
slow <- integer(length(names(working.corpus)))
terrible <- integer(length(names(working.corpus)))
waste <- integer(length(names(working.corpus)))
worst <- integer(length(names(working.corpus)))
wrong <- integer(length(names(working.corpus)))

reviews.tdm <- TermDocumentMatrix(working.corpus)

for(index.for.document in seq(along=names(working.corpus))) {
  amazing[index.for.document] <-
    sum(termFreq(working.corpus[[index.for.document]],
    control = list(dictionary = "amazing")))
  beautiful[index.for.document] <-
    sum(termFreq(working.corpus[[index.for.document]],
    control = list(dictionary = "beautiful")))
  classic[index.for.document] <-
    sum(termFreq(working.corpus[[index.for.document]],
    control = list(dictionary = "classic")))
  enjoy[index.for.document] <-
    sum(termFreq(working.corpus[[index.for.document]],
    control = list(dictionary = "enjoy")))
  enjoyed[index.for.document] <-
    sum(termFreq(working.corpus[[index.for.document]],
    control = list(dictionary = "enjoyed")))
  entertaining[index.for.document] <-
    sum(termFreq(working.corpus[[index.for.document]],
    control = list(dictionary = "entertaining")))
  excellent[index.for.document] <-
    sum(termFreq(working.corpus[[index.for.document]],
    control = list(dictionary = "excellent")))
  fans[index.for.document] <-
    sum(termFreq(working.corpus[[index.for.document]],
    control = list(dictionary = "fans")))
  favorite[index.for.document] <-
    sum(termFreq(working.corpus[[index.for.document]],
    control = list(dictionary = "favorite")))
  fine[index.for.document] <-
```

309

```
    sum(termFreq(working.corpus[[index.for.document]],
    control = list(dictionary = "fine")))
  fun[index.for.document] <-
    sum(termFreq(working.corpus[[index.for.document]],
    control = list(dictionary = "fun")))
  humor[index.for.document] <-
    sum(termFreq(working.corpus[[index.for.document]],
    control = list(dictionary = "humor")))
  lead[index.for.document] <-
    sum(termFreq(working.corpus[[index.for.document]],
    control = list(dictionary = "lead")))
  liked[index.for.document] <-
    sum(termFreq(working.corpus[[index.for.document]],
    control = list(dictionary = "liked")))
  love[index.for.document] <-
    sum(termFreq(working.corpus[[index.for.document]],
    control = list(dictionary = "love")))
  loved[index.for.document] <-
    sum(termFreq(working.corpus[[index.for.document]],
    control = list(dictionary = "loved")))
  modern[index.for.document] <-
    sum(termFreq(working.corpus[[index.for.document]],
    control = list(dictionary = "modern")))
  nice[index.for.document] <-
    sum(termFreq(working.corpus[[index.for.document]],
    control = list(dictionary = "nice")))
  perfect[index.for.document] <-
    sum(termFreq(working.corpus[[index.for.document]],
    control = list(dictionary = "perfect")))
  pretty[index.for.document] <-
    sum(termFreq(working.corpus[[index.for.document]],
    control = list(dictionary = "pretty")))
  recommend[index.for.document] <-
    sum(termFreq(working.corpus[[index.for.document]],
    control = list(dictionary = "recommend")))
  strong[index.for.document] <-
    sum(termFreq(working.corpus[[index.for.document]],
    control = list(dictionary = "strong")))
  top[index.for.document] <-
    sum(termFreq(working.corpus[[index.for.document]],
    control = list(dictionary = "top")))
  wonderful[index.for.document] <-
    sum(termFreq(working.corpus[[index.for.document]],
    control = list(dictionary = "wonderful")))
  worth[index.for.document] <-
    sum(termFreq(working.corpus[[index.for.document]],
    control = list(dictionary = "worth")))
  bad[index.for.document] <-
    sum(termFreq(working.corpus[[index.for.document]],
    control = list(dictionary = "bad")))
  boring[index.for.document] <-
    sum(termFreq(working.corpus[[index.for.document]],
    control = list(dictionary = "boring")))
  cheap[index.for.document] <-
    sum(termFreq(working.corpus[[index.for.document]],
    control = list(dictionary = "cheap")))
  creepy[index.for.document] <-
    sum(termFreq(working.corpus[[index.for.document]],
    control = list(dictionary = "creepy")))
  dark[index.for.document] <-
    sum(termFreq(working.corpus[[index.for.document]],
    control = list(dictionary = "dark")))
  dead[index.for.document] <-
    sum(termFreq(working.corpus[[index.for.document]],
    control = list(dictionary = "dead")))
```

310

```
    death[index.for.document] <-
      sum(termFreq(working.corpus[[index.for.document]],
      control = list(dictionary = "death")))
    evil[index.for.document] <-
      sum(termFreq(working.corpus[[index.for.document]],
    control = list(dictionary = "evil")))
  hard[index.for.document] <-
    sum(termFreq(working.corpus[[index.for.document]],
    control = list(dictionary = "hard")))
  kill[index.for.document] <-
    sum(termFreq(working.corpus[[index.for.document]],
    control = list(dictionary = "kill")))
  killed[index.for.document] <-
    sum(termFreq(working.corpus[[index.for.document]],
    control = list(dictionary = "killed")))
  lack[index.for.document] <-
    sum(termFreq(working.corpus[[index.for.document]],
    control = list(dictionary = "lack")))
  lost[index.for.document] <-
    sum(termFreq(working.corpus[[index.for.document]],
    control = list(dictionary = "lost")))
  miss[index.for.document] <-
    sum(termFreq(working.corpus[[index.for.document]],
    control = list(dictionary = "miss")))
  murder[index.for.document] <-
    sum(termFreq(working.corpus[[index.for.document]],
    control = list(dictionary = "murder")))
  mystery[index.for.document] <-
    sum(termFreq(working.corpus[[index.for.document]],
    control = list(dictionary = "mystery")))
  plot[index.for.document] <-
    sum(termFreq(working.corpus[[index.for.document]],
    control = list(dictionary = "plot")))
  poor[index.for.document] <-
    sum(termFreq(working.corpus[[index.for.document]],
    control = list(dictionary = "poor")))
  sad[index.for.document] <-
    sum(termFreq(working.corpus[[index.for.document]],
    control = list(dictionary = "sad")))
  scary[index.for.document] <-
    sum(termFreq(working.corpus[[index.for.document]],
    control = list(dictionary = "scary")))
  slow[index.for.document] <-
    sum(termFreq(working.corpus[[index.for.document]],
    control = list(dictionary = "slow")))
  terrible[index.for.document] <-
    sum(termFreq(working.corpus[[index.for.document]],
    control = list(dictionary = "terrible")))
  waste[index.for.document] <-
    sum(termFreq(working.corpus[[index.for.document]],
    control = list(dictionary = "waste")))
  worst[index.for.document] <-
    sum(termFreq(working.corpus[[index.for.document]],
    control = list(dictionary = "worst")))
  wrong[index.for.document] <-
    sum(termFreq(working.corpus[[index.for.document]],
    control = list(dictionary = "wrong")))
  }
```

311

312

附录E

术　语　表

为了更好地理解 Web 与网络数据科学（包括在线研究方法），对各类专业术语进行学习非常重要，包括数据科学、社会科学与商业研究方法、计算机、编程语言、社交网络分析（网络科学）、通信网络以及互联网等领域。本术语表会对本书中提到的专业术语和缩写词进行定义。想要获得更多与计算机和互联网相关的术语，请参考 Hale and Scanlon (1999) 和 Downing, Covington, Covington, Barrett, and Covington (2012)。Moran and Hunt (2009) 对网站搜索性能方面的术语进行了概述。Hyslop (2010), Casciano (2011) 和 Gasston (2011; 2013) 对当前 HTML 和 CSS 的概念进行了概述。Robbins (2003) 为网页设计者介绍了一个入门的应对方法。Amor (2002) 和 Tannenbaum and Wetherall (2010) 对数据通信、互联网和万维网进行了讨论。以上这些文献是本术语表的资料来源。

万维网（World Wide Web）通过将超文本与网络技术进行嫁接而产生。作为万维网的网络基础，互联网则源于早期的广域网，如被广泛应用于国防部门、高等院校及技术研究中心的 ARPANET 和 Usenet。ARPANET 最初为国防而建立，后来成为很多高等院校的研究人员使用的广域网。Usenet 将多台计算机用简单的电话拨号方式连接起来，用于交换电子邮件和文件传输。基于标准的网络协议使互联网通信成为可能。底层协议定义了数据如何通过通信链路可靠传输，高层协议则用于各类通信程序的开发，如用于终端与计算机通信的 Telnet 协议和用于文件传输的 ftp 协议。

313

万维网的概念于 1989 年提出，而超文本的构想则可以追溯到 20 世纪 60 年代。超文本是一种文本中存在指向其他文本的链接的文本，是一种开发动态文档的技术，由用户来确定文档中不同段落文本之间的路径。我们现在使用的超文本标记语言（HTML）原来只是一种文本系统。为了使文本内容能够按某种格式显示在计算机屏幕上，开发人员在 ASCII 文本文件中添加了标记代码。当然，软件开发者也对这些系统进行了扩展，使其能够像文本一样支持图形。万维网上使用的超文本链接包括统一资源定位符（URL），它不仅可以用于定位网络中的远程文件和 Web 网页，也可以用于定位用户计算机内的本地文件和 Web 网页。可扩展标记语言（XML）实现了作为数据交换标准的承诺，为进程间的通信和 Web 服务提供了基础。要了解万维网的历史，我们推荐去看万维网的发明者 Tim Berners-Lee (2000)，有关 Web

近期的发展信息可以从万维网联盟（World Wide Web Consortium）www.w3.org 获得。

Qualitative Research Online (Miller and Walkowski 2004) 有本书中的部分术语，在此使用得到了 Research Publishers LLC 的许可。

---

adjacency matrix（邻接矩阵） 考虑一个拥有节点以及节点之间连接的网络。如果节点 i 和节点 j 相连接，则 $X_{ij}=1$，否则 $X_{ij}=0$。无向网络图会得出对称的邻接矩阵，有向网络图则得出非对称邻接矩阵。

agent（代理） 又被称为“人工代理”。是指一个能够为计算机用户或程序自动执行任务的程序，其行为就像是一个有限操作领域内的人。

ARPANET 由军队资助开发的广域网，是互联网的前身。

ASP 是应用服务提供商（Application Service Provider）的英文缩写，是一个提供外包应用服务的公司。在微软的系统中，ASP 也代表动态服务器页面（Active Server Page）。

asynchronous focus group（异步焦点小组） 详见 bulletin board（公告栏）。

bandwidth（带宽） 与信息在网络中传输速度相关联的一个表达方式。“低带宽”指较低的速度，“高带宽”则指较高的速度。

betweenness centrality（介数中心性） 通常针对节点来计算，这个指标代表某个节点位于其他节点间最短路径中的程度。此指标可反映节点的重要性。对于一个网络连接，介数性也可以用类似方法得出，它是指某个连接位于其他节点间最短路径中的程度。 314

blog（博客） “网络日志（weblog）”的简称。

boundary (of a network)[边界（网络）] 包含节点和连接的网络的覆盖度。边界定义了网络的起点和终点，即感兴趣的布局范围。

bps（比特率）。用于度量通信链路的传输速度，有时也称为“波特率（baud rate）”。

browser launch（浏览器启动） 某些虚拟设备允许协调器打开（启动）一个新的浏览器窗口。

bulletin board（公告栏） 也称为留言板或异步焦点小组。这是一个让参与者（主持人、受访者、观察者）阅读和发布消息的在线应用和特定网站。通信可以持续数天。作为一个异步小组，它不要求所有参与者同时在线。讨论则按照讨论的主题来组织。

chat room（聊天室） 互联网上的一个虚拟会议室，让参与者聚集在一起互相交流。它可以是公共的（只要空间允许，对任何人开放）或私有的（需要密码才能进入）。

client（客户端） 就是用户用于与网络或互联网进行交互的一台计算机。Web 客户端与 Web 服务器进行通信。“客户（Client）”也用于表示那些为此而开展研究的个人或机构，还表示研究与信息服务的使用者。

client-server application（客户端 - 服务器应用） 一个实现客户端与服务器通过互联网进行通信的计算机应用。

closeness centrality（接近中心性） 是一个用于表示节点重要性的度量，反映网络中的某一个节点与其他节点的接近程度，而接近程度是指两个节点之间的跳数或连接的个数。

collage（拼图） 是一种在定性研究中常用的映射技术。针对一个给定的主题，受访者被要求构建出一个由图片、文字、图画或其他艺术品构成的拼图，能够表达受访者对主题的回应。这个拼图然后作为一个工具来帮助受访者表达出他们对不同的主题、品牌和问题的看法、意见、相信程度以及态度。

cookie 存储在用户本地终端机上的数据，用于识别用户身份并保存过去交互的信息。

cost per click (CPC)[单位点击成本（CPC）] Google AdWords 中使用的定价模式。广告商只有在用户点击广告后才支付费用，单位点击成本通过一个竞标过程来确定。

content analysis（内容分析） 从定性研究的角度对文本的含义进行分析。有时也被称为主题、语义和网络文本分析，内容分析往往从简单的单词计数开始。另请参见 text measure（文本度量）。

corpus（语料库） 文本分析和文本挖掘中使用的一个文档集合（通常是一组经过整理的副本）。此单词的复数是“corpora”。

crawler (Web crawler)[爬虫（Web 爬虫）] 一个能够从 Web 中自动抓取数据的计算机程序。又称为网页蜘蛛、网络机器人或机器人。

degree（度） 每一个无向图中的节点都有度或与其他节点相连接的连接个数。这是一个反映节点重要性的指标。

315

degree distribution（度分布） 一个网络中所有节点的分布或度值集。

degree centrality（度中心性） 对于一个节点来说另外一个关于度的术语。表示面向一个网络中所有节点的平均度，是网络的一个总体统计值。

density (of a network)[密度（网络密度）] 实际连接数占所有可能连接数的比例。一个完全无连接或断开连接的节点集拥有的连接数为 0，而一个具有 n 个节点的完全连接网络中的连接数为 $\frac{n(n-1)}{2}$。如果一个网络具有 l 条连接，那么该网络的密度计算公式为 $\frac{l}{n(n-1)/2}$。

discussion thread（讨论主线） 一个讨论或会话的主线。请参见 thread（主线）。

Document Object Model (DOM) [文档对象模型（DOM）] 用于定义 Web 页面上 HTML 中节点的正确放置或层次结构（树结构）的规则。

DSL（数字用户线路）。是由电话公司和经销商提供的一种高速互联网接入方式。

dyad（二人小组）。也用于描述有一个主持人和两个参与者的焦点小组。

eigenvector centrality（特征向量中心性） 对于一个节点来说，我们认为如果它接近其他重要的节点，那么这个节点就变得同样重要，相应地，其他这些节点又接近其他重要的节点，以此类推。就像第一性原则表示的是一组变量具有共同可变性一样，特征向量中心性表示一个节点对于网络中所有节点来说的中心性程度。

ethnography（民族志） 在现实生活的场景中对社会与消费者行为进行观察性或现场研究。

E-mail（电子邮件） 使用储存与转发方法通过互联网发送消息。有时在采访和调查中使用。

emoticons（表情符号） 在线通信时使用键盘上的字符来表达情绪。大写字母可用于表达强调或表明提高音调。一些比较流行的特殊字符串如下：-) 表示微笑或愉快的响应，;-) 表示眨眼或开玩笑，:-(表示不愉快或不赞同。

focused conversation（焦点讨论） 一场有多位参与者的开放式讨论，讨论的主题在会话指南中阐明。由于在讨论过程中没有访谈者或主持人对讨论进行引导，焦点讨论由参与者们自行导向。

frame（帧） 网页设计中的一个元素。某些 Web 页面使用帧来构成，也就是用户在一个页面中可以浏览的子块，浏览的同时保持其他所有帧不变。举例来说，一些在线设施会使用多个帧，其中一个帧用于聊天信息流、一个帧作为文本输入区域，另外一个帧则用于显示激励信息。

ftp 文件传输协议，是互联网中使用的众多网络标准之一。

game theory（博弈论） 数学和经济学中涉及对弈双方的特殊用语，对弈者为达到目的而采用的知识、目的和策略。存在有竞争性和合作性博弈。

generative grammar（生成语法） 是语言学中的一个术语，指形成有意义论述的规则。是形态学、语法学和语义学中的一个通用术语。

grounded theory（扎根理论） 是一种分析和解析定性数据的方法学。理论来源于数据。分析涉及识别类型以及按照类别对数据进行编码。共同主题与概念从类型之间所观察到的关系中浮现出来。

HTML（超文本标记语言，Hypertext markup language）。是对 Web 页面进行格式化的标准编码。

316

HTTP（超文本传输协议，Hypertext transfer protocol）。Web 中客户端 – 服务器之间进行通信的主要方法。

ICQ（我找你，I Seek You），一款即时通信软件。

IMHO 在线通信者使用的速记术语，意思是“依我拙见”(in my humble opinion)。

Internet（互联网） 一个由全球众多网络相互连接而形成的公共数据通信网络。

intranet（内部网） 一个机构内部的数据通信网络。

IRC（互联网中继聊天，Internet Relay Chat），一款聊天室软件。

ISP（互联网服务提供商，Internet Services Provider）。是一个提供互联网接入服务的机构。这些机构通常也会提供运行在互联网之上的软件应用，如浏览器和搜索引擎。

IT “信息技术”(information technology) 的缩写。

Java 是一款由 Sun Microsystems 公司开发的通用编程语言，由 Oracle 公司提供更新与技术支持。它特别适合开发在线应用。

JavaScript 一种用于 Web 的面向对象的脚本语言。广泛应用于客户端事件处理。嵌入在 Web 浏览器中。在 Node.js 中作为一种单独的语言来实现，也可以应用在服务器端。

JavaScript Object Notation, JSON（JavaScript 对象表示法） JavaScript 数据结构与无序的名称 / 值对的结合。为数组、字符串、数字、逻辑值（二进制 true/false）以及特殊值

null提供存储。在制作Web页面时，可以直接在JavaScript代码中使用。

JPEG 针对图像的图形比特位映射文件格式。

kbps（每秒传送的千比特位数，kilobits per second）。是对通信链路速度的度量。

keyword（关键字） 在搜索查询时使用的一个或一组单词，由搜索引擎提供商进行索引。

LAMP 一个用于开发在线应用的公共域软件套件，指Linux操作系统、Apache Web服务器、MySQL数据库以及Perl、PHP或Python编程语言。

listserv（邮件列表） 一个自动的电子邮件群。所有参与者（受访者和主持人）的电子邮件地址都被编入一个邮件列表中。主持人提出讨论主线。任何人对消息的回复都自动发送给邮件列表中的所有人。

LOL 网络通信者使用的速记术语，意思是“大声地笑”或“许多欢笑”。

Luddite（勒德分子） 反对技术革新的人。在19世纪的英国，很多工人中一个名叫Ned Ludd的工人毁坏了代替了人工的机器，以抗议工人们失去了工作。

MEG 有主持人的电子邮件组。是电子邮件组或邮件列表组的另一个名称。

message board（留言板） 描述异步焦点小组的另外一个术语。

modem 调制器，解调器。是一种对通过信息网络传输的数据进行编码和解码的通信设备。

morphology（形态学） 语言学的一个分支。用于形成复杂词汇的规则。另请参见generative grammar（生成语法）。

317

natural language processing（自然语言处理） 使用语法规则去模仿人类交流并将自然语言转换成结构化的文本进行分析。自然语言是指能够形成有意义论述所使用的单词及规则。

netiquette（网上礼节） 适合互联网环境的社会行为。例如，输入“都使用大写字母”（USING ALL CAPITAL LETTERS）就被认为是一种不好的礼节，因为这被看作与在网上大喊大叫相当。

network（网络） 相互连接的物体，即节点和连接这些节点的边。一个数据通信网络则包含计算机系统节点和数据通信链路。这些链路是有线或无线的电子线路。

offline（离线） 对于定性研究而言，是指任何不使用互联网的研究方法，如面对面焦点小组或电话焦点小组。

online（在线） 对于定性研究而言，是指任何使用互联网的研究方法，如实时聊天、公告板（留言板）和邮件列表（MEG）。

online community（在线社区） 通过互联网进行通信的个体群组，即一组在一段时间内进行在线交流的个体。也称作虚拟社区。在初步研究中，一个在线社区可能是由50～200招募来的人组成的一个小组，参与一项针对相关主题、持续3～12个月的研究。研究可能会用到各种各样的方法，包括在线调查、实时焦点小组和公告板。

organic search（有机搜索） 也称为自然搜索，就是搜索引擎找到与用户请求最相关的结果。到目前为止，相关性中最重要的部分就是一个Web页面接收到来自其他Web页面的链接。与有机搜索不同，付费搜索的结果是广告和广告商付费的链接。

page view（页面访问） 顾名思义，是一个用户对一个 Web 网页的一次访问。一段时间内的页面访问次数是对 Web 流量的一个常用度量。

PageRank 计算在 Web 页面搜索中链接的重要性、信任度或相关性的算法。与特征向量中心性相似。

paid search（付费搜索） 搜索结果是广告和广告商付费的链接。与有机搜索或自然搜索的结果不同。

panel（座谈小组） 招募来参与一段时间各类研究的一组受访者。

parser (text parser)[解析器（文本解析器）] 一个为后续分析而对文本进行预处理的计算机程序，会将一个字符串用另外一个字符串来替换。通常使用正则表达式来实现。

Perl（“实用提取与报表语言，Practical Extraction and Report Language)”。一种在 Web 应用中广泛使用的文本处理与脚本语言。

PHP（个人主页面，Personal home page)。是与 Perl 相似的另外一种脚本语言，广泛使用在 Web 应用中。

post（帖子）（名词，帖子）一个参与者（主持人、受访者或观察者）发布在留言板讨论区里的信息。一个参与者在实时在线小组里发表的看法或提出的问题。(动词，贴）发送一个消息。

provider（提供商） 代表客户或用户提供研究和信息服务的个人或组织。 318

psychographics（心理统计学） 与心理因素相关的人口统计学。

Python 通用编程语言。一种用于 Web 应用的与 Perl 和 PHP 相似的脚本语言。

real-time focus group（实时焦点小组） 一个主持人和参与者同时在线的焦点小组。有时也称为“实时聊天”。

random network (random graph)[随机网络（随机图）] 将一组节点完全随机地使用边进行连接。

regular expressions（正则表达式） 高效搜索与文本处理所需的识别字符串的专门语法。在操作系统的 shell 搜索工具以及如 Perl、Java、JavaScript、Python 及 R 语言的各类主机语言中实现。

scraper (Web scraper)[抓取程序（Web 抓取程序）] 在 Web 中识别特定数据的一个计算机程序。应用于 Web 的自动数据获取（Web 爬虫）场景中，通常使用 XPath 语法实现。

semantics（语义） 语言学的一个分支。研究通过语言表达出的含义。另请详见 generative grammar（生成语法）。

semantic web（语义网） 是一个将基于 Web 的数据通过计算机和人类都能够使用的方式加以定义和连接的想法。这就需要使用已被广泛接受的信息编码标准，如 XML。

stemming (word stemming)[词干提取（词干）] 在文本分析学中，通过去掉词尾而得到一个词的基础部分或词干。

synchronous focus group（同步焦点小组） 请详见 real-time focus group（实时焦点

小组）。

syntax（语法） 语言学的一个分支。形成短语与语句的规则。另请详见 generative grammar（生成语法）。

TCP/IP 传输控制协议/互联网协议（Transmission Control Protocol/Internet Protocol）。互联网中计算机与计算机之间通信时使用的基本协议。

telnet 互联网中终端与计算机之间的通信协议。

text analysis（文本分析） 用于描述文本分析与解析方法的通用术语，包括内容分析、扎根理论构建和文本挖掘。

text measure（文本度量） 情感分析中对文本进行描述的属性的分值。与调查工具的相似之处在于，文本度量也可以用于评估个性化、消费者偏好以及政治观点。不同之处在于，文本度量以非结构化文本（文档、副本）作为输入数据开始，而不是从强迫回复问卷开始。

text mining（文本挖掘） 自动或半自动地处理文本。它涉及将结构强加在文本上，并从文本中抽取出有价值的信息。与数据挖掘相似，它通常与大型数据库或文档集（语料库）分析有关。

thread（主线） 一个讨论或会话的主线。与一个特定题目相关的在线消息序列。引发其他问题和陈述的讨论和陈述。是一个广泛使用在异步焦点小组（公告板或留言板）而不是在实时焦点小组中的术语。

transcript（副本） 一场讨论的文本日志，例如通过一个在线焦点小组或焦点讨论而产生。

transitivity（传递性） 对一个网络连通性的度量。也称为网络的平均聚类系数。通常根据邻接矩阵计算出完全连接三元组（三合会）的比例。三元组或三合会是指有三个节点的集，而一个封闭的三合会则是两两相连接的三个节点的集。

319

triad（三合会） 三个节点、三人组或三倍数。在定性研究中用来描述有一个主持人和三个参与者的焦点小组。

Usenet 互联网的广域网前身。

URL 统一资源定位符，是互联网信息资源的地址。

virtual community（虚拟社区） 在线社区的另一个术语。一群通过互联网进行通信的个人。

virtual facility（虚拟设施） 一个描述在线的用于模拟一个实际的物理研究设施的形容词。例如，一款在线焦点小组应用软件提供一个虚拟设施，来模仿现实生活中的一个面对面焦点小组。

Web 指万维网（World Wide Web），建立在互联网之上。

web board（Web 留言板） 异步焦点小组的另一个术语。

weblog（博客） 一个具有以标识时间的日志格式登出在线信息的网站。用于个人和组织发布有关自己的最新消息。

web presence（Web 存在状况） 一个机构、品牌、网站或登录页面在 Web 上得到认可的程度。通常所称的“搜索引擎优化”（search engine optimization, SEO）是 Web 存在状况的一部分。一种评估 Web 存在状况的方法是根据网站在有机搜索结果列表中的排名来进行评估。

web server（Web 服务器） 作为 Web 应用主机的网络计算机，如电子邮件、邮件列表以及在线焦点小组。与之相对应的是用户与网络进行交互的客户端或计算机。

web service（web 服务） Web 服务使一台计算机上的进程与另一台计算机上的进程进行交互，提供进程间的网络通信。信息使用 XML 标准相互交换。

white board（白板） 用于显示图像（视觉激励、概念、Web 页面）的框架（虚拟设施窗口）。一个白板的虚拟相似体，在会议室中使用干清除标识笔进行书写。

Wiki Web 上的多人协作表达实践。用于快速生成一个可公开编辑网站的软件。Wiki Web 页面允许任何人在任何时候使用任何设备通过一个 Web 浏览器的访问进行编辑。

WWW 万维网（World Wide Web）的缩写。

XML 可扩展标记语言（Extensible markup language）。与 HTML 的结构和外观相似，XML 却有另一个不同的目的。它是通过标记数据来描述数据特性的一种方式。

XPath 是一种特殊语法，用于对文档对象模型（Document Object Model, DOM）中的节点和属性进行遍历，从而提取出有效的数据。 320

参考文献

Ackland, R. 2013. *Web Social Science: Concepts, Data and Tools for Social Scientists in the Digital Age*. Los Angeles: Sage.

Adler, J. 2010. *R in a Nutshell: A Desktop Quick Reference*. Sebastopol, Calif.: O'Reilly.

Agrawal, R., H. Mannila, R. Srikant, H. Toivonen, and A. I. Verkamo 1996. Fast discovery of association rules. In U. M. Fayyad, G. Piatetsky-Shapiro, P. Smyth, and R. Uthurusamy (eds.), *Handbook of Data Mining and Knowledge Discovery*, Chapter 12, pp. 307–328. Menlo Park, Calif. and Cambridge, Mass.: American Association for Artificial Intelligence and MIT Press.

Agresti, A. 2013. *Categorical Data Analysis* (third ed.). New York: Wiley.

Airoldi, E. M., D. Blei, E. A. Erosheva, and S. E. Fienberg (eds.) 2014. *Handbook of Mixed Membership Models and Their Applications*. Boca Raton, Fla.: CRC Press.

Albert, J. 2009. *Bayesian Computation with R*. New York: Springer. 244

Albert, R. and A.-L. Barabási 2002, January. Statistical mechanics of complex networks. *Reviews of Modern Physics* 74(1):47–97. 99

Alexa Internet, I. 2014. The top 500 sites on the web. Retrieved from the World Wide Web, October 21, 2014, at `http://www.alexa.com/topsites`.

Alfons, A. 2014a. *cvTools: Cross-Validation Tools for Regression Models*. Comprehensive R Archive Network. 2014. `http://cran.r-project.org/web/packages/cvTools/cvTools.pdf`.

Alfons, A. 2014b. *simFrame: Simulation Framework*. Comprehensive R Archive Network. 2014. `http://cran.r-project.org/web/packages/simFrame/simFrame.pdf`. 250

Alfons, A., M. Templ, and P. Filzmoser 2014. *An Object-Oriented Framework for Statistical Simulation: The R Package simFrame*. Comprehensive R Archive Network. 2014. `http://cran.r-project.org/web/packages/simFrame/vignettes/simFrame-intro.pdf`. 250

Allemang, D. and J. Hendler 2007. *Semantic Web for the Working Ontologist: Effective Modeling in RDFS and OWL* (second ed.). Boston: Morgan Kaufmann.

Allison, P. D. 2010. *Survival Analysis Using SAS: A Practical Guide* (second ed.). Cary, N.C.: SAS Institute Inc.

Amor, D. 2002. *The E-business (R)evolution: Living and Working in an Interconnected World* (second ed.). Upper Saddle River, N.J.: Prentice Hall.

Andersen, P. K., Ø. Borgan, R. D. Gill, and N. Keiding 1993. *Statistical Models Based on Counting Processes*. New York: Springer.

Appleton, D. R. 1995. May the best man win? *The Statistician* 44(4):529–538. 103

Asur, S. and B. A. Huberman 2010. Predicting the future with social media. In *Proceedings of the 2010 IEEE/WIC/ACM International Conference on Web Intelligence and Intelligent Agent Technology - Volume 01*, WI-IAT '10, pp. 492–499. Washington, DC, USA: IEEE Computer Society. http://dx.doi.org/10.1109/WI-IAT.2010.63. 133, 134

Auguie, B. 2014. *Package gridExtra: Functions in grid graphics*. Comprehensive R Archive Network. 2014. `http://cran.r-project.org/web/packages/gridExtra/`

gridExtra.pdf.

Baayen, R. H. 2008. *Analyzing Linguistic Data: A Practical Introduction to Statistics with R*. Cambridge, UK: Cambridge University Press. 259

Bache, K. and M. Lichman 2013. UCI machine learning repository. http://archive.ics.uci.edu/ml. 296

Baeza-Yates, R. and B. Ribeiro-Neto 1999. *Modern Information Retrieval*. New York: ACM Press.

Barabási, A.-L. 2003. *Linked: How Everything is Connected to Everything Else*. New York: Penguin/Plume.

Barabási, A.-L. 2010. *Bursts: The Hidden Pattern Behind Everything We Do*. New York: Penguin/Dutton.

Barabási, A.-L. and R. Albert 1999. Emergence of scaling in random networks. *Science* 286 (5439):509–512. 224

Bates, D. and M. Maechler 2014. *Matrix: Sparse and Dense Matrix Classes and Methods*. Comprehensive R Archive Network. 2014. http://cran.r-project.org/web/packages/support.CEs/support.CEs.pdf. 210

Bates, D. M. and D. G. Watts 2007. *Nonlinear Regression Analysis and Its Applications*. New York: Wiley.

Beazley, D. M. 2009. *Python Essential Reference* (fourth ed.). Upper Saddle River, N.J.: Pearson Education.

Beazley, D. M. and B. K. Jones 2013. *Python Cookbook* (third ed.). Sebastopol, Calif.: O'Reilly.

Becker, R. A. and W. S. Cleveland 1996. *S-Plus Trellis^{TM} Graphics User's Manual*. Seattle: MathSoft, Inc. 252

Behnel, S. 2014, September 10. lxml. Supported at http://lxml.de/. Documentation retrieved from the World Wide Web, October 26, 2014, at http://lxml.de/3.4/lxmldoc-3.4.0.pdf.

Belew, R. K. 2000. *Finding Out About: A Cognitive Perspective on Search Engine Technology and the WWW*. Cambridge: Cambridge University Press.

Belsley, D. A., E. Kuh, and R. E. Welsch 1980. *Regression Diagnostics: Identifying Influential Data and Sources of Collinearity*. New York: Wiley.

Berk, R. A. 2008. *Statistical Learning from a Regression Perspective*. New York: Springer.

Berners-Lee, T. 2000. *Weaving the Web: The Original Design and Ultimate Destiny of the World Wide Web*. New York: HarperBusiness.

Berry, M. and M. Browne 2005, October. Email surveillance using non-negative matrix factorization. *Computational and Mathematical Organization Theory* 11(3):249–264. 81

Berry, M. W. and M. Browne 1999. *Understanding Search Engines: Mathematical Modeling and Text Retrieval*. Philadelphia: Society for Industrial and Applied Mathematics.

Berthold, M. R., N. Cebron, F. Dill, T. R. Gabriel, T. Kötter, T. Meinl, P. Ohl, C. Sieb, K. Thiel, and B. Wiswedel 2007. KNIME: The Konstanz Information Miner. In *Studies in Classification, Data Analysis, and Knowledge Organization (GfKL 2007)*. New York: Springer. ISSN 1431-8814. ISBN 978-3-540-78239-1.

Bilton, N. 2013. *Hatching Twitter: A True Story of Money, Power, Friendship, and Betrayal*. New York: Portfolio/Penguin.

Bird, S., E. Klein, and E. Loper 2009. *Natural Language Processing with Python: Analyzing Text with the Natural Language Toolkit*. Sebastopol, Calif.: O'Reilly. http://www.nltk.org/book/. 172

Bishop, Y. M. M., S. E. Fienberg, and P. W. Holland 1975. *Discrete Multivariate Analysis: Theory and Practice*. Cambridge: MIT Press.

Bizer, C., J. Lehmann, G. Kobilarov, S. Auer, C. Becker, R. Cyganiak, and S. Hellmann 2009. DBpedia—A crystallization point for the web of data. *Web Semantics* 7:154–165.

Blei, D., A. Ng, and M. Jordan 2003. Latent Dirichlet allocation. *Journal of Machine Learning Research* 3:993–1022. 176, 251

Bogdanov, P. and A. Singh 2013. Accurate and scalable nearest neighbors in large networks based on effective importance. *CIKM '13*:1–10. ACM CIKM '13 conference paper retrieved from the World Wide Web at `http://www.cs.ucsb.edu/~petko/papers/bscikm13.pdf`.

Bojanowski, M. 2014. *Package intergraph: Coercion Routines for Network Data Objects in R*. Comprehensive R Archive Network. 2014. `http://cran.r-project.org/web/packages/intergraph/intergraph.pdf`.

Bollen, J., H. Mao, and X. Zeng 2011. Twitter mood predicts the stock market. *Journal of Computational Science* 2:1–8.

Borg, I. and P. J. F. Groenen 2010. *Modern Multidimensional Scaling: Theory and Applications* (second ed.). New York: Springer.

Borgatti, S. P., M. G. Everett, and J. C. Johnson 2013. *Analyzing Social Networks*. Thousand Oaks, Calif.: Sage. 102

Boser, B. E., I. M. Guyon, and V. N. Vapnik 1992. A training algorithm for optimal margin classifiers. In *Proceedings of the Fifth Conference on Computational Learning Theory*, pp. 144–152. Association for Computing Machinery Press. 129

Bourne, W. 2013, February 27. 2 reasons to keep and eye on github. *Inc.com*. Retrieved from the World Wide Web, November 6, 2014, at `http://www.inc.com/magazine/201303/will-bourne/2-reasons-to-keep-an-eye-on-github.html`. 235

Box, G. E. P. and D. R. Cox 1964. An analysis of transformations. *Journal of the Royal Statistical Society, Series B (Methodological)* 26(2):211–252.

Boztug, Y. and T. Reutterer 2008. A combined approach for segment-specific market basket analysis. *European Journal of Operational Research* 187(1):294–312. 210

Bradley, R. A. 1976, June. Science, statistics, and paired comparisons. *Biometrics* 32(2): 213–239.

Bradley, R. A. and M. E. Terry 1952, December. Rank analysis of incomplete block designs: I. the method of paired comparisons. *Biometrika* 39(3/4):324–345.

Bradley, S. P., A. C. Hax, and T. L. Magnanti 1977. *Applied Mathematical Programming*. Reading, Mass.: Addison-Wesley. 225

Brandes, U. and T. Erlebach (eds.) 2005. *Network Analysis: Methodological Foundations*. New York: Springer.

Brandes, U., L. C. Freeman, and D. Wagner 2014. Social networks. In R. Tamassia (ed.), *Handbook of Graph Drawing and Visualization*, Chapter 26, pp. 805–840. Boca Raton, Fla.: CRC Press/Chapman & Hall.

Breese, J. S., D. Heckerman, and C. M. Kadie 1998, May. Empirical analysis of predictive algorithms for collaborative filtering. Technical Report MSR-TR-98-12, Microsoft Research. 18 pp. `http://research.microsoft.com/apps/pubs/default.aspx?id=69656`. 210, 296

Breiman, L. 2001a. Random forests. *Machine Learning* 45(1):5–32.

Breiman, L. 2001b. Statistical modeling: The two cultures. *Statistical Science* 16(3):199–215.

Breiman, L., J. H. Friedman, R. A. Olshen, and C. J. Stone 1984. *Classification and Regression Trees*. New York: Chapman & Hall.

Brin, S. and L. Page 1998. The anatomy of a large-scale hypertextual web search engine. *Computer Networks and ISDN Systems* 30:107–117. 46, 103

Brownlee, J. 2011. *Clever Algorithms: Nature-Inspired Programming Recipies*. Melbourne, Australia: Creative Commons. `http://www.CleverAlgorithms.com`. 251

Bruzzese, D. and C. Davino 2008. Visual mining of association rules. In S. Simoff, M. H. Böhlen, and A. Mazeika (eds.), *Visual Data Mining: Theory, Techniques and Tools for*

Visual Analytics, pp. 103–122. New York: Springer.

Buchanan, M. 2002. *Nexus: Small Worlds and the Groundbreaking Science of Networks*. New York: W.W. Norton and Company.

Bühlmann, P. and S. van de Geer 2011. *Statistics for High-Dimensional Data: Methods, Theory and Applications*. New York: Springer. 249

Burt, J. M. and M. B. Garman 1971. Conditional Monte Carlo: A simulation technique for stochastic network analysis. *Management Science* 18(3):207–217. 225

Butts, C. T. 2008a, May 5. network: A package for managing relational data in R. *Journal of Statistical Software* 24(2):1–36. `http://www.jstatsoft.org/v24/i02`. Updated version available at `http://cran.r-project.org/web/packages/network/vignettes/networkVignette.pdf`.

Butts, C. T. 2008b, May 8. Social network analysis with sna. *Journal of Statistical Software* 24(6):1–51. `http://www.jstatsoft.org/v24/i06`.

Butts, C. T. 2014a. *Package network: Classes for Relational Data*. Comprehensive R Archive Network. 2014. `http://cran.r-project.org/web/packages/network/network.pdf`.

Butts, C. T. 2014b. *Package snad: Statistical Analysis of Network Data with R*. Comprehensive R Archive Network. 2014. `http://cran.r-project.org/web/packages/sand/sand.pdf`.

Butts, C. T. 2014c. *Package sna: Tools for Social Network Analysis*. Comprehensive R Archive Network. 2014. `http://cran.r-project.org/web/packages/sna/sna.pdf`. Also see the statnet website at `http://www.statnetproject.org/`.

Buvač, V. and P. J. Stone 2001, April 2. The General Inquirer user's guide. Software developed with the support of Harvard University and The Gallup Organization.

Cairo, A. 2013. *The Functional Art: An Introduction to Information Graphics and Visualization*. Berkeley, Calif: New Riders.

Cami, A. and N. Deo 2008. Techniques for analyzing dynamic random graph models of web-like networks: An overview. *Networks* 51(4):211–255.

Campbell, S. and S. Swigart 2014. *Going beyond Google: Gathering Internet Intelligence* (5 ed.). Oregon City, Oreg.: Cascade Insights. `http://www.cascadeinsights.com/gbgdownload`. 59

Cantelon, M., M. Harter, T. J. Holowaychuk, and N. Rajlich 2014. *Node.js in Action*. Shelter Island, N.Y.: Manning. 3

Carley, K. M. and D. Skillicorn 2005, October. Special issue on analyzing large scale networks: The Enron corpus. *Computational and Mathematical Organization Theory* 11(3): 179–181. 291

Carlin, B. P. 1996, February. Improved NCAA basketball tournament modeling via point spread and team strength information. *The American Statistician* 50(1):39–43. 103

Carlin, B. P. and T. A. Louis 1996. *Bayes and Empirical Bayes Methods for Data Analysis*. London: Chapman & Hall. 244

Carr, D., N. Lewin-Koh, and M. Maechler 2014. *hexbin: Hexagonal Binning Routines*. Comprehensive R Archive Network. 2014. `http://cran.r-project.org/web/packages/hexbin/hexbin.pdf`. 252

Carrington, P. J., J. Scott, and S. Wasserman (eds.) 2005. *Models and Methods in Social Network Analysis*. Cambridge, UK: Cambridge University Press.

Casciano, C. 2011. *The CSS Pocket Guide*. Upper Saddle River, N.J.: Pearson/Peachpit.

Cass, S. 2014, July. The top 10 programming languages: Spectrum's 2014 ranking. *IEEE Spectrum* 51(7):68. 298

Chakrabarti, S. 2003. *Mining the Web: Discovering Knowledge from Hypertext Data*. San Francisco: Morgan Kaufmann.

Chambers, J. M. and T. J. Hastie (eds.) 1992. *Statistical Models in S*. Pacific Grove, Calif.: Wadsworth & Brooks/Cole. Champions of S, S-Plus, and R call this "the white book." It introduced statistical modeling syntax using S3 classes.

Chan, E. P. 2009. *Quantitative Trading: How to Build Your Own Algorithmic Trading Business*. New York: Wiley.

Chan, E. P. 2013. *Algorithmic Trading: Winning Strategies and Their Rationale*. New York: Wiley.

Chang, W. 2013. *R Graphics Cookbook*. Sebastopol, Calif.: O'Reilly.

Chapanond, A., M. S. Krishnamoorthy, and B. Yener 2005, October. Graph theoretic and spectral analysis of Enron email data. *Computational and Mathematical Organization Theory* 11(3):265–281. 81

Charmaz, K. 2002. Qualitative interviewing and grounded theory analysis. In J. F. Gubrium and J. A. Holstein (eds.), *Handbook of Interview Research: Context and Method*, Chapter 32, pp. 675–694. Thousand Oaks, Calif.: Sage. 279

Charniak, E. 1993. *Statistical Language Learning*. Cambridge: MIT Press.

Chatterjee, S. and A. S. Hadi 2012. *Regression Analysis by Example* (fifth ed.). New York: Wiley.

Chau, M. and H. Chen 2003. Personalized and focused web spiders. In N. Zhong, J. Liu, and Y. Yao (eds.), *Web Intelligence*, Chapter 10, pp. 198–217. New York: Springer. 59

Cherny, L. 1999. *Conversation and Community: Chat in a Virtual World*. Stanford, Calif.: CSLI Publications. 265

Chodorow, K. 2013. *MongoDB: The Definitive Guide* (second ed.). Sebastopol, Calif.: O'Reilly. 3

Christakis, N. A. and J. H. Fowler 2009. *Connected: The Surprising Power of Our Social Networks and How They Shape Our Lives*. New York: Little Brown and Company.

Christensen, R. 1997. *Log-Linear Models and Logistic-Regression* (second ed.). New York: Springer.

Chun, W. J. 2007. *Core Python Programming* (second ed.). Upper Saddle River, N.J.: Pearson Education.

Cleveland, W. S. 1993. *Visualizing Data*. Murray Hill, N.J.: AT&T Bell Laboratories. Initial documentation for trellis graphics in S-Plus. 252

Clifton, B. 2012. *Advanced Web Metrics with Google Analytics* (third ed.). New York: Wiley. 19

Cohen, J. 1960, April. A coefficient of agreement for nominal data. *Educational and Psychological Measurement* 20(1):37–46. 248

Conway, D. and J. M. White 2012. *Machine Learning for Hackers*. Sebastopol, Calif.: O'Reilly.

Cook, R. D. 1998. *Regression Graphics: Ideas for Studying Regressions through Graphics*. New York: Wiley.

Cook, R. D. 2007. Fisher lecture: Dimension reduction in regression. *Statistical Science* 22: 1–26.

Cook, R. D. and S. Weisberg 1999. *Applied Regression Including Computing and Graphics*. New York: Wiley.

Copeland, R. 2013. *MongoDB Applied Design Patterns* (second ed.). Sebastopol, Calif.: O'Reilly. 3

Cox, T. F. and M. A. A. Cox 1994. *Multidimensional Scaling*. London: Chapman & Hall.

Cranor, L. F. and B. A. LaMacchia 1998. Spam! *Communications of the ACM* 41(8):74–83.

Cristianini, N. and J. Shawe-Taylor 2000. *An Introduction to Support Vector Machines and Other Kernel-Based Learning Methods*. Cambridge: Cambridge University Press.

Crnovrsanin, T., C. W. Mueldera, R. Faris, D. Felmlee, and K.-L. Ma 2014. Visualization techniques for categorical analysis of social networks with multiple edge sets. *Social Networks* 37:56–64.

Crockford, D. 2008. *JavaScript: The Good Parts*. Sebastopol, Calif.: O'Reilly.

Croll, A. and S. Power 2009. *Complete Web Monitoring: Watching Your Visitors, Performance, Communities & Competitors*. Sebastopol, Calif.: O'Reilly.

Csardi, G. 2014a. *Package igraphdata: A Collection of Network Data Sets for the igraph Package*. Comprehensive R Archive Network. 2014. `http://cran.r-project.org/web/packages/igraphdata/igraphdata.pdf`.

Csardi, G. 2014b. *Package igraph: Network Analysis and Visualization*. Comprehensive R Archive Network. 2014. `http://cran.r-project.org/web/packages/igraph/igraph.pdf`.

Daconta, M. C., L. J. Obrst, and K. T. Smith 2003. *The Semantic Web: A Guide to the Future of XML, Web Services, and Knowledge Management*. New York: Wiley.

Dale, J. and S. Abbott 2014. *Qual-Online The Essential Guide: What Every Researcher Needs to Know about Conducting and Moderating Interviews via the Web*. Ithaca, N.Y.: Paramount Market Publishing.

Datta, A., S. Shulman, B. Zheng, S.-D. Lin, A. Sum, and E.-P. Lim (eds.) 2011. *Social Informatics: Proceedings of the Third International Conference SocInfo 2011*. New York: Springer.

Davenport, T. H. and J. G. Harris 2007. *Competing on Analytics: The New Science of Winning*. Boston: Harvard Business School Press. 239

Davenport, T. H., J. G. Harris, and R. Morison 2010. *Analytics at Work: Smarter Decisions, Better Results*. Boston: Harvard Business School Press. 239

David, H. A. 1963. *The Method of Paired Comparisons*. London: Charles Griffin & Company Limited.

Davidson, R. R. and P. H. Farquhar 1976, June. A bibliography on the method of paired comparisons. *Biometrics* 32(2):241–252.

Davis, B. H. and J. P. Brewer 1997. *Electronic Discourse: Linguistic Individuals in Virtual Space*. Albany, N.Y.: State University of New York Press. 265

Davison, M. L. 1992. *Multidimensional Scaling*. Melbourne, Fla.: Krieger.

de Nooy, W., A. Mrvar, and V. Batagelj 2011. *Exploratory Social Network Analysis with Pajek*. Cambridge, U.K.: Cambridge University Press.

Dean, J. and S. Ghemawat 2004. MapReduce: Simplifed Data Processing on Large Clusters. Retrieved from the World Wide Web at `http://static.usenix.org/event/osdi04/tech/full_papers/dean/dean.pdf`.

Dehuri, S., M. Patra, B. B. Misra, and A. K. Jagadev (eds.) 2012. *Intelligent Techniques in Recommendation Systems*. Hershey, Pa.: IGI Global.

Delen, D., R. Sharda, and P. Kumar 2007. Movie forecast guru: A Web-based DSS for Hollywood managers. *Decision Support Systems* 43(4):1151–1170. 133

Demšar, J. and B. Zupan 2013. Orange: Data mining fruitful and fun—A historical perspective. *Informatica* 37:55–60. 250, 299

Dey, I. 1999. *Grounding Grounded Theory: Guidelines for Qualitative Inquiry*. New York: Wiley. 279

Di Battista, G., P. Eades, R. Tomassia, and I. G. Tollis 1999. *Graph Drawing: Algorithms for the Visualization of Graphs*. Upper Saddle River, N.J.: Prentice Hall.

Di Battista, G. and M. Rimondini 2014. Computer networks. In R. Tamassia (ed.), *Handbook of Graph Drawing and Visualization*, Chapter 25, pp. 763–804. Boca Raton, Fla.: CRC Press/Chapman & Hall.

Diesner, J., T. L. Frantz, and K. M. Carley 2005, October. Communication networks from the Enron email corpus. *Computational and Mathematical Organization Theory* 11(3): 201–228. 81

Dippold, K. and H. Hruschka 2013. Variable selection for market basket analysis. *Computational Statistics* 28(2):519–539. `http://dx.doi.org/10.1007/s00180-012-0315-3`. ISSN 0943-4062. 210

Dover, D. 2011. *Search Engine Optimization Secrets: Do What You Never Thought Possible with SEO*. New York: Wiley.

Downing, D., M. Covington, M. Covington, C. A. Barrett, and S. Covington 2012. *Dictionary of Computer and Internet Terms* (eleventh ed.). Hauppauge, N.Y.: Barron's Educational Series.

Draper, N. R. and H. Smith 1998. *Applied Regression Analysis* (third ed.). New York: Wiley.

DuCharme, B. 2013. *Learning SPARQL*. Sebastopol, Calif.: O'Reilly. 235

Duda, R. O., P. E. Hart, and D. G. Stork 2001. *Pattern Classification* (second ed.). New York: Wiley.

Dumais, S. T. 2004. Latent semantic analysis. In B. Cronin (ed.), *Annual Review of Information Science and Technology*, Volume 38, Chapter 4, pp. 189–230. Medford, N.J.: Information Today.

Easley, D. and J. Kleinberg 2010. *Networks, Crowds, and Markets: Reasoning about a Highly Connected World*. Cambridge, UK: Cambridge University Press.

Efron, B. 1986. Why isn't everyone a Bayesian (with commentary). *The American Statistician* 40(1):1–11.

Eichenwald, K. 2005. *Conspiracy of Fools: A True Story*. New York: Broadway Books/ Random House.

Ellis, D., R. Oldridge, and A. Vasconcelos 2004. Community and virtual community. In B. Cronin (ed.), *Annual Review of Information Science and Technology*, Volume 38, Chapter 3, pp. 145–186. Medford, N.J.: Information Today.

Engelbrecht, A. P. 2007. *Computational Intelligence: An Introduction* (second ed.). New York: Wiley. 251

Erdős, P. and A. Rényi 1959. On random graphs. *Publicationes Mathematicae* 6:290–297.

Erdős, P. and A. Rényi 1960. On the evolution of random graphs. *Publications of the Mathematical Institute of the Hungarian Academy of Sciences* 5:17–61.

Erdős, P. and A. Rényi 1961. On the strength of connectedness of a random graph. *Acta Mathematica Scientia Hungary* 12:261–267.

Erwig, M. 2000. The graph Voronoi diagram with applications. *Networks* 36(3):156–163.

Esipova, N., T. W. Miller, M. D. Zarnecki, J. Elzaurdia, and S. Ponnaiya 2004. Exploring the possibilities of online focus groups. In T. W. Miller and J. Walkowski (eds.), *Qualitative Research Online*, Chapter A, pp. 107–128. Manhattan Beach, Calif.: Research Publishers.

European Parliament 2002. Directive on privacy and electronic communications. Retrieved from the World Wide Web, October 21, 2014, at `http://eur-lex.europa.eu/LexUriServ/LexUriServ.do?uri=CELEX:32002L0058:en:HTML`. 19

Everitt, B. and G. Dunn 2001. *Applied Multivariate Data Analysis* (second ed.). New York: Wiley. 176

Everitt, B. and S. Rabe-Hesketh 1997. *The Analysis of Proximity Data*. London: Arnold.

Everitt, B. S., S. Landau, M. Leese, and D. Stahl 2011. *Cluster Analysis* (fifth ed.). New York: Wiley.

Fawcett, T. 2003, January 7. ROC graphs: Notes and practical considerations for researchers. http://www.hpl.hp.com/techreports/2003/HPL-2003-4.pdf.

Feinerer, I. 2012. *Introduction to the wordnet Package*. Comprehensive R Archive Network. 2012. `http://cran.r-project.org/web/packages/wordnet/vignettes/wordnet.pdf`. 259

Feinerer, I. 2014. *Introduction to the tm Package*. Comprehensive R Archive Network. 2014. `http://cran.r-project.org/web/packages/tm/vignettes/tm.pdf`.

Feinerer, I. and K. Hornik 2014a. *tm: Text Mining Package*. Comprehensive R Archive Network. 2014. `http://cran.r-project.org/web/packages/tm/tm.pdf`.

Feinerer, I. and K. Hornik 2014b. *wordnet: WordNet Interface*. Comprehensive R Archive Network. 2014. `http://cran.r-project.org/web/packages/wordnet/wordnet.pdf`. 259

Feinerer, I., K. Hornik, and D. Meyer 2008, 3 31. Text mining infrastructure in R. *Journal* of *Statistical Software* 25(5):1–54. http://www.jstatsoft.org/v25/i05. ISSN 1548-7660.

Feldman, R. 2002. Text mining. In W. Klösgen and J. M. Żytkow (eds.), *Handbook of Data Mining and Knowledge Discovery*, Chapter 38, pp. 749–757. Oxford: Oxford University Press.

Fellbaum, C. 1998. *WordNet: An Electronic Lexical Database*. Cambridge, Mass.: MIT Press. 259

Fellows, I. 2014a. wordcloud makes words less cloudy. Retrieved from the World Wide Web at http://blog.fellstat.com/. 176, 259

Fellows, I. 2014b. *wordcloud: Word Clouds*. Comprehensive R Archive Network. 2014. http://cran.r-project.org/web/packages/wordcloud/wordcloud.pdf. 176, 259

Fenzel, D., J. Hendler, H. Lieberman, and W. Wahlster (eds.) 2003. *Spinning the Semantic Web: Bringing the World Wide Web to Its Full Potential*. Cambridge: MIT Press.

Ferrucci, D., E. Brown, J. Chu-Carroll, J. Fan, D. Gondek, A. A. Kalyanpur, A. Lally, J. W. Murdock, E. Nyberg, J. Prager, N. Schlaefer, and C. Welty 2010, Fall. Building Watson: An overview of the DeepQA project. *AI Magazine* 31(3):59–79.

Few, S. 2009. *Now You See It: Simple Visualization Techniques and Quantitative Analysis*. Oakland, Calif.: Analytics Press.

Fielding, N. G. and R. M. Lee 1998. *Computer Analysis and Qualitative Research*. London: Sage. 279

Fienberg, S. E. 2007. *Analysis of Cross-Classified Categorical Data* (second ed.). New York: Springer.

Firth, D. 1991. Generalized linear models. In D. Hinkley and E. Snell (eds.), *Statistical Theory and Modeling: In Honour of Sir David Cox, FRS*, Chapter 3, pp. 55–82. London: Chapman and Hall.

Fisher, R. A. 1970. *Statistical Methods for Research Workers* (fourteenth ed.). Edinburgh: Oliver and Boyd. First edition published in 1925. 244

Fisher, R. A. 1971. *Design of Experiments* (ninth ed.). New York: Macmillan. First edition published in 1935. 244

Flam, F. 2014, September 30. The odds, continually updated. *The New York Times*:D1. Retrieved from the World Wide Web at http://www.nytimes.com/2014/09/30/science/the-odds-continually-updated.html?_r=0. 244

Flanagan, D. 2011. *JavaScript: The Definitive Guide* (sixth ed.). Sebastopol, Calif.: O'Reilly. 3

Fox, J. 2002, January. Robust regression: Appendix to an R and S-PLUS companion to applied regression. Retrieved from the World Wide Web at http://cran.r-project.org/doc/contrib/Fox-Companion/appendix-robust-regression.pdf. 249

Fox, J. 2014. *car: Companion to Applied Regression*. Comprehensive R Archive Network. 2014. http://cran.r-project.org/web/packages/car/car.pdf.

Fox, J. and S. Weisberg 2011. *An R Companion to Applied Regression* (second ed.). Thousand Oaks, Calif.: Sage. 128

Franceschet, M. 2011, June. PageRank: Standing on the shoulders of giants. *Communications of the ACM* 54(6):92–101.

Franks, B. 2012. *Taming the Big Data Tidal Wave: Finding Opportunities in Huge Data Streams with Advanced Analytics*. Hoboken, N.J.: Wiley. 239

Freeman, L. C. 2004. *The Development of Social Network Analysis: A Study in the Sociology of Science*. Vancouver, B.C.: Empirical Press.

Freeman, L. C. 2005. Graphic techniques for exploring social network data. In P. J. Carrington, J. Scott, and S. Wasserman (eds.), *Models and Methods in Social Network Analysis*, Chapter 12, pp. 248–269. Cambridge, UK: Cambridge University Press.

Friedl, J. E. F. 2006. *Mastering Regular Expressions* (third ed.). Sebastopol, Calif.: O'Reilly. 259

Fruchterman, T. M. J. and E. M. Reingold 1991. Graph drawing by force-directed placement. *Software—Practice and Experience* 21(11):1129–1164.

Fry, B. 2008. *Visualizing Data: Exploring and Explaining Data with the Processing Environment*. Sebastopol, Calif.: O'Reilly.

Gabriel, K. R. 1971. The biplot graphical display of matrices with application to principal component analysis. *Biometrika* 58:453–467.

Garcia-Molina, H., J. D. Ullman, and J. Widom 2009. *Database Systems: The Complete Book* (second ed.). Upper Saddle River, N.J.: Prentice-Hall.

Gasston, P. 2011. *The Book of CSS3: A Developer's Guide to the Future of Web Design*. San Francisco: No Starch Press.

Gasston, P. 2013. *The Modern Web: Multi-Device Web Development with HTML5, CSS3, and JavaScript*. San Francisco: No Starch Press.

Geisser, S. 1993. *Predictive Inference: An Introduction*. New York: Chapman & Hall. 244

Gelman, A., J. B. Carlin, H. S. Stern, and D. B. Rubin 1995. *Bayesian Data Analysis*. London: Chapman & Hall. 244

Gelman, A., J. Hill, Y.-S. Su, M. Yajima, and M. G. Pittau 2014. *mi: Missing Data Imputation and Model Checking*. Comprehensive R Archive Network. 2014. `http://cran.r-project.org/web/packages/mi/mi.pdf`.

Georgakopoulou, A. and D. Goutsos 1997. *Discourse Analysis: An Introduction*. Edinburgh, U.K.: Edinburgh University Press. 279

Gibson, H., J. Faith, and P. Vickers 2013, 07. A survey of two-dimensional graph layout techniques for information visualisation. *Information Visualization* 12(3-4):324–357.

Glaser, B. G. and A. L. Strauss 1967. *The Discovery of Grounded Theory: Strategies for Qualitative Research*. Hawthorne, N.Y.: Aldine de Gruyter. 279

Gnanadesikan, R. 1997. *Methods for Statistical Data Analysis of Multivariate Observations* (second ed.). New York: Wiley. 176

Goldenberg, A., A. X. Zheng, S. E. Fienberg, and E. M. Airoldi 2009, January. A survey of statistical network models. *Foundations and Trends in Machine Learning* 2(2):129–233.

Golombisky, K. and R. Hagen 2013. *White Space is Not Your Enemy: A Beginner's Guide to Communicating Visually through Graphic, Web, & Multimedia Design* (second ed.). Burlington, Mass.: Focal Press/Taylor & Francis.

Goodreau, S. M., M. S. Handcock, D. R. Hunter, C. T. Butts, and M. Morris 2008, 5 8. A statnet tutorial. *Journal of Statistical Software* 24(9):1–26. `http://www.jstatsoft.org/v24/i09`. CODEN JSSOBK. ISSN 1548-7660.

Gower, J. C. and D. J. Hand 1996. *Biplots*. London: Chapman & Hall.

Graybill, F. A. 1961. *Introduction to Linear Statistical Models, Volume 1*. New York: McGraw-Hill.

Graybill, F. A. 2000. *Theory and Application of the Linear Model*. Stamford, Conn.: Cengage Learning.

Gries, S. T. 2009. *Quantitative Corpus Linguistics with R: A Practical Introduction*. New York: Routledge.

Gries, S. T. 2013. *Statistics for Linguistics with R: A Pratical Introduction* (second revised ed.). Berlin: De Gruyter Mouton. 259

Grieve, T. 2003, October 14. The decline and fall of the Enron empire. *Salon*:165–178. Retrieved from the World Wide Web at `http://www.salon.com/2003/10/14/enron_22/`.

Groeneveld, R. A. 1990, November. Ranking teams in a league with two divisions of *t* teams. *The American Statistician* 44(4):277–281. 103

Grolemund, G. and H. Wickham 2014. *lubridate: Make Dealing with Dates a Little Easier*. Comprehensive R Archive Network. 2014. `http://cran.r-project.org/web/packages/lubridate/lubridate.pdf`. 62

Gross, J. L., J. Yellen, and P. Zhang (eds.) 2014. *Handbook of Graph Theory* (second ed.). Boca Raton, Fla.: CRC Press/Chapman & Hall.

Grothendieck, G. 2014a. *gsubfn: Utilities for Strings and Function Arguments*. Comprehensive R Archive Network. 2014. `http://cran.r-project.org/web/packages/gsubfn/gsubfn.pdf`.

Grothendieck, G. 2014b. *gsubfn: Utilities for Strings and Function Arguments (Vignette)*. Comprehensive R Archive Network. 2014. `http://cran.r-project.org/web/packages/gsubfn/vignettes/gsubfn.pdf`.

Groves, R. M., F. J. Fowler, Jr., M. P. Couper, J. M. Lepkowski, E. Singer, and R. Tourangeau 2009. *Survey Methodology* (second ed.). New York: Wiley.

Guilford, J. P. 1936. *Psychometric Methods*. New York: McGraw-Hill.

Gunning, D., V. K. Chaudhri, P. Clark, K. Barker, S.-Y. Chaw, M. Greaves, B. Grosof, A. Leung, D. McDonald, S. Mishra, J. Pacheco, B. Porter, A. Spaulding, D. Tecuci, and J. Tien 2010, Fall. Project Halo update—Progress toward digital Aristotle. *AI Magazine* 31(3):33–58.

Hagbert, A. and D. Schult 2014. *NetworkX: Python Software for Complex Networks*. NetworkX Development Team. 2014. Retrieved from the World Wide Web at `https://github.com/networkx/`.

Hahsler, M. 2014a. *recommenderlab: A Framework for Developing and Testing Recommendation Algorithms*. Comprehensive R Archive Network. 2014. `http://cran.r-project.org/web/packages/recommenderlab/vignettes/recommenderlab.pdf`.

Hahsler, M. 2014b. *recommenderlab: Lab for Developing and Testing Recommender Algorithms*. Comprehensive R Archive Network. 2014. `http://cran.r-project.org/web/packages/recommenderlab/recommenderlab.pdf`.

Hahsler, M., C. Buchta, B. Grün, and K. Hornik 2014a. *arules: Mining Association Rules and Frequent Itemsets*. Comprehensive R Archive Network. 2014. `http://cran.r-project.org/web/packages/arules/arules.pdf`.

Hahsler, M., C. Buchta, B. Grün, and K. Hornik 2014b. *Introduction to arules: A Computational Environment for Mining Association Rules and Frequent Itemsets*. Comprehensive R Archive Network. 2014. `http://cran.r-project.org/web/packages/arules/vignettes/arules.pdf`.

Hahsler, M., C. Buchta, and K. Hornik 2008. Selective association rule generation. *Computational Statistics* 23:303–315. 210

Hahsler, M. and S. Chelluboina 2014a. *arulesViz: Visualizing Association Rules and Frequent Itemsets*. Comprehensive R Archive Network. 2014. `http://cran.r-project.org/web/packages/arulesViz/arulesViz.pdf`.

Hahsler, M. and S. Chelluboina 2014b. *Visualizing Association Rules: Introduction to the R-extension Package arulesViz*. Comprehensive R Archive Network. 2014. `http://cran.r-project.org/web/packages/arulesViz/vignettes/arulesViz.pdf`.

Hahsler, M., S. Chelluboina, K. Hornik, and C. Buchta 2011. The arules R-package ecosystem: Analyzing interesting patterns from large transaction data sets. *Journal of Machine Learning Research* 12:2021–2025.

Hahsler, M., B. Grün, and K. Hornik 2005, September 29. arules: A computational environment for mining association rules and frequent item sets. *Journal of Statistical Software* 14(15):1–25. `http://www.jstatsoft.org/v14/i15`.

Hale, C. and J. Scanlon 1999. *Wired Style: Principles of English Usage in the Digital Age*. New York: Broadway Books.

Hand, D. J. 1997. *Construction and Assessment of Classification Rules*. New York: Wiley.

Handcock, M. S. and K. Gile 2007, April 28. Modeling social networks with sampled or missing data. `http://www.csss.washington.edu/Papers/wp75.pdf`. 109

Handcock, M. S., D. R. Hunter, C. T. Butts, S. M. Goodreau, and M. Morris 2008, May 5. statnet: Software tools for the representation, visualization, analysis and simulation of network data. *Journal of Statistical Software* 24(1):1–11. `http://www.jstatsoft.org/v24/i01`.

Hanneman, R. A. and M. Riddle 2005. *Introduction to Social Network Methods*. Riverside, Calif.: University of California Riverside. `http://www.faculty.ucr.edu/~hanneman/nettext/`.

Harrell, Jr., F. E. 2001. *Regression Modeling Strategies: With Applications to Linear Models, Logistic Regression, and Survival Analysis*. New York: Springer.

Harrington, P. 2012. *Machine Learning in Action*. Shelter Island, N.Y.: Manning. 210

Hart, R. P. 2000a. *Campaign Talk: Why Elections Are Good for Us*. Princeton, N.J.: Princeton University Press.

Hart, R. P. 2000b. *DICTION 5.0: The Text Analysis Program*. Thousand Oaks, Calif.: Sage.

Hart, R. P. 2001. Redeveloping Diction: theoretical considerations. In M. D. West (ed.), *Theory, Method, and Practice in Computer Content Analysis*, Chapter 3, pp. 43–60. Westport, Conn.: Ablex.

Hastie, T., R. Tibshirani, and J. Friedman 2009. *The Elements of Statistical Learning: Data Mining, Inference, and Prediction* (second ed.). New York: Springer.

Hausser, R. 2001. *Foundations of Computational Linguistics: Human-Computer Communication in Natural Language* (second ed.). New York: Springer-Verlag.

Hearst, M. A. 1997. TextTiling: Segmenting text into multi-paragraph subtopic passages. *Computational Linguistics* 23(1):33–64. 259

Heer, J., M. Bostock, and V. Ogievetsky 2010, May 1. A tour through the visualization zoo: A survey of powerful visualization techniques, from the obvious to the obscure. *acmqueue: Association for Computing Machinery*:1–22. Retrieved from the World Wide Web at `http://queue.acm.org/detail.cfm?id=1805128`.

Hemenway, K. and T. Calishain 2004. *Spidering Hacks: 100 Industrial-Strength Tips & Tools*. Sabastopol, Calif.: O'Reilly. 59

Herman, A. and T. Swiss (eds.) 2000. *The World Wide Web and Contemporary Cultural Theory*. New York: Routledge. 265

Hinkley, D. V., N. Reid, and E. J. Snell (eds.) 1991. *Statistical Theory and Modeling*. London: Chapman and Hall. 244

Ho, Q. and E. P. Xing 2014. Analyzing time-evolving networks using an evolving cluster mixed membership blockmodel. In E. M. Airoldi, D. Blei, E. A. Erosheva, and S. E. Fienberg (eds.), *Handbook of Mixed Membership Models and Their Applications*, Chapter 23, pp. 489–525. Boca Raton, Fla.: CRC Press.

Hoberman, S. 2014. *Data Modeling for MongoDB: Building Well-Designed and Supportable MongoDB Databases*. Basking Ridge, N.J.: Technics Publications. 3

Hoerl, A. E. and R. W. Kennard 2000. Ridge regression: biased estimation for nonorthogonal problems. *Technometrics* 42(1):80–86. Reprinted from *Technometrics*, volume 12. 249

Hoffman, P. and Scrapy Developers 2014, June 26. Scrapy documentation release 0.24.0. Supported at `http://scrapy.org/`. Documentation retrieved from the World Wide Web, October 26, 2014, at `https://media.readthedocs.org/pdf/scrapy/0.24/scrapy.pdf`.

Holden, D. 2006. Hierarchical edge bundles: Visualization of adjacency relations in hierarchical data. *IEEE Transactions on Visualization and Computer Graphics* 12(5):741–748.

Holeton, R. (ed.) 1998. *Composing Cyberspace: Identity, Community, and Knowledge in the Electronic Age*. Boston: McGraw-Hill. 265

Holland, P. W., K. B. Laskey, and S. Leinhardt 1983. Stochastic blockmodels: First steps. *Social Networks* 5:109–138.

Honaker, J., G. King, and M. Blackwell 2014. *Amelia II: A Program for Missing Data*. Comprehensive R Archive Network. 2014. `http://cran.r-project.org/web/packages/Amelia/Amelia.pdf`.

Hornik, K. 2014a. *RWeka Odds and Ends*. Comprehensive R Archive Network. 2014. `http://cran.r-project.org/web/packages/RWeka/vignettes/RWeka.pdf`.

Hornik, K. 2014b. *RWeka: R/Weka Interface*. Comprehensive R Archive Network. 2014. http://cran.r-project.org/web/packages/RWeka/RWeka.pdf.

Hosmer, D. W., S. Lemeshow, and S. May 2013. *Applied Survival Analysis: Regression Modeling of Time to Event Data* (second ed.). New York: Wiley.

Hosmer, D. W., S. Lemeshow, and R. X. Sturdivant 2013. *Applied Logistic Regression* (third ed.). New York: Wiley. 128

Houston, B., L. Bruzzese, and S. Weinberg 2002. *The Investigative Reporter's Handbook: A Guide to Documents, Databases and Techniques* (fourth ed.). Boston: Bedford/St. Martin's.

Hu, M. and B. Liu 2004, August 22–25. Mining and summarizing customer reviews. *Proceedings of the ACM SIGKDD International Conference on Knowledge Discovery & Data Mining (KDD-2004)*. Full paper available from the World Wide Web at http://www.cs.uic.edu/~liub/publications/kdd04-revSummary.pdf Original source for opinion and sentiment lexicon, available from the World Wide Web at http://www.cs.uic.edu/~liub/FBS/sentiment-analysis.html#lexicon.

Huberman, B. A. 2001. *The Laws of the Web: Patterns in the Ecology of Information*. Cambridge: MIT Press. 265

Huet, S., A. Bouvier, M.-A. Poursat, and E. Jolivet 2004. *Statistical Tools for Nonlinear Regression: A Practical Guide with S-Plus and R Examples* (second ed.). New York: Springer.

Hughes-Croucher, T. and M. Wilson 2012. *Node Up and Running*. Sebastopol, Calif.: O'Reilly. 3

Huisman, M. 2010. Imputation of missing network data: Some simple procedures. *Journal of Social Structure* 10. http://www.cmu.edu/joss/content/articles/volume10/huisman.pdf. 109

Hyslop, B. 2010. *The HTML Pocket Guide*. Upper Saddle River, N.J.: Pearson/Peachpit.

Ihaka, R., P. Murrell, K. Hornik, J. C. Fisher, and A. Zeileis 2014. *colorspace: Color Space Manipulation*. Comprehensive R Archive Network. 2014. http://cran.r-project.org/web/packages/colorspace/colorspace.pdf.

Indurkhya, N. and F. J. Damerau (eds.) 2010. *Handbook of Natural Language Processing* (second ed.). Boca Raton, Fla.: Chapman and Hall/CRC.

Ingersoll, G. S., T. S. Morton, and A. L. Farris 2013. *Taming Text: How to Find, Organize, and Manipulate It*. Shelter Island, N.Y.: Manning. 176, 251

Izenman, A. J. 2008. *Modern Multivariate Statistical Techniques: Regression, Classification, and Manifold Learning*. New York: Springer. 176

Jackson, M. O. 2008. *Social and Economic Networks*. Princeton, N.J.: Princeton University Press.

James, G., D. Witten, T. Hastie, and R. Tibshirani 2013. *An Introduction to Statistical Learning with Applications in R*. New York: Springer.

Janssen, J., M. Hurshman, and N. Kalyaniwalla 2012. Model selection for social networks using graphlets. *Internet Mathematics* 8(4):338-363.

Johnson, K. 2008. *Quantitative Methods in Linguistics*. Malden, Mass.: Blackwell Publishing. 259

Johnson, R. A. and D. W. Wichern 1998. *Applied Multivariate Statistical Analysis* (fourth ed.). Upper Saddle River, N.J.: Prentice Hall. 176

Johnson, S. 1997. *Interface Culture*. New York: Basic Books. 265

Jones, S. G. (ed.) 1997. *Virtual Culture: Identity & Communication in Cybersociety*. Thousand Oaks, Calif.: Sage. 265

Jones, S. G. (ed.) 1999. *Doing Internet Research: Critical Issues and Methods for Examining the Net*. Thousand Oaks, Calif.: Sage. 265

Jonscher, C. 1999. *The Evolution of Wired Life: From the Alphabet to the Soul-Catcher Chip—How Information Technologies Change Our World*. New York: Wiley. 265

Joula, P. 2008. *Authorship Attribution*. Hanover, Mass.: Now Publishers. 259

Jung, C. G. 1968. *The Archetypes of the Collective Unconscious (Collected Works of C. G. Jung, Vol. 9, Part 1)* (second ed.). Princeton, N.J.: Princeton University Press.

Jünger, M. and P. Mutzel (eds.) 2004. *Graph Drawing Software*. New York: Springer. 82

Jurafsky, D. and J. H. Martin 2009. *Speech and Language Processing: An Introduction to Natural Language Processing, Computational Linguistics, and Speech Recognition* (second ed.). Upper Saddle River, N.J.: Prentice Hall.

Kadushin, C. 2012. *Understanding Social Networks*. New York: Oxford University Press.

Kamada, T. and S. Kawai 1989. An algorithm for drawing general undirected graphs. *Information Processing Letters* 31(1):7–15.

Kaufman, L. and P. J. Rousseeuw 1990. *Finding Groups in Data: An Introduction to Cluster Analysis*. New York: Wiley.

Kaushik, A. 2010. *Web Analytics 2.0: The Art of Online Accountability & Science of Customer Centricity*. New York: Wiley/Sybex.

Kay, M. 2008. *XSLT 2.0 and XPath 2.0 Programmer's Reference* (fourth ed.). New York: Wiley/Wrox. 29

Keener, J. P. 1993, March. The Perron-Frobenius theorem and the ranking of football teams. *SIAM Review* 35(1):80–93.

Keila, P. and D. B. Skillicorn 2005, October. Structure of the Enron email dataset. *Computational and Mathematical Organization Theory* 11(3):183–199. 81

Kelle, U. (ed.) 1995. *Computer-Aided Qualitative Data Analysis: Theory, Methods and Practice*. Thousand Oaks, Calif.: Sage. 279

Kendall, M. G. and B. B. Smith 1940, March. On the method of paired comparisons. *Biometrika* 31(3/4):324–345.

Kirk, R. E. 2013. *Experimental Design: Procedures for the Behavioral Sciences* (fourth ed.). Thousand Oaks, Calif.: Sage.

Klimmt, B. and Y. Yang 2004. Introducing the Enron corpus. `http://ceas.cc/2004/168.pdf`.

Kolaczyk, E. D. 2009. *Statistical Analysis of Network Data: Methods and Models*. New York: Springer. 224

Kolaczyk, E. D. and G. Csárdi 2014. *Statistical Analysis of Network Data with R*. New York: Springer.

Koller, M. 2014. *Simulations for Sharpening Wald-type Inference in Robust Regression for Small Samples*. Comprehensive R Archive Network. 2014. `http://cran.r-project.org/web/packages/robustbase/vignettes/lmrob_simulation.pdf`. 249

Koller, M. and W. A. Stahel 2011. Sharpening Wald-type inference in robust regression for small samples. *Computational Statistics and Data Analysis* 55(8):2504–2515. 249

Kossinets, G. 2006. Effects of missing data in social networks. *Social Networks* 28(3):247–268. 109

Kratochvíl, J. (ed.) 1999. *Graph Drawing: 7th International Symposium / Proceedings GD'99*. New York: Springer.

Krippendorff, K. H. 2012. *Content Analysis: An Introduction to Its Methodology* (third ed.). Thousand Oaks, Calif.: Sage. 133

Krug, S. 2014. *Don't Make Me Think: A Common Sense Approach to Web Usability* (third ed.). Upper Saddle River, N.J.: Pearson/New Riders.

Kuhn, M. 2014. *caret: Classification and Regression Training*. Comprehensive R Archive Network. 2014. `http://cran.r-project.org/web/packages/caret/caret.pdf`.

Kuhn, M. and K. Johnson 2013. *Applied Predictive Modeling*. New York: Springer.

Kutner, M. H., C. J. Nachtsheim, J. Neter, and W. Li 2004. *Applied Linear Statistical Models* (fifth ed.). Boston: McGraw-Hill.

Kvale, S. 1996. *InterViews: An Introduction to Qualitative Research Interviewing*. Thousand Oaks, Calif.: Sage. 275

Lacy, L. W. 2005. *Owl: Representing Information Using the Web Ontology Language*. Victoria, B.C.: Trafford. 235

Lamp, G. 2014. *ggplot for Python*. GitHub. 2014. `https://github.com/yhat/ggplot`. 252

Landauer, T. K., P. W. Foltz, and D. Laham 1998. An introduction to latent semantic analysis. *Discourse Processes* 25:259–284. Retrieved from the World Wide Web, October 31, 2014, at `http://lsa.colorado.edu/papers/dp1.LSAintro.pdf`. 176

Landauer, T. K., D. S. McNamara, S. Dennis, and W. Kintsch (eds.) 2014. *Handbook of Latent Semantic Analysis* (reprint ed.). New York: Psychology Press. 176

Lander, J. P. 2014. *R for Everyone: Advanced Analytics and Graphics*. Upper Saddle River, N.J.: Pearson Education.

Lang, D. T. 2014a. *Package RCurl: General Network (HTTP/FTP/...) Client Interface for R*. Comprehensive R Archive Network. 2014. `http://cran.r-project.org/web/packages/RCurl/RCurl.pdf`. 62

Lang, D. T. 2014b. *Package XML: Tools for Parsing and Generating XML within R and S-Plus*. Comprehensive R Archive Network. 2014. `http://cran.r-project.org/web/packages/XML/XML.pdf`. 62

Langville, A. N. and C. D. Meyer 2006. *Google's Page Rank and Beyond: The Science of Search Engine Rankings*. Princeton, N.J.: Princeton University Press.

Langville, A. N. and C. D. Meyer 2012. *Who's 1?: The Science of Rating and Ranking*. Princeton, N.J.: Princeton University Press.

Lantz, B. 2013. *Machine Learning with R*. Birmingham, U.K.: Packt Publishing.

Laursen, G. H. N. and J. Thorlund 2010. *Business Analytics for Managers: Taking Business Intelligence Beyond Reporting*. Hoboken, N.J.: Wiley. 239

Law, A. M. 2014. *Simulation Modeling and Analysis* (fifth ed.). New York: McGraw-Hill.

Le, C. T. 1997. *Applied Survival Analysis*. New York: Wiley.

Leber, J. 2013. The immortal life of the Enron e-mails. *MIT Technology Review*. `http://www.technologyreview.com/news/515801/the-immortal-life-of-the-enron-e-mails/`. 291

Ledoiter, J. 2013. *Data Mining and Business Analytics with R*. New York: Wiley.

Leetaru, K. 2011. *Data Mining Methods for Content Analysis: An Introduction to the Computational Analysis of Content*. New York: Routledge. 133

Lefkowitz, M. 2014. Twisted documentation. Contributors listed as Twisted Matrix Laboratories at `http://twistedmatrix.com/trac/wiki/TwistedMatrixLaboratories`. Documentation available at `http://twistedmatrix.com/trac/wiki/Documentation`.

Lehmann, J., R. Isele, M. Jakob, A. Jentzsch, D. Kontokostas, P. N. Mendes, S. Hellmann, M. Morsey, P. van Kleef, S. Auer, and C. Bizer 2014. DBpedia—A large-scale, multilingual knowledge base extracted from wikipedia. *Semantic Web Journal*:in press. Retrieved from the World Wide Web at `http://svn.aksw.org/papers/2013/SWJ_DBpedia/public.pdf`.

Leisch, F. and B. Gruen 2014. *CRAN Task View: Cluster Analysis & Finite Mixture Models*. Comprehensive R Archive Network. 2014. `http://cran.r-project.org/web/views/Cluster.html`.

Lenat, D., M. Witbrock, D. Baxter, E. Blackstone, C. Deaton, D. Schneider, J. Scott, and B. Shepard 2010, Fall. Harnessing Cyc to answer clinical researchers' ad hoc queries. *AI Magazine* 31(3):13–32.

Lengauer, T. and R. E. Tarjan 1979. A fast algorithm for finding dominators in a flowgraph. *ACM Transactions on Programming Languages and Systems (TOPLAS)* 1(1):121–141. 82

Leskovec, J. and C. Faloutsos 2006. Sampling from large graphs. Retrieved from the World Wide Web at `http://cs.stanford.edu/people/jure/pubs/sampling-kdd06.pdf`. 109

Leskovec, J., D. Huttenlocher, and J. Kleinbert 2010a. Predicting positive and negative links in online social networks. Retrieved from the World Wide Web at http://cs.stanford.edu/people/jure/pubs/triads-chi10.pdf.

Leskovec, J., D. Huttenlocher, and J. Kleinbert 2010b. Signed networks in social media. Retrieved from the World Wide Web at http://cs.stanford.edu/people/jure/pubs/triads-chi10.pdf.

Leskovec, J., K. J. Lang, A. Dasgupta, and M. W. Mahoney 2009. Community structure in large networks: Natural cluster sizes and the absence of large well-defined clusters. *Internet Mathematics* 6(1):29–123. http://www.technologyreview.com/news/515801/the-immortal-life-of-the-enron-e-mails/.

Lévy, P. 1997. *Collective Intelligence: Mankind's Emerging World in Cyberspace*. Cambridge, Mass.: Perseus Books. 265

Lévy, P. 2001. *Cyberculture*. Minneapolis: University of Minnesota Press. 265

Lewin-Koh, N. 2014. *Hexagon Binning: an Overview*. Comprehensive R Archive Network. 2014. http: //cran.r-project.org/web/packages/hexbin/vignettes/hexagon_binning.pdf. 252

Lewis, T. G. 2009. *Network Science: Theory and Applications*. New York: Wiley. 224

Liaw, A. and M. Wiener 2014. *randomForest: Breiman and Cutler's Random Forests for Classification and Regression*. Comprehensive R Archive Network. 2014. http://cran.r-project.org/web/packages/randomForest/randomForest.pdf.

Lindahl, C. and E. Blount 2003. Weblogs: simplifying web publishing. *Computer* 36(11): 114–116.

Little, R. J. A. and D. B. Rubin 1987. *Statistical Analysis with Missing Data*. New York: Wiley.

Liu, B. 2010. Sentiment analysis and subjectivity. In N. Indurkhya and F. J. Damerau (eds.), *Handbook of Natural Language Processing* (second ed.)., pp. 627–665. Boca Raton, Fla.: Chapman and Hall/CRC.

Liu, B. 2011. *Web Data Mining: Exploring Hyperlinks, Contents, and Usage Data*. New York: Springer. 251

Liu, B. 2012. *Sentiment Analysis and Opinion Mining*. San Rafael, Calif.: Morgan & Claypool.

Lopez, V., C. Unger, P. Cimiano, and E. Motta 2013. Evaluating question answering over linked data. *Web Semantics* 21:3-13.

Lord, F. M. and M. R. Novick 1968. *Statistical Theories of Mental Test Scores*. Reading, Mass.: Addison-Wesley.

Lovink, G. (ed.) 2011. *Networks Withoug a Cause: A Critique of Social Media*. Cambridge, UK: Polity Press. 134

Luangkesorn, L. 2014. *Simulation Programming with Python*. University of Pittsburgh. 2014. Retrieved from the World Wide Web at http://users.iems.northwestern.edu/~nelsonb/IEMS435/PythonSim.pdf. Translation of chapter 4 of Nelson (2014).

Luger, G. F. 2008. *Artificial Intelligence: Structures and Strategies for Complex Problem Solving* (sixth ed.). Boston: Addison-Wesley. 251

Lumley, T. 2010. *Complex Surveys: A Guide to Analysis Using R*. New York: Wiley.

Lumley, T. 2014. *mitools: Tools for Multiple Imputation of Missing Data*. Comprehensive R Archive Network. 2014. http://cran.r-project.org/web/packages/mitools/mitools.pdf.

Maas, A. L., R. E. Daly, P. T. Pham, D. Huang, A. Y. Ng, and C. Potts 2011, June. Learning word vectors for sentiment analysis. In *Proceedings of the 49th Annual Meeting of the Association for Computational Linguistics: Human Language Technologies*, pp. 142–150. Portland, Ore.: Association for Computational Linguistics. Retrieved from the World Wide Web at http://ai.stanford.edu/~amaas/papers/wvSent_acl2011.pdf.

Macal, C. M. and M. J. North 2006. *Introduction to Agent-Based Modeling and Simulation*.

http://www.docstoc.com/docs/36015647/Introduction-to-Agent-based-Modeling-and-Simulation.

Maechler, M. 2014a. *Package cluster*. Comprehensive R Archive Network. 2014. http://cran.r-project.org/web/packages/cluster/cluster.pdf.

Maechler, M. 2014b. *robustbase: Basic Robust Statistics*. Comprehensive R Archive Network. 2014. http://cran.r-project.org/web/packages/robustbase/robustbase.pdf. 249

Maisel, L. S. and G. Cokins 2014. *Predictive Business Analytics: Forward-Looking Capabilities to Improve Business Performance*. New York: Wiley. 239

Manly, B. F. J. 1994. *Multivariate Statistical Methods: A Primer* (second ed.). London: Chapman & Hall. 176

Mann, C. and F. Stewart 2000. *Internet Communication and Qualitative Research: A Handbook for Researching Online*. London: Sage. 265

Mann, C. and F. Stewart 2002. Internet interviewing. In J. F. Gubrium and J. A. Holstein (eds.), *Handbook of Interview Research: Context and Method*, Chapter 29, pp. 603–627. Thousand Oaks, Calif.: Sage.

Manning, C. D. and H. Schütze 1999. *Foundations of Statistical Natural Language Processing*. Cambridge: MIT Press.

Markham, A. N. 1998. *Life Online: Researching Real Experience in Virtual Space*. Walnut Creek, Calif.: AltaMira Press. 265

Maronna, R. A., D. R. Martin, and V. J. Yohai 2006. *Robust Statistics Theory and Methods*. New York: Wiley. 249

Matloff, N. 2011. *The Art of R Programming*. San Francisco: no starch press.

Maybury, M. T. (ed.) 1997. *Intelligent Multimedia Information Retrieval*. Menlo Park, Calif./ Cambridge: AAAI Press / MIT Press.

McArthur, R. and P. Bruza 2001. The ABCs of online community. In N. Zhong, Y. Yao, J. Liu, and S. Ohsuga (eds.), *Web Intelligence: Research and Development*, pp. 141–147. New York: Springer.

McCallum, Q. E. (ed.) 2013. *Bad Data Handbook*. Sebastopol, Calif.: O'Reilly.

McCullagh, P. and J. A. Nelder 1989. *Generalized Linear Models* (second ed.). New York: Chapman and Hall.

McGrayne, S. B. 2011. *The Theory that Would Not Die: How Bayes' Rule Cracked the Enigma Code, Hunted Down Russian Submarines and Emerged Triumphant from Two Centuries of Controversy*. New Haven, Conn.: Yale University Press. 244

McKellar, J. and A. Fettig 2013. *Twisted Network Programming Essentials: Event-driven Network Programming with Python* (second ed.). Sebastopol, Calif.: O'Reilly.

McLean, B. and P. Elkind 2003. *The Smartest Guys in the Room: The Amazing Rise and Scandalous Fall of Enron*. New York: Penguin.

McTaggart, R. and G. Daroczi 2014. *Package Quandl: Quandl Data Connection*. Comprehensive R Archive Network. 2014. http://cran.r-project.org/web/packages/Quandl/Quandl.pdf with additional online documentation available at https://www.quandl.com/. 62

Meadow, C. T., B. R. Boyce, and D. H. Kraft 2000. *Text Information Retrieval Systems* (second ed.). San Diego: Academic Press.

Merkl, D. 2002. Text mining with self-organizing maps. In W. Klösgen and J. M. Z̊ytkow (eds.), *Handbook of Data Mining and Knowledge Discovery*, Chapter 46.9, pp. 903–910. Oxford: Oxford University Press.

Mertz, D. 2002. *Charming Python: SimPy Simplifies Complex Models*. IBM developerWorks. 2002. Retrieved from the World Wide Web at http://www.ibm.com/developerworks/linux/library/l-simpy/l-simpy-pdf.pdf.

Meyer, D. 2014a. *Proximity Measures in the proxy Package for R*. Comprehensive R Archive Network. 2014. http://cran.r-project.org/web/packages/proxy/vignettes/overview.pdf.

Meyer, D. 2014b. *proxy: Distance and Similarity Measures*. Comprehensive R Archive Network. 2014. http://cran.r-project.org/web/packages/proxy/proxy.pdf.

Meyer, D., E. Dimitriadou, K. Hornik, A. Weingessel, and F. Leisch 2014. *e1071: Misc Functions of the Department of Statistics (e1071), TU Wien*. Comprehensive R Archive Network. 2014. http://cran.r-project.org/web/packages/e1071/e1071.pdf.

Michalawicz, Z. and D. B. Fogel 2004. *How to Solve It: Modern Heuristics*. New York: Springer. 251

Mikowski, M. S. and J. C. Powell 2014. *Single Page Web Applications JavaScript End-to-End*. Shelter Island, N.Y.: Manning. 3

Milborrow, S. 2014. *rpart.plot: Plot rpart models. An Enhanced Version of plot.rpart*. Comprehensive R Archive Network. 2014. http://cran.r-project.org/web/packages/rpart.plot/rpart.plot.pdf.

Milgram, S. 1967. The small world problem. *Psychology Today* 1(1):60–67. 99

Miller, D. and D. Slater 2000. *The Internet: An Ethnographic Approach*. Oxford: Berg. 265

Miller, G. A. 1995. Wordnet: A lexical database for English. *Communications of the ACM* 38 (11):39–41. 259

Miller, J. H. and S. E. Page 2007. *Complex Adaptive Systems: An Introduction to Computational Models of Social Life*. Princeton, N. J.: Princeton University Press.

Miller, J. P. 2000a. *Millennium Intelligence: Understanding and Conducting Competitive Intelligence in the Digital Age*. Medford, N.J.: Information Today, Inc. 59

Miller, T. W. 2000b. Marketing research and the information industry. *CASRO Journal 2000*:21–26.

Miller, T. W. 2001, Summer. Can we trust the data of online research. *Marketing Research*: 26–32. Reprinted as "Make the call: Online results are a mixed bag," *Marketing News*, September 24, 2001:20–25.

Miller, T. W. 2002. Propensity scoring for multimethod research. *Canadian Journal of Marketing Research* 20(2):46–61.

Miller, T. W. 2005. *Data and Text Mining: A Business Applications Approach*. Upper Saddle River, N.J.: Pearson Prentice Hall.

Miller, T. W. 2015a. *Modeling Techniques in Predictive Analytics with Python and R: A Guide to Data Science*. Upper Saddle River, N.J.: Pearson Education. http://www.ftpress.com/miller.

Miller, T. W. 2015b. *Modeling Techniques in Predictive Analytics: Business Problems and Solutions with R* (revised and expanded ed.). Upper Saddle River, N.J.: Pearson Education. http://www.ftpress.com/miller.

Miller, T. W. and P. R. Dickson 2001. On-line market research. *International Journal of Electronic Commerce* 5(3):139–167.

Miller, T. W., D. Rake, T. Sumimoto, and P. S. Hollman 2001. Reliability and comparability of choice-based measures: Online and paper-and-pencil methods of administration. *2001 Sawtooth Software Conference Proceedings*:123–130. Published by Sawtooth Software, Sequim, WA.

Miller, T. W. and J. Walkowski (eds.) 2004. *Qualitative Research Online*. Manhattan Beach, Calif.: Research Publishers LLC. 261, 313

Miniwatts Marketing Group 2014. Internet users of the world. Retrieved from the World Wide Web, October 21, 2014, at http://www.internetworldstats.com/stats.htm. 45

Mitchell, M. 1996. *An Introduction to Genetic Algorithms*. Cambridge: MIT Press. 251

Mitchell, M. 2009. *Complexity: A Guided Tour*. Oxford, U.K.: Oxford University Press. 227

Moran, M. and B. Hunt 2009. *Search Engine Marketing, Inc.: Driving Search Traffic to Your Company's Web Site* (second ed.). Boston: IBM Press/Pearson Education.

Moreno, J. L. 1934. Who shall survive?: Foundations of sociometry, group psychotherapy, and sociodrama. Reprinted in 1953 (second edition) and in 1978 (third edition) by Beacon House, Inc., Beacon, N.Y. 96

Mosteller, F. and D. L. Wallace 1984. *Applied Bayesian and Classical Inference: The Case of "The Federalist" Papers* (second ed.). New York: Springer. Earlier edition published in 1964 by Addison-Wesley, Reading, Mass. The previous title was *Inference and Disputed Authorship: The Federalist*.

Müller, K., T. Vignaux, and C. Chui 2014. *SimPy: Discrete Event Simulation for Python*. SimPy Development Team. 2014. Retrieved from the World Wide Web at `https://simpy.readthedocs.org/en/latest/` with code available at `https://bitbucket.org/simpy/simpy/`.

Murphy, K. P. 2012. *Machine Learning: A Probabilistic Perspective*. Cambridge, Mass.: MIT Press. 176, 244, 251

Murrell, P. 2011. *R Graphics* (second ed.). Boca Raton, Fla.: CRC Press.

Nair, V. G. 2014. *Getting Started with Beautiful Soup: Build Your Own Web Scraper and Learn All About Web Scraping with Beautiful Soup*. Birmingham, UK: PACKT Publishing.

Nelson, B. L. 2013. *Foundations and Methods of Stochastic Simulation: A First Course*. New York: Springer. Supporting materials available from the World Wide Web at `http://users.iems.northwestern.edu/~nelsonb/IEMS435/`.

Nelson, W. B. 2003. *Recurrent Events Data Analysis for Product Repairs, Disease Recurrences, and Other Applications*. Series on Statistics and Applied Probability. Philadelphia and Alexandria, Va.: ASA-SIAM.

Neuendorf, K. A. 2002. *The Content Analysis Guidebook*. Thousand Oaks, Calif.: Sage.

Neuwirth, E. 2014. *Package RColorBrewer: ColorBrewer palettes*. Comprehensive R Archive Network. 2014. `http://cran.r-project.org/web/packages/RColorBrewer/RColorBrewer.pdf`.

Newman, M. E. J. 2010. *Networks: An Introduction*. Oxford, UK: Oxford University Press. 224

Nolan, D. and D. T. Lang 2014. *XML and Web Technologies for Data Sciences with R*. New York: Springer. 29

North, M. J. and C. M. Macal 2007. *Managing Business Complexity: Discovering Strategic Solutions with Agent-Based Modeling and Simulation*. Oxford, U.K.: Oxford University Press. 226

Nunnally, J. C. 1967. *Psychometric Theory*. New York: McGraw-Hill.

O'Hagan, A. 2010. *Kendall's Advanced Theory of Statistics: Bayesian Inference*, Volume 2B. New York: Wiley. 244

Osborne, J. W. 2013. *Best Practices in Data Cleaning: A Complete Guide to Everything You Need to Do Before and After Collecting Your Data*. Los Angeles: Sage.

Osgood, C. 1962. Studies in the generality of affective meaning systems. *American Psychologist* 17:10–28. 133

Osgood, C., G. Suci, and P. Tannenbaum (eds.) 1957. *The Measurement of Meaning*. Urbana, Ill.: University of Illinois Press. 133

Pang, B. and L. Lee 2008. Opinion mining and sentiment analysis. *Foundations and Trends in Information Retrieval* 2(1–2):1–135.

Pedregosa, F., G. Varoquaux, A. Gramfort, V. Michel, B. Thirion, O. Grisel, M. Blondel, P. Prettenhofer, R. Weiss, V. Dubourg, J. Vanderplas, A. Passos, D. Cournapeau, M. Brucher, M. Perrot, and E. Duchesnay 2011. Scikit-learn: Machine learning in Python. *Journal of Machine Learning Research* 12:2825–2830.

Penenberg, A. L. and M. Barry 2000. *Spooked: Espionage in Corporate America*. Cambridge, Mass.: Perseus Publishing. 56

Pentland, A. 2014. *Social Physics: How Good Ideas Spread—The Lessons from a New Science*. New York: Penguin.

Peters, G. and J. T. Woolley 2014. State of the Union Addresses and Messages. Online papers available from The American Presidency Project. Retrieved from the World Wide Web, November 1–14, 2014 at `http://www.presidency.ucsb.edu`. Documentation provided at `http://www.presidency.ucsb.edu/sou.php`. 295

Pinker, S. 1994. *The Language Instinct*. New York: W. Morrow and Co.

Pinker, S. 1997. *How the Mind Works*. New York: W.W. Norton & Company.

Pinker, S. 1999. *Words and Rules: The Ingredients of Language*. New York: HarperCollins.

Popping, R. 2000. *Computer-Assisted Text Analysis*. Thousand Oaks, Calif.: Sage. 133, 279

Porter, D. (ed.) 1997. *Internet Culture*. New York: Routledge. 265

Potts, C. 2011. On the negativity of negation. In *Proceedings of Semantics and Linguistic Theory 20*, pp. 636–659. CLC Publications. Retrieved from the World Wide Web at `http://elanguage.net/journals/salt/article/view/20.636/1414`.

Powazak, D. M. 2001. *Design for Community: The Art of Connecting Real People in Virtual Places*. Indianapolis: New Riders.

Powers, S. 2003. *Practical RDF*. Sebastopol, Calif.: O'Reilly. 235

Preece, J. 2000. *Online Communities: Designing Usability, Supporting Sociability*. New York: Wiley.

Priebe, C. E., J. M. Conroy, D. J. Marchette, and Y. Park 2005, October. Scan statistics on Enron graphs. *Computational and Mathematical Organization Theory* 11(3):229–247.

Priebe, J. 2009, April 29. A study of Internet users' cookie and JavaScript settings. Retrieved from the World Wide Web, October 21, 2014, at `http://smorgasbork.com/component/content/article/84-a-study-of-internet-users-cookie-and-javascript-settings`. 16

Provost, F. and T. Fawcett 2014. *Data Science for Business: What You Need to Know About Data Mining and Data-Analytic Thinking*. Sebastopol, Calif.: O'Reilly. 239

Pustejovsky, J. and A. Stubbs 2013. *Natural Language Annotation for Machine Learning*. Sebastopol, Calif.: O'Reilly.

Putler, D. S. and R. E. Krider 2012. *Customer and Business Analytics: Applied Data Mining for Business Decision Making Using R*. Boca Raton, Fla: Chapman & Hall/CRC.

Radcliffe-Brown, A. R. 1940. On social structure. *Journal of the Royal Anthropological Society of Great Britain and Ireland* 70:1–12. 96

Rajaraman, A. and J. D. Ullman 2012. *Mining of Massive Datasets*. Cambridge, UK: Cambridge University Press. 210

Reingold, E. M. and J. S. Tilford 1981. A fast algorithm for finding dominators in a flowgraph. *IEEE Transactions on Software Engineering* 7:223–228. 82

Reitz, K. 2014a, October 15. Python guide documentation (release 0.0.1). Retrieved from the World Wide Web, October 23, 2014, at `https://media.readthedocs.org/pdf/python-guide/latest/python-guide.pdf`.

Reitz, K. 2014b. Requests: HTTP for humans. Documentation available from the World Wide Web at `http://docs.python-requests.org/en/latest/`.

Rencher, A. C. and G. B. Schaalje 2008. *Linear Models in Statistics* (second ed.). New York: Wiley.

Resig, J. and BearBibeault 2013. *Secrets of the JavaScript Ninja*. Shelter Island, N.Y.: Manning. 3

Resnick, M. 1998. *Turtles, Termites, and Traffic Jams: Explorations in Massively Parallel Microworlds*. Cambridge, Mass.: MIT Press. 227

Rheingold, H. 2000. *The Virtual Community: Homesteading on the Electronic Frontier* (revised ed.). Cambridge: The MIT Press. 265

Ricci, F., L. Rokach, B. Shapira, and P. B. Kantor (eds.) 2011. *Recommender Systems Handbook*. New York: Springer.

Richardson, L. 2014. Beautiful soup documentation. Available from the World Wide Web at `http://www.crummy.com/software/BeautifulSoup/bs4/doc/`.

Robbins, J. N. 2003. *Learning Web Design: A Beginners Guide to HTML, CSS, JavaScript, and Web Graphics* (fourth ed.). Sebastopol, Calif.: O'Reilly.

Robert, C. P. 2007. *The Bayesian Choice: From Decision Theoretic Foundations to Computational Implementation* (second ed.). New York: Springer. 244

Robert, C. P. and G. Casella 2009. *Introducing Monte Carlo Methods with R*. New York: Springer. 244

Roberts, C. W. (ed.) 1997. *Text Analysis for the Social Sciences: Methods for Drawing Statistical Inferences from Texts and Transcripts*. Mahwah, N.J.: Lawrence Erlbaum Associates. 133, 279

Robinson, I., J. Webber, and E. Eifrem 2013. *Graph Databases*. Sebastopol, Calif.: O'Reilly.

Rosen, J. 2001. *The Unwanted Gaze: The Destruction of Privacy in America*. New York: Vintage. 134

Rossant, C. 2014. *IPython Interactive Computing and Visualization Cookbook*. Birmingham, UK: Packt Publishing.

Rousseeuw, P. J. 1987. Silhouettes: A graphical aid to the interpretation and validation of cluster analysis. *Journal of Computational and Applied Mathematics* 20:53–65.

Rubin, D. B. 1987. *Multiple Imputation for Nonresponse in Surveys*. New York: Wiley.

Russell, M. A. 2014. *Mining the Social Web: Data Mining Facebook, Twitter, LinkedIn, Google+, GitHub, and More*. Sebastopol, Calif.: O'Reilly. 61

Russell, S. and P. Norvig 2009. *Artificial Intelligence: A Modern Approach* (third ed.). Upper Saddle River, N.J.: Prentice Hall. 251

Rusu, A. 2014. Tree drawing algorithms. In R. Tamassia (ed.), *Handbook of Graph Drawing and Visualization*, Chapter 5, pp. 155–192. Boca Raton, Fla.: CRC Press/Chapman & Hall. 82

Ryan, G. W. and H. R. Bernard 2000. Data management and analysis methods. In N. K. Denzin and Y. S. Lincoln (eds.), *Handbook of Qualitative Research: Context and Method* (second ed.), Chapter 29, pp. 769–802. Thousand Oaks, Calif.: Sage. 279

Ryan, J. A. 2014. *Package quantmod: Quantitative Financial Modelling Framework*. Comprehensive R Archive Network. 2014. `http://cran.r-project.org/web/packages/quantmod/quantmod.pdf`. 62

Ryan, J. A. and J. M. Ulrich 2014. *Package xts: eXtensible Time Series*. Comprehensive R Archive Network. 2014. `http://cran.r-project.org/web/packages/xts/xts.pdf`. 62

Ryan, T. P. 2008. *Modern Regression Methods* (second ed.). New York: Wiley. 128

Šalamon, T. 2011. *Design of Agent-Based Models: Developing Computer Simulations for a Better Understanding of Social Processes*. Czech Republic: Tomáš Bruckner, Řepín-Živonín.

Samara, T. 2007. *Design Elements: A Graphic Style Manual*. Gloucester, Mass.: Rockport Publishers.

Sarkar, D. 2008. *Lattice: Multivariate Data Visualization with R*. New York: Springer.

Sarkar, D. 2014. *lattice: Lattice Graphics*. Comprehensive R Archive Network. 2014. `http://cran.r-project.org/web/packages/lattice/lattice.pdf`.

Sarkar, D. and F. Andrews 2014. *latticeExtra: Extra Graphical Utilities Based on Lattice*. Comprehensive R Archive Network. 2014. `http://cran.r-project.org/web/packages/latticeExtra/latticeExtra.pdf`.

Savage, D., X. Zhang, X. Yu, P. Chou, and Q. Wang 2014. Anomaly detection in online social networks. *Social Networks* 39:62–70.

Schafer, J. L. 1997. *Analysis of Incomplete Multivariate Data*. London: Chapman & Hall.

Schauerhuber, M., A. Zeileis, D. Meyer, and K. Hornik 2008. Benchmarking open-source tree learners in R/RWeka. In C. Preisach, H. Burkhardt, L. Schmidt-Thieme, and R. Decker (eds.), *Data Analysis, Machine Learning, and Applications*, pp. 389–396. New York: Springer.

Schaul, J. 2014. *ComplexNetworkSim Package Documentation*. pypi.python.org. 2014. Retrieved from the World Wide Web at `https://pythonhosted.org/Complex-NetworkSim/` with code available at `https://github.com/jschaul/Complex-ComplexNetworkSim`.

Schnettler, S. 2009. A structured overview of 50 years of small-world research. *Social Networks* 31:165–178. 99

Schrott, P. R. and D. J. Lanoue 1994. Trends and perspectives in content analysis. In I. Borg and P. Mohler (eds.); *Trends and Perspectives in Empirical Social Research*, pp. 327–345. Berlin: Walter de Gruyter.

Scott:, J. 2013. *Social Network Analysis* (third ed.). Thousand Oaks, Calif.: Sage.

Seale, C. 2002. Computer-assisted analysis of qualitative interview data. In J. F. Gubrium and J. A. Holstein (eds.), *Handbook of Interview Research: Context and Method*, Chapter 31, pp. 651–670. Thousand Oaks, Calif.: Sage. 279

Sebastiani, F. 2002. Machine learning in automated text categorization. *ACM Computing Surveys* 34(1):1–47.

Seber, G. A. F. 2000. *Multivariate Observations*. New York: Wiley. Originally published in 1984. 176

Segaran, T., C. Evans, and J. Taylor 2009. *Programming the Semantic Web*. Sebastopol, Calif.: O'Reilly. 235

Sen, R. and M. H. Hansen 2003, March. Predicting web user's next access based on log data. *Journal of Computational and Graphical Statistics* 12(1):143–155.

Senkul, P. and S. Salin 2012, March. Improving pattern quality in web usage mining by using semantic information. *Knowledge and Information Systems* 30(3):527–541.

SEOMoz, Inc. 2013. Search engine ranking factors: survey and correlation data. Retrieved from the World Wide Web, November 1, 2014, at `http://moz.com/search-ranking-factors`.

Shakhnarovich, G., T. Darrell, and P. Indyk (eds.) 2006. *Nearest-Neighbor Methods in Learning and Vision: Theory and Practice*. Cambridge, Mass.: MIT Press.

Sharda, R. and D. Delen 2006. Predicting box office success of motion pictures with neural networks. *Expert Systems with Applications* 30:243–254. 133

Sharma, S. 1996. *Applied Multivariate Techniques*. New York: Wiley. 176

Sherry, Jr., J. F. and R. V. Kozinets 2001. Qualitative inquiry in marketing and consumer research. In D. Iacobucci (ed.), *Kellogg on Marketing*, Chapter 8, pp. 165–194. New York: Wiley.

Shmueli, G. 2010. To explain or predict? *Statistical Science* 25(3):289–310.

Shroff, G. 2013. *The Intelligent Web: Search, Smart Algorithms, and Big Data*. Oxford, U.K.: Oxford University Press.

Siegel, E. 2013. *Predictive Analytics: The Power to Predict Who Will Click, Buy, Lie, or Die*. Hoboken, N.J.: Wiley. 239

Silverman, D. 1993. *Interpreting Qualitative Data: Methods for Analysing Talk, Text, and Interaction*. London: Sage. 279

Silverman, D. 2000a. Analyzing talk and text. In N. K. Denzin and Y. S. Lincoln (eds.), *Handbook of Qualitative Research: Context and Method* (second ed.), Chapter 31, pp. 821–834. Thousand Oaks, Calif.: Sage. 279

Silverman, D. 2000b. *Doing Qualitative Research: A Practical Approach*. Thousand Oaks, Calif.: Sage.

Silverman, D. 2001. *Interpreting Qualitative Data: Methods for Analysing Talk, Text, and Interaction* (second ed.). Thousand Oaks, Calif.: Sage.

Simpson, J. 2002. *XPath and XPointer: Locating Content in XML Documents*. Sebastopol, Calif.: O'Reilly. 29

Sing, T., O. Sander, N. Beerenwinkel, and T. Lengauer 2005. ROCR: Visualizing classifier performance in R. *Bioinformatics* 21(20):3940–3941.

Slater, J. B., P. V. M. Broekman, M. Corris, A. Iles, B. Seymour, and S. Worthington (eds.) 2013. *Proud to Be Flesh—A Mute Magazine Anthology of Cultural Politics After the Net*. London: Mute Publishing.

Smith, J. A. and J. Moody 2013. Structural effects of network sampling coverage I: Nodes missing at random. *Social Networks* 35:652–668. 109

Smith, M. A. and P. Kollock (eds.) 1999. *Communities in Cyberspace*. New York: Routledge. 265

Snedecor, G. W. and W. G. Cochran 1989. *Statistical Methods* (eighth ed.). Ames, Iowa: Iowa State University Press. First edition published by Snedecor in 1937. 244

Socher, R., J. Pennington, E. H. Huang, A. Y. Ng, and C. D. Manning 2011. Semi-Supervised Recursive Autoencoders for Predicting Sentiment Distributions. In *Proceedings of the 2011 Conference on Empirical Methods in Natural Language Processing (EMNLP)*.

Srivastava, A. N. and M. Sahami (eds.) 2009. *Text Mining: Classification, Clustering, and Applications*. Boca Raton, Fla.: CRC Press.

StatCounter 2014. StatCounter global stats. Retrieved from the World Wide Web, October 21, 2014, at `http://gs.statcounter.com/#browser-ww-monthly-200807-201410`.

Stefanov, S. 2010. *JavaScript Patterns: Building Better Applications with Coding and Design Patterns*. Sebastopol, Calif.: O'Reilly. 3

Stone, P. J. 1997. Thematic text analysis: New agendas for analyzing text content. In C. W. Roberts (ed.), *Text Analysis for the Social Sciences: Methods for Drawing Statistical Inferences from Texts and Transcripts*, Chapter 2, pp. 35–54. Mahwah, N.J.: Lawrence Erlbaum Associates. 279

Stone, P. J., D. C. Dunphy, M. S. Smith, and D. M. Ogilvie 1966. *The General Inquirer: A Computer Approach to Content Analysis*. Cambridge: MIT Press.

Strauss, A. and J. Corbin 1998. *Basics of Qualitative Research: Techniques and Procedures for Developing Grounded Theory* (second ed.). Thousand Oaks, Calif.: Sage. 279

Stuart, A., K. Ord, and S. Arnold 2010. *Kendall's Advanced Theory of Statistics: Classical Inference and the Linear Model*, Volume 2A. New York: Wiley. 244

Suess, E. A. and B. E. Trumbo 2010. *Introduction to Probability Simulation and Gibbs Sampling with R*. New York: Springer. 244

Sullivan, D. 2001. *Document Warehousing and Text Mining: Techniques for Improving Business Operations, Marketing, and Sales*. New York: Wiley.

Supowit, K. J. and E. M. Reingold 1983. The complexity of drawing trees nicely. *Acta Informatica* 18:177–392. 82

Taddy, M. 2013a. Measuring political sentiment on Twitter: factor-optimal design for multinomial inverse regression. Retrieved from the World Wide Web at `http://arxiv.org/pdf/1206.3776v5.pdf`.

Taddy, M. 2013b. Multinomial inverse regression for text analysis. Retrieved from the World Wide Web at `http://arxiv.org/pdf/1012.2098v6.pdf`.

Taddy, M. 2014. *textir: Inverse Regression for Text Analysis*. 2014. `http://cran.r-project.org/web/packages/textir/textir.pdf`.

Tamassia, R. (ed.) 2014. *Handbook of Graph Drawing and Visualization*. Boca Raton, Fla.: CRC Press/Chapman & Hall.

Tan, P.-N., M. Steinbach, and V. Kumar 2006. *Introduction to Data Mining*. Boston: Addison-Wesley. 210

Tang, W., H. He, and X. M. Tu 2012. *Applied Categorical and Count Data Analysis*. Boca Raton, Fla.: Chapman & Hall/CRC.

Tannenbaum, A. S. and D. J. Wetherall 2010. *Computer Networks* (fifth ed.). Upper Saddle River, N.J.: Pearson/Prentice Hall.

Tanner, M. A. 1996. *Tools for Statistical Inference: Methods for the Exploration of Posterior Distributions and Likelihood Functions* (third ed.). New York: Springer. 244

Tennison, J. 2001. *XSLT and XPath On The Edge*. New York: Wiley. 29

Therneau, T. 2014. *survival: Survival Analysis*. Comprehensive R Archive Network. 2014. http://cran.r-project.org/web/packages/survival/survival.pdf.

Therneau, T., B. Atkinson, and B. Ripley 2014. *rpart: Recursive Partitioning*. Comprehensive R Archive Network. 2014. http://cran.r-project.org/web/packages/rpart/rpart.pdf.

Therneau, T. and C. Crowson 2014. *Using Time Dependent Covariates and Time Dependent Coefficients in the Cox Model*. Comprehensive R Archive Network. 2014. http://cran.r-project.org/web/packages/survival/vignettes/timedep.pdf.

Therneau, T. M. and P. M. Grambsch 2000. *Modeling Survival Data: Extending the Cox Model*. New York: Springer.

Thiele, J. C. 2014. *Package RNetLogo: Provides an Interface to the Agent-Based Modelling Platform NetLogo*. Comprehensive R Archive Network. 2014. User manual: http://cran.r-project.org/web/packages/RNetLogo/RNetLogo.pdf. Additional documentation available at http://rnetlogo.r-forge.r-project.org/.

Thompson, M. 1975. On any given Sunday: Fair competitor orderings with maximum likelihood methods. *Journal of the American Statistical Association* 70(351):536–541. 103

Thompson, T. (ed.) 2000. *Writing about Business* (second ed.). New York: Columbia University Press.

T.Hothorn, F. Leisch, A. Zeileis, and K. Hornik 2005, September. The design and analysis of benchmark experiments. *Journal of Computational and Graphical Statistics* 14(3):675–699.

Thurstone, L. L. 1927. A law of comparative judgment. *Psychological Review* 34:273–286.

Tibshirani, R. 1996. Regression shrinkage and selection via the lasso. *Journal of the Royal Statistical Society, Series B* 58:267–288. 249

Tidwell, J. 2011. *Designing Interfaces: Patterns for Effective Interactive Design* (second ed.). Sebastopol, Calif.: O'Reilly.

Toivonen, R., L. Kovanen, J.-P. O. Mikko Kivelä, J. SaramŁki, and K. Kaski 2009. A comparative study of social network models: Network evolution models and nodal attribute models. *Social Networks* 31:240–254.

Tong, S. and D. Koller 2001. Support vector machine active learning with applications to text classification. *Journal of Machine Learning Research* 2:45–66.

Torgerson, W. S. 1958. *Theory and Methods of Scaling*. New York: Wiley.

Travers, J. and S. Milgram 1969, December. Experimental study of small world problem. *Sociometryy* 32(4):425–443. 99

Trybula, W. J. 1999. Text mining. In M. E. Williams (ed.), *Annual Review of Information Science and Technology*, Volume 34, Chapter 7, pp. 385–420. Medford, N.J.: Information Today, Inc.

Tsvetovat, M. and A. Kouznetsov 2011. *Social Network Analysis for Startups: Finding Connections on the Social Web*. Sebastopol, Calif.: O'Reilly.

Tufte, E. R. 1990. *Envisioning Information*. Cheshire, Conn.: Graphics Press.

Tufte, E. R. 1997. *Visual Explanations: Images and Quantities, Evidence and Narrative*. Cheshire, Conn.: Graphics Press.

Tufte, E. R. 2004. *The Visual Display of Quantitative Information* (second ed.). Cheshire, Conn.: Graphics Press.

Tufte, E. R. 2006. *Beautiful Evidence*. Cheshire, Conn.: Graphics Press.

Tukey, J. W. 1977. *Exploratory Data Analysis*. Reading, Mass.: Addison-Wesley.

Tukey, J. W. and F. Mosteller 1977. *Data Analysis and Regression: A Second Course in Statistics*. Reading, Mass.: Addison-Wesley.

Turney, P. D. 2002, July 8–10. Thumbs up or thumbs down? Semantic orientation applied to unsupervised classification of reviews. *Proceedings of the 40th Annual Meeting of the Association for Computational Linguistics (ACL '02)*:417–424. Available from the National Research Council Canada publications archive. 134

Turow, J. 2013. *The Daily You: How the New Advertising Industry Is Defining Your Identity and Your Worth*. New Haven, Conn.: Yale University Press. 134

Uddin, S., J. Hamra, and L. Hossain 2013. Exploring communication networks to understand organizational crisis using exponential random graph models. *Computational and Mathematical Organization Theory* 19:25–41.

Uddin, S., A. Khan, L. Hossain, M. Piraveenan, and S. Carlsson 2014. A topological framework to explore longitudinal social networks. *Computational and Mathematical Organization Theory*:21 pages. Manuscript published online by Springer Science+Business Media, New York.

Unwin, A., M. Theus, and H. Hofmann (eds.) 2006. *Graphics of Large Datasets: Visualizing a Million*. New York: Springer. 252

Vanderbei, R. J. 2014. *Linear Programming: Foundations and Extensions* (fourth ed.). New York: Springer. 225

Vapnik, V. N. 1998. *Statistical Learning Theory*. New York: Wiley. 129

Vapnik, V. N. 2000. *The Nature of Statistical Learning Theory* (second ed.). New York: Springer. 129

Venables, W. N. and B. D. Ripley 2002. *Modern Applied Statistics with S* (fourth ed.). New York: Springer-Verlag. Champions of S, S-Plus, and R call this *the mustard book*.

Vine, D. 2000. *Internet Business Intelligence: How to Build a Big Company System on a Small Company Budget*. Medford, N.J.: Information Today, Inc. 59

W3Techs 2014. Usage of traffic analysis tools for websites. Retrieved from the World Wide Web, October 21, 2014, at `http://w3techs.com/technologies/overview/traffic_analysis/all`. 3, 14

Walther, J. B. 1996. Computer-mediated communication: impersonal, interpersonal, and hyperpersonal interaction. *Communication Research* 23(1):3–41. 265

Wanderschneider, M. 2013. *Learning Node.js: A Hands-On Guide to Building Web Applications in JavaScript*. Upper Saddle River, N.J.: Pearson Education/Addison-Wesley. 3

Wang, X., H. Tao, Z. Xie, and D. Yi 2013. Mining social networks using wave propogation. *Computational and Mathematical Organization Theory* 19:569–579.

Wasserman, L. 2010. *All of Statistics: A Concise Course in Statistical Inference*. New York: Springer. 244

Wasserman, S. and C. Anderson 1987. Stochastic *a posteriori* blockmodels: Construction and assessment. *Social Networks* 9:1–36.

Wasserman, S. and K. Faust 1994. *Social Network Analysis: Methods and Applications*. Cambridge, UK: Cambridge University Press. Chapter 4: Graphs and Matrices contributed by D. Iacobucci.

Wasserman, S. and D. Iacobucci 1986. Statistical analysis of discrete relational data. *British Journal of Mathematical and Statistical Psychology* 39:41–64.

Wasserman, S. and P. Pattison 1996, September. Logit models and logistic regression for social networks: I An introduction to markov graphs and p^*. *Psychometrika* 61(3):401–425.

Watts, D. J. 1999. *Small Worlds: The Dynamics of Networks between Order and Randomness*. Princeton, N.J.: Princeton University Press. 99

Watts, D. J. 2003. *Six Degrees: The Science of a Connected Age*. New York: W.W. Norton.

Watts, D. J. and S. H. Strogatz 1998. Collective dynamics of 'small-world' networks. *Nature* 393(6684):440–442. 99, 224

Wei, T. H. 1952. The algebraic foundations of ranking theory. Ph.D. thesis, Cambridge University, Cambridge, UK.

Weiner, J. 2014. *Package riverplot: Sankey or ribbon plots*. Comprehensive R Archive Network. 2014. http://cran.r-project.org/web/packages/riverplot/riverplot.pdf.

Weiss, S. M., N. Indurkhay, and T. Zhang 2010. *Fundamentals of Predictive Text Mining*. New York: Springer.

Weiss, S. M., N. Indurkhya, and T. Zhang 2010. *Fundamentals of Predictive Text Mining*. New York: Springer.

Weitzman, E. A. 2000. Software and qualitative research. In N. K. Denzin and Y. S. Lincoln (eds.), *Handbook of Qualitative Research: Context and Method* (second ed.)., Chapter 30, pp. 803–820. Thousand Oaks, Calif.: Sage. 279

Weitzman, E. A. and M. B. Miles 1995. *Computer Programs for Qualitative Data Analysis*. Thousand Oaks, Calif.: Sage. 279

Werry, C. and M. Mowbray (eds.) 2001. *Online Communities: Commerce, Community Action, and the Virtual University*. Upper Saddle River, N.J.: Prentice-Hall.

West, B. T. 2006. A simple and flexible rating method for predicting success in the NCAA basketball tournament. *Journal of Quantitative Analysis in Sports* 2(3):1–14. 103

West, M. D. (ed.) 2001. *Theory, Method, and Practice in Computer Content Analysis*. Westport, Conn.: Ablex. 133, 279

White, T. 2011. *Hadoop: The Definitive Guide* (second ed.). Sebastopol, Calif.: O'Reilly.

Wickham, H. 2010. stringr: Modern, consistent string processing. *The R Journal* 2(2):38–40.

Wickham, H. 2014a. *Advanced R*. Boca Raton, Fla.: Chapman & Hall/CRC.

Wickham, H. 2014b. *stringr: Make It Easier to Work with Strings*. Comprehensive R Archive Network. 2014. http://cran.r-project.org/web/packages/stringr/stringr.pdf. 259

Wickham, H. and W. Chang 2014. *ggplot2: An Implementation of the Grammar of Graphics*. Comprehensive R Archive Network. 2014. http://cran.r-project.org/web/packages/ggplot2/ggplot2.pdf. 62, 252

Wikipedia 2014a. 2011 Virginia earthquake—Wikipedia, the free encyclopedia. Retrieved from the World Wide Web, September 18, 2014, at url = http://en.wikipedia.org/w/index.php?title=2011_Virginia_earthquake&oldid=625563344.

Wikipedia 2014b. Directory on privacy and electronic communications. Retrieved from the World Wide Web, October 21, 2014, at http://en.wikipedia.org/wiki/Directive_on_Privacy_and_Electronic_Communication. 19

Wikipedia 2014c. History of Wikipedia—Wikipedia, the free encyclopedia. Retrieved from the World Wide Web, September 23, 2014, at http://en.wikipedia.org/wiki/History_of_Wikipedia.

Wilensky, U. 1999. Netlogo. NetLogo 5.1.0 User Manual and documentation available from the Center for Connected Learning and Computer-Based Modeling, Northwestern University, Evanston, IL at http://ccl.northwestern.edu/netlogo/.

Wilkinson, L. 2005. *The Grammar of Graphics* (second ed.). New York: Springer.

Williams, H. P. 2013. *Model Building in Mathematical Programming* (fifth ed.). New York: Wiley. 225

Witten, I. H., E. Frank, and M. A. Hall 2011. *Data Mining: Practical Machine Learning Tools and Techniques*. Burlington, Mass.: Morgan Kaufmann. 210, 250

Witten, I. H., A. Moffat, and T. C. Bell 1999. *Managing Gigabytes: Compressing and Indexing Documents and Images* (second ed.). San Francisco: Morgan Kaufmann.

Wolcott, H. F. 1994. *Transforming Qualitative Data: Description, Analysis, and Interpretation*. Thousand Oaks, Calif.: Sage.

Wolcott, H. F. 1999. *Ethnography: A Way of Seeing*. Walnut Creek, Calif.: AltaMira.

Wolcott, H. F. 2001. *The Art of Fieldwork*. Walnut Creek, Calif.: AltaMira.

Wood, D., M. Zaidman, and L. Ruth 2014. *Linked Data: Structured Data on the Web*. Shelter

Island, N.Y.: Manning.

Yau, N. 2011. *Visualize This: The FlowingData Guide to Design, Visualization, and Statistics*. New York: Wiley.

Yau, N. 2013. *Data Points: Visualization That Means Something*. New York: Wiley.

Youmans, G. 1990. Measuring lexical style and competence: The type-token vocabulary curve. *Style* 24(4):584–599. 259

Youmans, G. 1991. A new tool for discourse analysis: The vocabulary management profile. *Language* 67(4):763–789. 259

Zeileis, A., G. Grothendieck, and J. A. Ryan 2014. *Package zoo: S3 Infrastructure for Regular and Irregular Time Series (Z's ordered observations)*. Comprehensive R Archive Network. 2014. `http://cran.r-project.org/web/packages/zoo/zoo.pdf`. 62

Zeileis, A., K. Hornik, and P. Murrell 2009, July. Escaping RGBland: Selecting colors for statistical graphics. *Computational Statistics and Data Analysis* 53(9):3259–3270.

Zeileis, A., K. Hornik, and P. Murrell 2014. *HCL-Based Color Palettes in R*. Comprehensive R Archive Network. 2014. `http://cran.r-project.org/web/packages/colorspace/vignettes/hcl-colors.pdf`.

Zubcsek, P. P., I. Chowdhury, and Z. Katona 2014. Information communities: The network structure of communication. *Social Networks* 38:50–62.

索　引

索引中所标页码为英文版原书页码，与页边栏中的页码一致。

A

adjacency matrix（邻接矩阵），70，314

agent（代理），314

Alteryx，250，299

ARPANET，313，314

ASP（应用服务提供商、动态服务器页面），314

association rule（关联规则），203-210

antecedent（前驱），203

confidence（信心），203，204，209

consequent（后继），203

item set（条目集），203

lift（提升度），204，209

support（支持），203，204，209

B

bandwidth（带宽），314

bar chart, *see* data visualization, bar chart（柱状图，参见数据可视化，柱状图）

Bayesian statistics（贝叶斯统计），243，244

Bayes' theorem（贝叶斯定理），243

betweenness centrality（介数中心性），71，102，314

big data（大数据），240

biologically-inspired methods（生物启发方法），251

black box model（黑盒模型），250

bot, *see* crawler (web crawler) [机器人，参见爬虫（Web 爬虫）]

boundary (of a network) [（网络）边界]，315

bps（比特率），315

browser launch（浏览器启动），315

browser usage（浏览器用处），5，6，15，17，18，28

bulletin board（公告栏），264，271-273，314，315

advantages（优势），273

applications（应用，应用程序），271

C

case study（案例分析）

Anonymous Microsoft Web Data（匿名微软网络数据），204，205，211，296

E-Mail or Spam?（电子邮件还是垃圾邮件？），281-283

Enron E-Mail Corpus and Network（安然电子邮件语料库与网络），75-82，91，108，291

Keyword Games（关键字游戏），47，51，288，290

POTUS Speeches（美国总统的演讲），172，176，178-183，192，197，295

Quake Talk（地震谈话），294

ToutBay Begins（开启 ToutBay 之旅），14-17，20，284-287

Wikipedia Votes（维基百科选举），109，115，292

chat room（聊天室），264，315

circle network（环形网络），71

classical statistics（经典的统计），242，244

null hypothesis（零假设），242

power（幂），243

statistical significance（统计上的重要性），242，243

classification（分类），121，129，238，246，248，250

predictive accuracy（预测精确度），247，248，300，303

client（客户，客户端），267，315

client-server application（客户端 – 服务器端应用程序），315

closeness centrality（接近中心性），102，315

cluster analysis（聚类分析），107，173，176，251
clustering coefficient (of a network), *see* transitivity [（网络）聚类系数]，参见传递性）
coefficient of determination（决定的系数），246
collaborative filtering（协同过滤），202
collage（拼图），315
competitive intelligence（竞争性情报），59
complexity, of model（模型的复杂度），249
content analysis（内容分析），315
Continuum Analytics（公司名）），299
cookie，315
corpus（语料库），315
correlation heat map, *see* data visualization, correlation heat map（关联热度图，参见数据可视化，关联热度图）
cost per click (CPC) [单位点击成本（CPC）]，315
CPC, *see* cost per click (CPC) [CPC，参见单位点击成本（CPC）]
crawler (web crawler) [爬虫（Web 爬虫）]，43，315
cross-validation（交叉验证），249

D

data preparation（数据准备），241
missing data（缺失数据），241
data science（数据科学），237，239
data visualization（数据可视化）
    bar chart（柱状图），15，206
    bubble chart（气泡图），208
    correlation heat map（关联热度图）104，307
    diagnostics（诊断，诊断学），248
    dot chart（点阵图），131
    histogram（方柱图），101
    lattice plot（格子图），252
    multidimensional scaling map（多维标度图），174，197
    network diagram（网络图），71-74，76-80，98-100
    Sankey diagram（桑基图），16，18，20
    scatter plot（散点图），207
    stacked area graph（堆叠区域图），5，10
    text map（文本图），174，197
    time line（时间线），269，272，277
    time series plot（时间序列图），17，62
    tree diagram（树形图），132
    word cloud（字云），176，178-182，259
database system（数据库系统），240
    non-relational（非关系型），240，241
    relational（关系型），240，241
degree（度，程度），70，315
degree centrality（度中心性），70，315
degree distribution（度分布），70，315
density (of a network) [（网络）密度]，316
discussion guide（讨论指南），264，267，275
Document Object Model (DOM) [文档对象模型]，26，316
DOM, *see* Document Object Model (DOM) [DOM，参见文档对象模型]
DSL（数字用户线路），316
dyad（二人小组），316

E

e-mail（电子邮件），263，316
eigenvector centrality（特征向量中心性），102，316
emoticons（表情符号），316
Enthought，299
ethnography（民族志），264，265，316
    digital ethnography（数字民族志），265
    netnography（网络志），265
    virtual ethnography（虚拟民族志），265
    event duration chart, *see* data visualization, time line（事件持续图，参见数据可视化，时间线）
explanatory model（解释性模型），238
explanatory variable（解释性变量），246

F

focused conversation（焦点讨论），276，277，316
frame（帧），316
Fruchterman-Reingold algorithm（Fruchterman-Reingold 算法），81
ftp（文件传输协议），316

G

game theory（博弈论），316
General Inquirer，133
generalized linear model（广义线性模型），246，249
generative grammar（生成语法），316
genetic algorithms（遗传算法），251
grounded theory（扎根理论），279，316

H

heuristics（探索法），251
histogram, *see* data visualization, histogram（方柱

图，参见数据可视化，方柱图）
HTML，314，316
HTTP，261，316

I

IBM（国际商用机器公司），250，299
ICQ，317
IMHO，317
interaction effect（交互效应），248
Internet（互联网），313，317
Internet Services Provider（互联网服务提供商），317
interview（访谈），267，268
intranet（内网），317
IRC，317
IT，317
item analysis, psychometrics（字条分析，心理学）

J

Java，317
JavaScript，3，298，317
JavaScript Object Notation (JSON)[JavaScript 对象表示法（JSON）]，317
JPEG，317
JSON, *see* JavaScript Object Notation (JSON) [JSON, 参见 JavaScript 对象表示法（JSON）]

K

Kamada-Kawai algorithm（Kamada-Kawai 算法），81
kbps（千比特率），317
keyword（关键字），317
keyword density（关键字密度），46
KNIME（KNIME），250

L

LAMP，317
line network（线性网），71
linear model（线性模型），246，249
linear predictor（线性预测器），246
listserv（邮件列表），263，317
log-linear models（对数线性模型），107
logistic regression（逻辑回归），128，246
LOL，317
Luddite（勒德分子），317

M

machine learning（机器学习），250，251
MEG，317
Microsoft（微软），299
modem（调制解调器），317
moderator（主持人），264，266，267，270，272，274
morphology（形态学），317
multidimensional scaling（多维标度），107，173，176
multidimensional scaling map, *see* data visualization, multidimensional scaling map（多维标度图，参见数据可视化，多维标度图）
multivariate methods（多变量方法），176

N

natural language processing（自然语言处理），317
nearest neighbor model（最近邻居模型），100，106，107，202
nearest-neighbor model（最近邻居模型），202
netiquette（网上礼节），318
network（网络），318
network diagram, *see* data visualization, network diagram（网络图，参见数据可视化，网络图）
network visualization（网络可视化），69-95

O

observer（观察者），270
online community（在线社区），265，318，320
online focus group（在线焦点小组），266，273
  differences with traditional focus group（与传统焦点小组的区别），266，269
  similarities with traditional focus group（与传统焦点小组的相似之处），266
online observation（在线观察），268
optimization（优化），251
organic search（有机搜索），46，318
over-fitting（过度拟合），248

P

page view（页面访问），318
PageRank，103，318
paid search（付费搜索），45，318
panel（座谈小组），318
parametric models（带有参数的模型），248
parser (text parser) [解析器（文本解析器）]，43，318
Perl，261，318
PHP，261，318
Poisson regression（泊松回归），245

post（发布），318
predictive model（预测模型），238
primary source (of information) [（信息的）主要来源]，58
principal component analysis（主成份分析），251，
psychographics（心理统计学），318
Python，3，4，297，298，319
Python package（Python 软件包）
    BeautifulSoup，30
    datetime，9
    fnmatch，183
    lxml，30
    matplotlib，9，83，135，183
    networkx，83
    nltk，135，183
    numpy，83，135，183，229
    os，33，36，135，183
    pandas，9，135，183，229
    patsy，135
    re，135，183
    requests，30
    scipy，183
    scrapy，33，36
    sklearn，135，183
    statsmodels，135，229

Q

qualitative research（定性研究），278

R

R（R 语言），3，298
R package（R 语言软件包）
    arules，211
    arulesViz，211
    car，211
    caret，151
    e1071，151
    ggplot2，10，20，63，151，197，231
    grid，151
    gridExtra，20
    igraph，87，91，110，115
    intergraph，91
    lattice，110，115
    latticeExtra，151
    lubridate，20，63
    network，91
    Quandl，63
    quantmod，63
    randomForest，151
    RColorBrewer，20，211
    RCurl，32，63
    riverplot，20
    RJSONIO，51
    RNetLogo，231
    rpart，151
    rpart.plot，151
    stringr，151
    tm，151
    wordcloud，192
    XML，32，63
    xts，63
    zoo，63
R-squared（R 平方），246
random forest（随机森林），129-131
random network (random graph) [随机网络（随机图）]，97，319
real-time focus group（实时焦点小组），268，271，319
    advantages（优势），270
    applications（应用，应用程序），270
    disadvantages（劣势），270
    system failures（系统故障），269
recommender systems（推荐系统），201-222
regression（回归），128，238，245，249
    nonlinear regression（非线性回归），249
    robust methods（稳健的方法），249
regular expressions（正则表达式），26，319
regularized regression（正则化回归），249
Reingold-Tilford algorithm（Reingold-Tilford 算法），82
response（响应、回复），245
robot, *see* crawler (web crawler) [机器人，参见爬虫（Web 爬虫）]
root mean-squared error（根均方差，RMSE），246
Rstudio（RStudio），299

S

sampling（抽样）
    sampling variability（抽样变异性），243
Sankey diagram, *see* data visualization, Sankey diagram（桑基图，参见数据可视化，桑基图）

SAS，250，299
scheduling（调度），251
scraper (web scraper) [抓取程序（web 抓取程序）]，43，319
search engine optimization, *see* web presence testing（搜索引擎优化，参见 web 存在测试）
secondary source (of information) [（信息的）次要来源]，58
segmentation（分段），210
semantic web（语义网），234，235，319
semantics（语义学），319
semi-supervised learning（半监督式学习），251
sentiment analysis（情感分析），119-171
SEO, *see* web presence testing（SEO，参见互动存在测试）
shrinkage estimators（收缩估算器），249
simulation（仿真），249
  benchmark study（基准研究），129，249，250
  what-if analysis（假设分析），238
small-world network（小世界网络），99
smoothing methods（平滑方法），249
  splines（样条函数），249
social network analysis（社交网络分析），95-118
sparse matrix（稀疏矩阵），201
spider, *see* crawler (web crawler) [蜘蛛，参见爬虫（Web 爬虫）]
star network（星形网络），70
statistic（统计）
  interval estimate（区间估计），242
  p-value，242
  point estimate（点估计），242
  test statistic（测试统计），242
stemming (word stemming) [词干提取（词干）]，319
stop words（停止词），172
Strategic and Competitive Intelligence Professionals（战略与竞争性情报专家），59
supervised learning（监督式学习），245，251，258
support vector machines（支持向量机），129
syntax（语法），319

T

TCP/IP，319
telnet，319
term frequency-inverse document frequency（词频率－文档频率倒数），173
terms-by-documents matrix（术语－文档矩阵），173
testing links（测试连接），47
text analysis（文本分析），267，319
text analytics（文本分析学），171-200，253-259
  content analysis（内容分析），133
  generative grammar（生成语法）253，254
  latent Dirichlet allocation（潜在狄利克雷分配），251
  latent semantic analysis（潜在语义分析），251
  morphology（形态学），254
  natural language processing（自然语言处理），134，253
  semantics（语义学），254
  stemming（词干提取），255
  syntax（语法），254
  terms-by-documents matrix（术语－文档矩阵），255，257
  text summarization（文本总结），258
  thematic analysis（主题分析），133，251
text map, *see* data visualization, text map（文本图，参见数据可视化，文本图）
text measure（文本度量），120，133，175，301
text measures（文本度量），319
text mining（文本挖掘），319
TF-IDF, *see* term frequency-inverse document-frequency（TF-IDF，参见考词频率－文档频率倒数）
thread, of discussion（讨论主线），273，275，316，319
time line, *see* data visualization, time line（时间线，参见数据可视化，时间线）
time series plot, *see* data visualization（时间序列图，参见数据可视化，时间序列图）
training-and-test regimen（训练－测试方案），129，238
transcript（副本），319
transitivity（传递性），109，319
tree network（树型网络），73
tree-structured model classification（树形结构模型分类），130，132
triad（三合会），109，320
triple, *see* triad（三倍数，参见三合会）

U

unsupervised learning（非监督式学习），251，256
URL，314，320
Usenet，320

V

variable transformation（变量变换），248
virtual facility（虚拟设施），320

W

web board（Web 留言板），320
web browser（Web 浏览器），267
web presence（Web 存在），47
web presence testing（Web 存在测试），320
web server（Web 服务器），267，320
web services（Web 服务），320
weblog（博客），264，265，320
Weka，210
Wiki，320
word stemming, *see* stemming (word stemming) [词干提取，参见词干提取（词干）]
World Wide Web（万维网），267，314

X

XML，320
XPath，26，320

推荐阅读

预测分析建模：Python与R语言实现

作者：[美] 托马斯 W. 米勒 ISBN：978-7-111-54887-4 定价：79.00元

在数据和算法统治的当下，我们只有通过强大的分析技术和信息交流才能获得稍纵即逝的竞争优势。本书将战略与管理、方法与模型、信息技术与代码三者完美结合，系统讲解如何使用横截面数据、时间序列、空间数据及时空数据解决我们面临的商务挑战，包括市场细分、品牌定位、产品选择建模、定价研究、财经分析、体育分析、文本分析、情感分析和社交网络分析等。

书中循序渐进地讲解了如何定义问题、识别数据、打造和优化模型、编写有效的Python和R代码、解释结果等。每章集中讨论一个预测分析的关键应用，介绍相关的预测分析模型知识、使用方法及优化策略。如果你从事数据分析工作，那么可以通过本书的案例，学习如何一步一步地分析问题、解决问题、找出问题的答案。

社交网站的数据挖掘与分析（原书第2版）

作者：[美] Matthew A. Russell ISBN：978-7-111-48699-2 定价：79.00元

社交网站数据如同深埋地下的“金矿”，如何利用这些数据来发现哪些人正通过社交媒介进行联系？他们正在谈论什么？或者他们在哪儿？本书第2版对上一版内容进行了全面更新和修订，它将揭示回答这些问题的方法与技巧。你将学到如何获取、分析和汇总散落于社交网站（包括Facebook、Twitter、LinkedIn、Google+、 GitHub、邮件、网站和博客等）的数据，以及如何通过可视化找到你一直在社交世界中寻找的内容和你闻所未闻的有用信息。

推荐阅读

R语言编程艺术

作者：[美] Norman Matloff ISBN：978-7-111-42314-0 定价：69.00元

R语言是世界上最流行的用于数据处理和统计分析的脚本语言。考古学家用它来跟踪古代文明的传播，医药公司用它来探索哪种药物更安全、更有效，精算师用它评估金融风险以保证市场的平稳运行。总之，在大数据时代，统计数据、分析数据都离不开计算机软件的支持，在这方面R语言尤其出色。

本书将带领你踏上R语言软件开发之旅，从最基本的数据类型和数据结构开始，到闭包、递归和匿名函数等高级主题，由浅入深，讲解细腻，读者完全不需要统计学的知识，甚至不需要编程基础。书中提到的很多高级编程技巧，是作者多年编程经验的总结，对有经验的开发者也大有裨益。本书精选了44个扩展案例，这些案例都源自于作者亲身参与过的咨询项目，都是与数据分析相关的，生动展示了R语言在统计学中的高效应用。

Python语言程序设计

作者：[美] 梁勇 ISBN：978-7-111-48768-5 定价：79.00元

Python易于学习，且编程有趣。Python的代码简单、短小，易读、直观，而且有丰富的库，使用极为方便，已经成为目前最流行的高级语言，特别是在数据分析领域，得到越来越多用户的青睐。本书是一本优秀的Python程序设计入门教材，沿袭了作者一贯的“以实例教，由实践学”的教学原则。书中采用了他所提出的已经经过实践检验的“基础先行”的方法，即在定义类之前，首先使用清晰简明的语言介绍基本程序设计概念，如选择语句、循环和函数；在介绍面向对象程序设计和GUI编程之前，首先介绍基本逻辑和程序设计概念。书中除了给出一些以游戏和数学为主的典型实例外，还在每章的开始使用简单的图形给出一两个例子，以激发学生的学习兴趣。通过阅读本书，读者可掌握利用Python解决实际问题的方法和能力。

推荐阅读

数据挖掘与商务分析：R语言

作者：约翰尼斯·莱道尔特 ISBN：978-7-111-54940-6 定价：69.00元

统计学习导论——基于R应用

作者：加雷斯·詹姆斯 等 ISBN：978-7-111-49771-4 定价：79.00元

数据科学：理论、方法与R语言实践

作者：尼娜·朱梅尔 等 ISBN：978-7-111-52926-2 定价：69.00元

商务智能：数据分析的管理视角（原书第3版）

作者：拉姆什·沙尔达 等 ISBN：978-7-111-49439-3 定价：69.00元